EXPERIMENTAL DESIGN IN PSYCHOLOGICAL RESEARCH

Books by ALLEN L. EDWARDS

EXPERIMENTAL DESIGN IN PSYCHOLOGICAL RESEARCH, *Fourth Edition*

STATISTICAL METHODS, *Second Edition*

STATISTICAL ANALYSIS, *Third Edition*

PROBABILITY AND STATISTICS

EXPECTED VALUES OF DISCRETE RANDOM VARIABLES AND ELEMENTARY STATISTICS

SOCIAL DESIRABILITY VARIABLE IN PERSONALITY ASSESSMENT AND RESEARCH

THE MEASUREMENT OF PERSONALITY TRAITS BY SCALES AND INVENTORIES

TECHNIQUES OF ATTITUDE SCALE CONSTRUCTION

FOURTH EDITION

EXPERIMENTAL DESIGN IN PSYCHOLOGICAL RESEARCH

ALLEN L. EDWARDS
Professor of Psychology, The University of Washington

Holt, Rinehart and Winston, Inc.
New York Chicago San Francisco Atlanta Dallas Montreal Toronto London Sydney

Library of Congress Catalog Card Number: 73-162648

ISBN: 0-03-086214-0

Printed in the United States of America
2345 038 987654321

To those students
who will some day make their contribution
to psychology and the behavioral sciences
by research and experimentation

To test a hypothesis for significance
is relatively easy;
to find a significant hypothesis to test
is much more difficult

PREFACE

In preparing this, the fourth, edition of *Experimental Design in Psychological Research,* I have been guided by the same principles I followed in writing the earlier editions. I have tried to write a book that can be understood by those who are familiar with elementary statistical analysis and who have a working knowledge of algebra.

I have included some proofs in Chapters 1–6. In almost all cases where a proof is not given in these chapters, it can be found in my book, *Probability and Statistics.* For those unfamiliar with the algebra of mathematical expectation, which I have used in the proofs in later chapters, I suggest that reading *Probability and Statistics* will be helpful.

A number of hypothetical sets of data are interspersed with the results of actual experiments. I have attempted to include in the problems at the end of each chapter some that illustrate by fairly easy computations the analyses described in the text. Some problems requiring more prolonged calculations are also included. The problems are in all cases modeled after the methods presented in the chapter. I have also included in the problems at various times a brief discussion of a particular point that the problem has been designed to illustrate. Answers to all problems involving calculations are given in the back of the book. I am indebted to Dr. Robert D. Abbott for his assistance in checking the answers to these problems.

Although I have eliminated in this edition the chapter on significance tests for correlation coefficients, I realize that some instructors who may use this book as a text may wish to discuss these tests. Consequently, I have retained in the appendix the table of significant values of the correlation coefficient and also the table of the z' transformation for the correlation coefficient so that students will not have to refer to another source for them.

I am indebted to the estate of the late Sir Ronald A. Fisher and to Oliver and Boyd, Ltd., Edinburgh, for permission to reprint Tables IV,

V, and VI from their book *Statistical Methods for Research Workers.* Table I is reproduced by permission of Messrs. Kendall and Smith and the Royal Statistical Society. Table VIII has been reproduced from Professor Snedecor's book, *Statistical Methods,* by permission of the author and his publisher, the Iowa State College Press. Additional values of t were also taken from Professor Snedecor's book by permission. Portions of Table II have been taken from *Handbook of Statistical Nomographs, Tables, and Formulas* by permission of Drs. J. W. Dunlap and A. K. Kurtz and their publishers, the World Book Company.

I am also indebted to Maxine Merrington and Catherine M. Thompson and *Biometrika* for permission to reproduce the entries in Table IX. H. Leon Harter kindly made Tables Xa and Xc available to me in advance of their publication in *Biometrics.* Tables XIIa to XIId were developed by Charles W. Dunnett and are reproduced by permission of the *Journal of the American Statistical Association.*

To the American Psychological Association, the American Statistical Association, the Biometric Society, the Royal Statistical Society, the British Psychological Society, the University College of London, Williams and Wilkins Company, and Warwick and York, Inc., I am indebted for permission to quote brief passages and to make use of data published in various professional journals.

Allen L. Edwards

Seattle, Washington

CONTENTS

10

FACTORIAL EXPERIMENTS: FACTORS WITH MORE THAN TWO LEVELS *184*

11

MODELS FOR FACTORIAL EXPERIMENTS WITH A RANDOMIZED GROUP DESIGN *201*

12

FACTORIAL EXPERIMENTS: FURTHER CONSIDERATIONS 215

13

RANDOMIZED BLOCK DESIGNS 231

14

RANDOMIZED BLOCK DESIGNS: REPEATED MEASURES 259

15

LATIN SQUARE DESIGNS *285*

16

LATIN SQUARE DESIGNS: REPEATED MEASURES *301*

17

TREND ANALYSIS DESIGNS *330*

18

THE ANALYSIS OF COVARIANCE FOR A RANDOMIZED GROUP DESIGN *369*

19

THE METHOD OF LEAST SQUARES *394*

EXPERIMENTAL DESIGN IN PSYCHOLOGICAL RESEARCH

CHAPTER I

THE NATURE OF RESEARCH

1.1 *Introduction*

Incidental observation is frequently important in initiating research for such observation may motivate us to formulate hypotheses, to state problems, and to ask questions. We may then seek answers to these questions by planned or systematic observation that characterizes research and experimentation. In research we do not haphazardly make observations of any and all kinds, but rather our attention is directed toward those observations that we believe to be relevant to questions we have previously formulated. The objective of research, as recognized by all sciences, is to use observation as a basis for answering questions of interest.[1]

Even when observations are made systematically, the raw unclassified observations are often of such a nature that in their original form they do not lend themselves to an obvious interpretation with respect to the questions we have posed. We therefore resort to techniques that will reduce the observations to a more manageable form. These techniques involve classifying and operating on the observations to reduce them to frequencies, proportions, means, variances, correlation coefficients, and other statistical measures. On the basis of these statistical measures we hope to be able to draw certain conclusions or inferences that will bear upon questions of interest.

[1] The brief discussion of research and experimentation in the behavioral sciences given in this chapter obviously does not do justice to the subject. For a much more complete treatment of research methods and techniques, see Lindzey (1954), Festinger and Katz (1953), Selltiz *et al.* (1964), Underwood (1957), Stollak *et al.* (1966), Megargee (1966), and Shontz (1965).

To consider a simple example, suppose that for some reason we have become interested in the question of whether a particular coin is unbiased. By "unbiased" we mean that if the coin is tossed in a random fashion it is equally probable that the coin will fall heads or tails. In an attempt to obtain an answer to the question raised, we may undertake systematic observation of the outcomes of a specified number of tosses of the coin. A single observation will consist of the result of a single toss; that is, we observe and record whether the coin falls heads or tails.

If we make 100 such observations, they may be recorded in the form $HTTHTTT \ldots H$, with each H and T being the record of a single observation. But this series of observations, in unreduced form, is not easy to interpret with respect to the question of interest. We can reduce the observations, however, by counting the number of H's and the number of T's, and these two frequencies will summarize succinctly the complete set of observations.

It may be intuitively obvious to us that, if the coin is unbiased, the two frequencies should be approximately equal and that any departure from equality would provide *some* evidence against the notion that the coin is unbiased. An important problem in research is how to evaluate objectively the evidence provided by any given set of observations. We shall see later that there are techniques for doing this and that these techniques require additional operations on the two observed frequencies.

The example cited characterizes fairly well the nature of much research. One or more questions are formulated. Systematic observation is then made of things believed to be relevant to the questions. Having made a series of observations, the observer then reduces these to a limited number of statistical measures that provide a summary description of the complete set. By means of further operations on the descriptive measures, the evaluation of the evidence provided by the observations, with respect to the questions of interest, is placed on an objective basis.

A number of additional aspects of research are also illustrated by the example. How many observations should be made? Obviously, only one observation would not provide us with adequate information in the present instance. If an insufficient number of observations are made, we shall not be much better off than with none. On the other hand, if we make more observations than are needed, we shall have wasted time and energy that might be more fruitfully spent in other endeavors. We shall later discuss techniques that are of value in estimating the number of observations needed for stated objectives.

We have also mentioned the relevance of the observations to the questions of interest. In the case at hand, the relevance of the observations made may seem obvious. We do not, for example, direct our attention to observing the position of the coin on the floor after each toss but rather to observing whether the coin falls heads or tails. We observe the latter because we believe these observations are pertinent to the question we have asked. In other research problems, the relevance of the observations made to the question of interest may not always be so clear-cut. Whether the observations to be made in a given research problem are relevant to the question they are supposed to answer must always be given serious consideration in planning the research.

The manner in which the coin is tossed is also important. If the coin is tossed in a systematic fashion, we may observe an excess of heads over tails or vice versa. The observations then made would bear not on the nature of the coin—the problem of interest—but rather on the nature of the toss. We have indicated that the method of tossing the coin should be such as to introduce randomness in the observations made so that, if the coin is unbiased, heads and tails may be assumed to be equally probable.

Furthermore, although we have indicated that techniques are available for making the evaluation of the evidence provided by a set of observations objective, this does not mean that we shall always be correct in the inference made simply because it is made on an objective basis. We may, for example, conclude that the coin is biased when, in fact, it is not. Or we may fail to conclude that the coin is biased when, in fact, it is. These two kinds of errors also require consideration in the planning of research.

In this book, we shall be concerned with various problems, some of which have been briefly described, involved in planning research and in designing experiments. We shall also be concerned with the analysis of observations made in the course of research. If a research study is well planned or an experiment is well designed, the methods of analysis to be applied to the observations will have been given consideration in the planning stage. Only in this way can one have some assurance that the observations to be made will enable one to obtain answers to the questions of interest. It would obviously be frustrating to any research worker if he were to make a large number of observations, only to find they could not be analyzed in any way that would answer the questions he was interested in. The only safe way to avoid this kind of frustration is to plan the analysis *before* the observations have been made.

1.2 Observations and Variables

We have stressed the importance of observations in research. The things that are observed are called *variates* or *variables.* In the example cited earlier, the variable, the thing observed, was the face of the coin. Any particular observation is called a *value* of the variable. For the face of a coin there are only two possible values of the variable:

heads or tails. The value of a variable indicates the class to which an observation is to be assigned. If a coin falls heads, we consider that observation and all others with the same value as belonging to the same class. In order for something we are observing to be considered a variable, we must have at least two possible classes of observations. The classes must also be mutually exclusive; that is, any given observation can be assigned to only one of the available classes. By a variable, therefore, we shall mean anything that we can observe and that is of such a nature that each single observation can be classified into one and only one of a number of mutually exclusive classes.

1.3 Behavioral Variables

By a *behavioral* variable we shall mean any variable that refers to some action or response of an organism. At one extreme, these actions may consist of relatively simple responses such as finger flexions, eye blinks, pressing a key when a light is flashed, marking True or False for an item on a test, pressing a bar in a Skinner box, and so forth. At the other extreme, we have such complex behavior patterns as those involved in problem solving, typewriting, or perhaps the even more complicated behavior involved in aggression, dominance, leadership, and social adaptability.

One commonly used behavioral variable in psychological research is the time required for some action or response to occur. A stimulus is presented to a subject and the time required for him to make a certain response is measured. A rat is placed in a starting box of a maze and the time required for the rat to run to the goal box is measured. A subject is presented with a problem and the time required for him to reach a solution is measured. Typewriting skill may be measured by determining how long it takes an individual to type a standard passage.

In other cases, we may hold the time constant and count the number of responses of a given kind that occur within a fixed period. For example, we may count the number of bar presses made by a rat in a fixed interval of time. Or we may count the number of problems solved in a fixed interval of time. We may count the number of "aggressive" responses made by a child during a play period or the number of times the child withdraws in response to the advances of another child.

In still other cases, we may simply count the number of responses of a given kind without regard to the time required to make these responses. For example, we may count the number of errors made by a rat running a maze without regard to the time required by the rat to run the maze. Or we may count the number of correct responses made to a fixed number of problems without concern for the time required by the subject to make his responses.

Instead of counting the number of responses of a given kind made by a child, we may ask judges to observe the behavior of the child in a specific situation and then to rate on a 5-point scale the degree to which they believe the behavior of the child reflects a variable of interest. Rating scales are frequently used in psychological research in an attempt to obtain measures of behavioral variables that are not otherwise easily quantified.

1.4 Stimulus Variables

Although psychology has, on occasion, been defined as the science of behavior, it is obvious that behavior does not occur in a vacuum but always in a particular setting or environment. The general class of things we observe that relate to the environment, situation, or conditions of stimulation, we shall refer to as *stimulus* variables. The stimulus variables in a psychological experiment may consist of relatively simple things, such as electric shock, light, sound, or pressure. These may be quantified by measuring the physical intensity of the stimulus.

There are other stimulus variables of interest to the psychologist for which we have no measures corresponding to physical intensity. These may consist of problem-solving situations, motor conflict situations, social situations, and so forth, and they are relatively more difficult to quantify. Indeed, in much research we can only say that the variations in stimulation in which we are interested consist of complex combinations of stimuli differing in kind rather than degree. We shall refer to any such differences in the conditions of stimulation, in a given experiment, as differences in *treatments*. The term "treatment" will thus be used to refer to a particular set of stimulus or experimental conditions.

In a given experiment, for example, we may be interested in the behavior of certain subjects when they are involved with an "authoritarian" leader and in the behavior of other subjects when they are involved with a "democratic" leader. The behavior of the leader will vary, but we may attempt to keep all other aspects of the stimulus situation constant. Assuming we are successful in this respect, we must still recognize that the differential role-playing activities of the leader may be expected to result in complex differences in the two situations. The difference, in other words, may not be in one dimension or variable but rather in a set of many differences in many variables. The term "treatment" is a useful one because it can be used in connection with differences between complex sets of experimental or stimulus conditions, and also in reference to differences in experimental conditions where a single variable is involved.

In many research problems the situation is varied by having some aspect of the situation either present or absent. Thus, in a learning experiment, the behavior of rats in a maze may be observed under conditions where food is present in the goal box and under conditions where food is absent. One group of subjects may study a programmed text under conditions where they have knowledge of results; that is, they are required to respond to material in the text and are immediately provided with information as to whether their response is correct or incorrect. Another group of subjects may study the same programmed text but without knowledge of results; that is, they are not provided with information as to whether their response is correct. The group of subjects for which some aspect or condition of the situation is present is often referred to as the *experimental* group, and the group of subjects for which the aspect or condition of the situation is absent is often referred to as the *control* group.

1.5 Organismic Variables

Organismic variables arise from ways in which organisms may be classified and from the observations and measurements of physical, physiological, and psychological characteristics of organisms. For example, we may measure the heights or weights of a group of individuals. These observations do not correspond to response variables or stimulus variables, but they may be conveniently described as *organismic* variables. They are characteristic ways in which the particular group of organisms under observation vary. Similarly, organisms may be classified according to the color of their hair or eyes. Or they may be classified according to whether they are males or females, or in terms of their age.

In research frequent use is also made of *response-inferred* organismic variables. By a "response-inferred organismic variable" is meant a classification based on prior observations of response. A person's IQ, for example, is determined by observing his response to a standardized testing situation. In many cases, however, it is convenient to regard IQ as something that is associated with the organism, that is, as an organismic variable. As another example of a response-inferred organismic variable, it is not uncommon to refer to one group of subjects in a given research problem as the "anxious" group and another group as the "nonanxious" group. This classification is often based on prior observation of response to some test of anxiety.

1.6 Discrete and Continuous Variables

In a discrimination experiment, a variable of interest may be the number of correct discriminations made by a subject in a set of ten trials. The only possible values of this variable are

0, 1, 2, . . ., 10, and this variable would be described as *discrete*. Values of a discrete variable are *always* exact. If we observe and record that a subject makes eight correct responses, there is no uncertainty with respect to this value. On the other hand, if we should measure the time required for a subject to make a discrimination, this variable would be described as *continuous*. Values of continuous variables are *never* exact, because no matter how accurately we measure the variable there is always uncertainty with respect to the observed or recorded value of the variable. For example, if we measure time in units of 0.001 second, we can never be sure that a recorded value of 1.213 seconds is actually 1.213 followed by an infinite number of zeros, that is, 1.213000. . .seconds. Because of the uncertainty associated with observed values of continuous variables, it is customary to regard such values as representing an interval ranging from one-half unit below up to one-half unit above the actual unit of measurement. For example, if time is measured in units of 0.001 second, then a recorded value of 1.213 seconds would be regarded as representing an interval ranging from 1.2125 up to 1.2135.

We shall take the same position with respect to variables that are, in fact, discrete. For example, it will be convenient to regard eight correct responses as representing an interval ranging from 7.5 up to 8.5. It will become obvious, in later chapters, that there are practical reasons for regarding measures of discrete variables as representing intervals, just as we regard measures of continuous variables as representing intervals. The fact is that all observed or recorded measurements are discrete, regardless of whether the variables are continuous or discrete.

1.7 Quantitative and Qualitative Variables

Variables for which the possible values represent differences in degree along a single dimension are often referred to as *quantitative* variables. All continuous variables are quantitative variables because the differences between the possible values of such variables are matters of degree. All discrete variables for which the possible values are counts of the number of responses of a given kind are also quantitative because the differences between the possible values are also matters of degree. A subject who makes eight correct responses, for example, has made one more correct response than a subject who makes only seven correct responses.

Quantitative variables are also described as *ordered* variables in that the possible different observed values of such variables, discrete or continuous, can be ordered on a continuum where the relationships of equal to, less than, or greater than, will hold true.

Unordered variables are variables in which differences in the possible values are matters of kind rather than degree. Such variables are often described as *qualitative* variables. The classification of subjects according to color of hair is an example of an organismic variable that is unordered or qualitative, as is also the classification of individuals as males or females.

1.8 Research and Experiments

It is the concern of psychologists and other scientists who are interested in the behavior of organisms to describe and study stimulus, response, and organismic variables. Much of psychological research is concerned with attempts to improve our methods of description of these variables by devising apparatus and developing techniques for more precise measurement of them.[2] In the hands of other psychologists these devices are used to make systematic observations of variables of interest.

As we have pointed out earlier, systematic observation is undertaken in an attempt to obtain answers to questions in which we are interested. In some cases the questions of interest have to do with the accurate description of a group of subjects with respect to one or more variables. An instructor, for example, may be interested in how the intelligence test scores for one of his classes are distributed. His systematic observations might consist of the score of each of his students on some standardized test of intelligence. These observations may be reduced (classified) so that he has available for each score the frequency with which it was observed. He may also be interested in finding out what the average score is for his class and something about the range or spread of scores.

In other cases, the questions of interest may concern the degree of association or relationship between two variables. For example, the same instructor may also be interested in determining whether the intelligence test scores of the students are in any way related to or associated with scores on a final examination. For example, do students with high intelligence test scores also tend to obtain high scores on the final examination, while those with low scores on the intelligence test tend to obtain low scores on the final examination?

In certain instances it is possible for an investigator to vary quantitatively one variable, usually a stimulus variable, and to study the behavior of groups of subjects under each value of the variable. As an example, we might vary the size of type in which a list of words is printed. Subjects are assigned to a given type size and the words for each type size are exposed at a constant rate. The observations obtained for each variation

[2] Sidowski (1966) covers a wide range of experimental techniques and instrumentation applicable to the study of both human and animal behavior.

in the stimulus conditions might be the number of words correctly recognized by each subject. If the average number of words correctly recognized for each type size is obtained, we may then determine the relationship between these averages and the type size. We might be interested, for example, in finding out whether the average number of words recognized increases as the type size is increased. Furthermore, we might be interested in determining whether the relationship between these two variables is linear or not.

In the case of the instructor interested in the relationship between scores on the intelligence test and scores on the final examination, the instructor is not able to control or manipulate the variables in which he is interested. For example, he has no control over the intelligence test scores or over the final examination scores of the subjects. The values of these variables are fixed or determined by each subject and cannot be manipulated or changed directly by the instructor. However, in the case just cited, one of the variables was directly under the control of the investigator. This variable was the stimulus variable, that is, the type size. The investigator can vary or alter this variable in the manner described. When certain variables can be controlled or manipulated directly in a research problem by the investigator, the research procedure is often described as an *experiment.*

The variables over which the investigator has control are called the *independent* variables. They are those which the investigator himself manipulates or varies. As the independent variables are changed or varied, the investigator observes other variables to see whether they are associated with or related to the changes introduced. These variables are called the *dependent* variables. In the case described, the dependent variable was the average number of words correctly recognized for each type size.

Many experiments are concerned with a comparison between what we have previously called various treatments. The questions asked in these experiments have to do with differences in the dependent variable under different treatments. Experiments of this kind have been described as *comparative* experiments. In a comparative experiment interest is directed toward the problem of discovering whether the different treatments result in differences in the observed values of the dependent variable. When the treatments represent a quantitative variable, then we may also be interested in studying the functional relationship between the quantitative independent variable and the dependent variable, that is, in determining whether the relationship is linear or of some other form.

The advantages of observations made under experimental or controlled conditions over observations made without such controls have been pointed out by Woodworth (1938, p. 2):

1. The experimenter makes the event happen at a certain time and place and so is fully *prepared* to make an accurate observation.

2. Controlled conditions being *known* conditions, the experimenter can set up his experiment a second time and repeat the observation; and—what is very important in view of the social nature of scientific investigation—he can report his conditions so that another experimenter can duplicate them and check the data.

3. The experimenter can systematically *vary* the conditions and note the concomitant variation in the results.

QUESTIONS AND PROBLEMS

1.1 Can you think of a case in the behavioral sciences in which incidental observation was instrumental in the formulation of some hypothesis which was then investigated by means of systematic observation?

1.2 Suppose someone is interested in the IQ's of 1000 fifth-grade children; that is, he has available 1000 such observations. Of what value would some form of data reduction be in this instance?

1.3 Make a list of five response variables, five stimulus variables, and five organismic variables other than the ones mentioned in the chapter. What available methods are there for quantifying each of the variables? How might those variables for which methods are not available be quantified?

1.4 Assume that one of the variables in a research problem is socioeconomic status. If you were constructing an index or a test for this variable, what factors, in addition to income, would you want to take into consideration?

1.5 A fortunate basketball coach at a small college once had five players trying for the position of center on the college team. The members of the coaching staff were unable to differentiate between the abilities of the five players. What situation tests might be developed to yield quantitative data concerning the ability of each player for the position of center?

1.6 A graduate department of psychology awards a number of research fellowships in psychology to students who are believed to be outstanding. Assume that the awards are to be based primarily on the potentiality of the students to do research in the field of psychology. What factors should be taken into consideration in making the awards? What methods might be devised for quantifying the variable of interest?

1.7 Comment on the following statement: "If a variable is truly continuous, then, in theory, no two observations could ever have the same value."

1.8 Select and read a research article in some journal. What question or questions was the research attempting to answer? What variables were involved in the research? What was the nature of the observations made?

1.9 In psychoanalytic theory, the *id* is said to be that aspect of the individual concerned with instinctual reactions for satisfying motives. According to Morgan (1956, p. 633), "The id seeks immediate gratification of motives with little regard for the consequences or for the realities of life." Regardless of whether this characteristic is called the id or something else, it seems reasonable that individuals do differ in the degree to which they manifest the characteristic. What observations might be made to obtain some measure of the "strength" of the id?

1.10 Psychologists who are interested in personality research make use of a large number of variables that refer to personality traits or characteristics (organismic variables?). Some examples are honesty, ego control, dependency, achievement motive, deference, and social introversion. What are some of the most frequently used methods of observing these variables?

1.11 Comment on the following statement: "Naming a variable may suggest the observations to be made but, in a very real sense, the observations actually made define the variable itself."

1.12 Suppose you want to know whether a die is biased. How would you go about obtaining an answer to this question?

1.13 How does an organismic variable differ from a stimulus and a behavioral variable?

1.14 What is the difference between a control group and an experimental group?

1.15 What is the difference between a dependent variable and an independent variable?

1.16 How does a quantitative variable differ from a qualitative variable?

1.17 What is the nature of a comparative experiment?

1.18 What is the difference between a continuous variable and a discrete variable?

CHAPTER 2

PRINCIPLES OF EXPERIMENTAL DESIGN

2.1 *The Farmer from Whidbey Island*

A farmer from nearby Whidbey Island visited the psychological laboratory of the University of Washington. He had with him a carved whalebone and claimed that in his hands the bone was an extremely powerful instrument capable of detecting the existence of even small quantities of water. To support his claim he said that several of his neighbors on Whidbey Island had tried unsuccessfully to bring in water wells. Finally they had called upon him for help. He had taken his whalebone, grasped one fork in each hand, and walked slowly over their ground. Suddenly, the point or apex of the bone had dipped sharply toward the ground. When his neighbors had drilled wells at the points he had located in this fashion, they had found water.[1]

The farmer added that he was unable to explain his peculiar power. His neighbors were unable to use the whalebone in locating water. It had to be in his hands before it would dip sharply to indicate the presence of water. He was somewhat disturbed by his ability and he thought that perhaps the psychologists at the university would be interested in examining him and telling him why it was that he was able to use the bone so effectively while others could not. He himself thought that it had something to do with "magnetism" that emanated from his body. Anyway, he would be willing to demonstrate his ability so that the psychologists could see for themselves. Perhaps then they could explain it to him.

[1] For an interesting account of water witching in the United States, see the article by Hyman and Vogt (1967).

At this point in his story, the farmer asked for a paper cup filled with water. When he was given the cup, he placed it on the floor. He then grasped the whalebone and held it stiffly in front of him as he moved slowly about the room. When the apex of the bone passed over the cup of water, his arms trembled slightly and the bone dipped toward the ground. The farmer showed signs of strain and remarked that the force was so powerful he was almost unable to keep the bone in his grip.

The psychologist thanked the farmer for his demonstration and said that he would like to test the farmer's ability to locate water under controlled conditions, but that this would require some preparation. Would the farmer agree to return for these tests next week? The farmer agreed and promised to return at the appointed time.

2.2 The Experiment

When the farmer returned to the psychological laboratory the next week, he was greeted by the psychologist and taken to one of the laboratory rooms. Spread around the floor of the room were ten pieces of plywood about 8 × 8 inches in size. Numbers from 1 to 10 had been marked on the top of each square. The pieces of plywood were resting upon tin cans, about No. 2 in size, with the labels removed. The psychologist explained that five of the cans had been filled with water and that five had been left empty. He had not used any systematic basis in determining which cans were to be filled and which were to be left empty, but rather, as he put it, "This was left to chance." As a matter of fact, he added, he himself did not know which of the cans contained water and which were empty, because he had left this task to a laboratory assistant. He was as much in the dark as the farmer, but he hoped that the farmer, with the aid of his whalebone, would soon be able to enlighten him. He again emphasized to the farmer that under five of the sections of plywood were cans with water and under five other sections were empty cans and that the arrangement of the empty and filled cans was purely a chance or random one.

The psychologist now wanted the farmer to take his whalebone and attempt to select the five sections of plywood covering cans filled with water. The farmer did not need to make his choice in any particular order; he was merely to select the set of five sections of plywood under which he believed cans filled with water would be found.

The observations to be made in this experiment consist of the choices the farmer makes. The outcome of the experiment is the particular group of choices the farmer makes. We shall examine this experiment in some detail. We shall pay particular attention to the set of all possible outcomes of the experiment, the question which the experimenter hopes to answer by the observations made, and the manner in which he proposes to arrive at this answer.

2.3 The Question of Interest

The question that motivated the experimenter to make the observations is not necessarily the one of interest to the farmer. The farmer, in his previous conversation, had indicated that he wanted to know *why* he could divine the presence of water. It is apparent that the farmer implicitly assumes that he can detect the presence of water. The question of interest to the psychologist, on the other hand, is whether the farmer is successful in doing what he believes he can do; that is, can he actually detect the presence of water? It would appear to be obvious that to ask why the farmer is successful in divining the presence of water is meaningful only if it can first be determined that he is, in fact, successful in doing so.

The psychologist may reason in this way: Let us assume that the farmer does *not* possess any particular powers that enable him to locate water with his whalebone; that the only factor operating in determining his choice is chance. More specifically, the question that the experimenter wishes to answer is: Can the farmer do any better in his choices than might be expected on the basis of chance?

2.4 Sample Space and Probability

We note that the task set for the farmer is to select five cans, those he believes contain water, from a set of ten. We shall regard the set actually selected by the farmer as a *sample* of $n = 5$ observations from a *finite population* in which $N = 10$. The particular sample selected by the farmer will represent one of the possible outcomes of the experiment. There will be other samples, not selected by the farmer, that represent other possible outcomes of the experiment.

The set, S, of all possible outcomes of an experiment is called the *sample space* of the experiment. The elements of S are called *sample points,* and each possible outcome of the experiment must correspond to one and only one sample point. Not only do we want to know the number of possible outcomes of the experiment, but we also want to be able, on some reasonable basis, to assign to each sample point or outcome a number, P, that gives the probability associated with the sample point. If the values of P assigned to each sample point are to be called probabilities, then they, in turn, must satisfy the following rules or axioms:

1. The values of P assigned to each sample point must be equal to or greater than 0 and equal to or less than 1; that is, $0 \leq P \leq 1$.

2. The sum of the values of P assigned to the sample points must be equal to 1; that is, we must have $\Sigma P = 1$.

3. If n subsets, $E_1, E_2, E_3, \ldots, E_n$ are defined on the sample space, and if the n subsets are mutually exclusive and exhaustive, then we must also have

$$P(E_1) + P(E_2) + P(E_3) + \cdots + P(E_n) = 1$$

2.5 Simulation of the Experiment

We can simulate the experiment with the farmer by placing ten disks in a box. We let each disk correspond to an observation without, for the moment, specifying the value of the observation. We identify the disks in the same manner in which the ten pieces of plywood are identified, that is, by the numbers 1 to 10. We now shake the box thoroughly and then let one disk fall through a small slit in the box. We shall assume that this procedure results in a random selection of a disk. By *random selection* we mean that we shall assume that the probability of any given disk in the box falling through the slit is the same for all disks. Having selected one disk, *without replacing it in the box*, we again shake the box and draw a second disk from the remaining nine. We again shake the box and draw a third disk from the remaining eight. We continue in this manner until we have a sample of $n = 5$ disks or observations from the $N = 10$ disks. We let this sample correspond to one of the sets of five choices that could be made by the farmer in the experiment.

If our method of sampling is random, then on each draw each of the disks in the box has an equal probability of being selected. For example, the probability of a particular one of the ten disks being selected on the first draw is $1/10$; on the second draw, with nine disks in the box, the probability of a particular one being drawn is $1/9$, and so on, until on the last draw the probability of a particular one being selected is $1/6$.

2.6 Permutations[2]

What is the probability that a sample, drawn in the manner described, will include the observations identified by the numbers 10, 8, 5, 4, and 1? Consider first the probability of obtaining these five observations in the order specified. The probability of obtaining 10 on the first draw is $1/10$. Given that we have drawn 10, the probability of 8 on

[2] If you have difficulty with this and the next few sections, then you need to review the topics of permutations and combinations in a text on probability and statistics. See, for example, Mosteller *et al.* (1961) or Edwards (1971). Both of these texts provide proofs of the equations given in this chapter.

the second draw is $\frac{1}{9}$. Given that we have obtained 10 on the first draw and 8 on the second, the probability of 5 on the third draw is $\frac{1}{8}$, and so on. The probability that the sample will be the particular set of five observations drawn in the order specified will therefore be

$$\frac{1}{10} \times \frac{1}{9} \times \frac{1}{8} \times \frac{1}{7} \times \frac{1}{6} = \frac{1}{30{,}240}$$

We note that the denominator in the above expression is simply the number of permutations of $N = 10$ objects taken $n = 5$ at a time. *Permutations* refer to the number of different orders in which a set of objects may be arranged. In general, the number of permutations of N objects taken all together is given by

$$_NP_N = N! \tag{2.1}$$

where $N!$ is called factorial N and represents $(N)(N-1)(N-2)\ldots(2)(1)$ or the product of all of the successive integers from N to 1. By definition $0!$ is always taken equal to 1. The number of permutations of N objects taken n at a time is given by

$$_NP_n = \frac{N!}{(N-n)!} \tag{2.2}$$

In the problem at hand, the number of permutations of $N = 10$ objects taken $n = 5$ at a time is

$$_{10}P_5 = \frac{10!}{(10-5)!} = 30{,}240$$

This number gives every possible set of five arranged in every possible order. That is, any one of the ten disks may be selected first; this selection may then be followed by any one of the remaining nine; this selection may be followed by any one of the remaining eight; and so on, until five have been selected.

2.7 Combinations

We have seen that the probability of selecting disks 10, 8, 5, 4, and 1, in that order, is 1/30,240. If we are not interested in the order in which these particular five observations are drawn, but simply in the probability that the sample will contain the specified five observations, then we note that the five observations may be arranged in $5! = 120$ different orders. Dividing 30,240 by 120, we obtain 252 ways in which five objects can be selected from ten objects, if the arrangement or order is ignored. In general, the number of *combinations* (arrangement or order ignored) of N distinct objects taken n at a time is given by

$$_NC_n = \frac{_NP_n}{_nP_n} = \frac{\frac{N!}{(N-n)!}}{n!} = \frac{N!}{n!(N-n)!} \tag{2.3}$$

or, in the present problem,

$$_{10}C_5 = \frac{10!}{5!(10-5)!} = 252$$

Then the probability that a sample selected in the manner described will contain the specified five observations, the order in which they are drawn being immaterial, will be $120/30{,}240 = 1/252$.

The probability we have just obtained will be exactly the same for any other specified set of five observations differing from the sample considered in one or more observations. For example, the probability that the sample will contain the observations 10, 4, 3, 2, and 1 is also $1/252$. Thus, as (2.3) shows, there are $10!/5!5! = 252$ different *unordered* samples, and each of these samples has a probability of $1/252$ of being selected.

2.8 Random Samples

The probability that any specified observation will be included in the sample of $n = 5$ observations is $\frac{1}{2}$. To see that this is so, we consider the probability that the specified observation will be the first one drawn. This probability is $\frac{1}{10}$. The probability that the observation will be the second one drawn will be equal to the probability that it is *not* obtained on the first draw times the probability that it will be selected from the remaining nine. Thus $(\frac{9}{10})(\frac{1}{9}) = \frac{1}{10}$ is the probability of obtaining the specified observation on the second draw. Similarly, the probability that the observation will be the third one selected will be given by $(\frac{9}{10})(\frac{8}{9})(\frac{1}{8}) = \frac{1}{10}$. In the same manner, we find that the probability of the observation being the fourth one drawn is $\frac{1}{10}$, and this is also the probability that it will be the fifth one drawn. Because these are *mutually exclusive events*, the probability that the observation will be included in the sample is $\frac{5}{10} = \frac{1}{2}$. This probability is the same for each of the ten observations.

We can thus say that every possible sample (order ignored) of five observations has the same probability ($\frac{1}{252}$) of being drawn and that every observation has the same probability ($\frac{1}{2}$) of being included in the set of five observations selected. These properties are used to define a *simple random sample* or, more briefly, a *random sample*. The use of the term "random" in connection with a sample should be considered as applying to the particular procedure or method used in selecting the observations. In other words, a random sample is one obtained by a particular method that we believe introduces randomness in the selection of the observations. In the present instance, we assumed that randomness in the selection of the disks was introduced by the use of the sampling box in which the disks were thoroughly mixed before one was selected. More useful procedures of random selection will be discussed later.

2.9 Probabilities of Possible Outcomes

Let us now assume that we have assigned, again by random methods, a value to each of the ten disks in such a way that five of the disks or observations have a value of W, corresponding to a filled or wet can, and that five have a value of D, corresponding to an empty or dry can. Our method of sampling remains the same, but we are now interested in the number of W's in each of the 252 possible samples of $n = 5$ observations. It is obvious that only one of the 252 samples can include the five observations with values of W, and we can therefore say that the probability of obtaining a sample with five W's is $1/252$.

What is the probability of obtaining a sample with four W's and one D? Specifically, let the first four observations drawn have the value of W and the last one the value of D. The probability of obtaining a W on the first draw is $5/10$; if this occurs, then there will be four observations with W's left in the box and five with D's, and the probability of W on the second draw will be $4/9$. If the first two draws are W's, then there will be three W's and five D's left in the box, and the probability of W on the third draw will be $3/8$. If we obtain a W on the third draw, then there will be two W's left and five D's. The probability of W on the fourth draw will then be $2/7$. If this occurs, then we have one W and five D's left in the box, and the probability that the fifth draw will be a D will be $5/6$. Therefore, the probability of the sample $WWWWD$, in the order specified, is

$$\frac{5}{10} \times \frac{4}{9} \times \frac{3}{8} \times \frac{2}{7} \times \frac{5}{6} = \frac{600}{30{,}240} = \frac{5}{252}$$

If we shift the D to any other position in the sequence, we could show, in the same manner, that the probability of this sequence is the same as when D is in the last position. It is clear that because D can appear in any one of five positions and because the five sequences are mutually exclusive that the probability of drawing a sample with four W's and one D is

$$5\left(\frac{5}{252}\right) = \frac{25}{252}$$

The probability of obtaining, in the order specified, $WWWDD$, is

$$\frac{5}{10} \times \frac{4}{9} \times \frac{3}{8} \times \frac{5}{7} \times \frac{4}{6} = \frac{1200}{30{,}240} = \frac{10}{252}$$

and again this probability remains unchanged for all possible permutations of three W's and two D's. We can find the number of permutations of three W's and two D's by considering the general formula for the number of permutations of n objects when the objects can be divided into k sets such that the objects within each set are alike. We let $r_1, r_2, \ldots, r_k$ be the number of objects in each of the respective sets, with $n = r_1 + r_2 + \cdots + r_k$. Then

$$_nP_{r_1,r_2,\ldots,r_k} = \frac{n!}{r_1!r_2!\ldots r_k!} \tag{2.4}$$

For the case at hand we have $r_1 = 3W$'s and $r_2 = 2D$'s. Then

$$_5P_{3,2} = \frac{5!}{3!2!} = 10$$

and we see that there are ten samples that will contain three W's and two D's. Because each of these samples has a probability of $10/252$ of being selected and because they are mutually exclusive, the probability of obtaining a sample with three W's and two D's is

$$10\left(\frac{10}{252}\right) = \frac{100}{252}$$

Using the same methods, we would find that the probability of obtaining a sample with two W's and three D's is $100/252$; the probability of obtaining a sample of one W and four D's is $25/252$; and the probability of obtaining a sample with no W's and five D's is $1/252$.

2.10 A Sample Space for the Experiment

Let T be the number of W's in a sample of $n = 5$. Then T is a variable that can, in the experiment described, take the possible values of 0, 1, 2, 3, 4, or 5. If the outcome of interest in the experiment is T, then as a sample space for the experiment we have $S = \{0, 1, 2, 3, 4, 5\}$. We note that each possible outcome, T, of the experiment is associated with one and only one sample point. Furthermore, each sample point has associated with it a probability, P. We also note that the values of P satisfy the previously stated axioms.[3]

Now the best that the farmer could possibly do in the present experiment would be to select the particular sample of five W's. There is only one way in which this could happen, and this particular sample would be 1 out of 252 possibilities. If the farmer's selections are being made solely on the basis of chance and if this experiment were repeated an indefinitely large number of times, then we would expect this particular sample to be selected with a theoretical relative frequency of $1/252$. We have $P(T = 5) = 1/252 = 0.004$ and only about 4 times in 1000 would the farmer be expected to have all of his choices correct on the basis of chance alone. Thus, if the outcome of the experiment is $T = 5$, then either a relatively improbable event has occurred by chance or else the farmer is *not* making his selections on the basis of chance.

2.11 Testing a Null Hypothesis

We constructed a sample space for this experiment by assuming that the farmer did *not* have the ability to detect water with his whalebone and that if this was, in fact, the case, then he must make his selections simply on the basis of chance. This assumption may be regarded as a hypothesis, often referred to as a *null hypothesis*, that the experiment is designed to test. Assuming the hypothesis to be true, it was possible to determine the probability of each possible outcome of the experiment.

Because the outcome $T = 5$ is relatively improbable on the basis of chance, but is *not* improbable if the farmer actually can detect the presence of water, many people would find it reasonable to reject the null hypothesis.

In testing a null hypothesis we must make some decisions as to how small the probability of a given outcome must be before we will decide to reject the hypothesis. The probability we choose to use in rejecting the null hypothesis is called the *significance level* of the test and is indicated by α. Judging on the basis of published research, many investigators appear to choose $\alpha = 0.05$ as the significance level of their tests; that is, they reject a null hypothesis whenever the outcome of the experiment has a probability equal to or less than 0.05, when the null hypothesis is true. Some people might regard this as a relatively lenient standard and might insist that α be much smaller than 0.05, say 0.001. In the present experiment, if we use $\alpha = 0.001$ as a standard, it would be impossible to reject the null hypothesis because, even when all five of the farmer's choices are correct,

[3] If we have a discrete variable X such that with each possible value of X there is an associated probability P_i that X is equal to X_i, and if the values of P_i satisfy the previously stated rules or axioms, then X is called a discrete random variable. It is obvious, in the present instance, that T is a discrete random variable.

the probability of this outcome is 0.004, a value that is larger than $\alpha =$ 0.001. If we choose α very small, we decrease the probability of rejecting the null hypothesis. But there is no sense in doing an experiment and testing the significance of the outcome if we have decided beforehand that we will not reject the null hypothesis no matter how improbable the outcome is under that null hypothesis.

It should also be clear that no single experiment can establish the absolute proof of the falsity of a null hypothesis, no matter how improbable the outcome of the experiment is under the null hypothesis. Improbable as $T = 5$ is in the present experiment, we have no way of knowing whether this outcome, if it occurs, is one that might be expected to occur by chance about 4 times in every 1000. As Fisher (1942) has pointed out, "In order to assert that a natural phenomenon is experimentally demonstrable we need, not an isolated record, but a reliable method of procedure. In relation to the test of significance, we may say that a phenomenon is experimentally demonstrable when we know how to conduct an experiment which will rarely fail to give us a statistically significant result" (pp. 13–14).

2.12 Type I and Type II Errors

As we have pointed out above, in making tests of significance we shall sometimes be in error in the inference drawn concerning the hypothesis tested. When the null hypothesis is true and the results of our test of significance lead us to decide that it is false, we describe this as a Type I error. When the null hypothesis is, in fact, false and we decide on the basis of a test of significance not to reject it, we describe this as a Type II error. The probability of a Type I error is set by α, the significance level of our test. If we always reject a hypothesis when the outcome of the experiment has probability of 0.05 or less, and if we consistently apply this standard, then we shall *incorrectly* reject 5 percent of the hypotheses we test; that is, we shall declare them false when they are in fact true. By choosing α small we can decrease the probability of a Type I error. But, at the same time, we will increase the probability of a Type II error.

Suppose, in the present experiment, we choose $\alpha = 0.05$. Then, obviously, if the outcome of the experiment is $T = 5$, the null hypothesis would be rejected. However, suppose that the outcome of the experiment is $T = 4$. Then $P(T \geq 4) = P(T = 4) + P(T = 5) = 0.099 + 0.004 = 0.103$ and, in terms of our standard, this is not a significant outcome. The outcome, however, is obviously better than chance; it just happens to be not *significantly* better than chance in terms of the standard we have chosen. If the outcome of the experiment is $T = 4$ and if we decide not to reject the null hypothesis, it is possible that we will be making a Type II error.

The *power* of a statistical test is defined as $1 - P(\text{Type II error})$ or, equivalently, as the probability of rejecting a null hypothesis when it is false and should be rejected. In general, if we hold α (the probability of a Type I error) constant, we can increase the power of a test by increasing the number of observations in the sample.

It would appear that in the experiment with the farmer, the number of observations is so small that the test of significance has relatively little power. The experimenter is well protected against a Type I error, but he has little protection against a Type II error. The fact is that the farmer may be able to do better than chance in locating water, but the experiment is not sufficiently sensitive to this ability. It could be made more sensitive, that is, the power of the test of significance could be increased by increasing the number of observations.

2.13 Experimental Controls

In the experiment described, if $T = 5$, we know that the probability of this outcome, under the null hypothesis, is 0.004. With $\alpha = 0.05$ and with $T = 5$, the null hypothesis would be rejected. If the null hypothesis is rejected, this means only that the experimenter is not willing to assume that chance alone determined the farmer's selections. It does not *prove* that the whalebone had any particular influence upon the farmer's selections. The test of the null hypothesis has nothing to do with *why* the farmer was able to choose correctly. The experimenter might be willing to assume or infer that the whalebone played some part in the farmer's selections, but he would undoubtedly do this only if other possible alternative explanations could be ruled out in terms of experimental controls. What are some of these alternative explanations?

Without the experimenter knowing about it, the farmer might use the toe of his foot to tap the cans under the board. Because in this manner the cans filled with water could be easily distinguished from the empty cans, this alone would account for a perfect selection on the part of the farmer. If this is the basis of the farmer's selections, then, obviously, the whalebone has nothing to do with his choices. The farmer might even deny that he is using this cue—the sound of the can when tapped with his foot—if questioned about it. But the psychologist knows that many of our choices and judgments are based on factors of which we are not aware. It would be the experimenter's responsibility to rule this possibility out by observation or by some other control.

The psychologist would also want to make sure that the farmer did not tap the tip of the whalebone on the tops of the plywood sections. If the farmer does this, his choice might be determined by the differences in sound of the sections covering the empty and filled cans. He might thus make a perfect selection and the experimenter would reject the null hypothesis of chance. But note again that the rejection of the hypothesis of chance does not establish the validity of the farmer's claim concerning the influence of the whalebone.

Another possible explanation of a perfect selection might be that the experimenter's assistant had spilled some of the water on the floor in filling the cans. The water might have been carefully mopped up, but slight cues may have remained. The absence of dust or the cleanliness of the floor under the sections of plywood containing water, as a result of the mopping, might provide cues for the farmer's choices.

If the experimenter, rather than his assistant, had filled the cans, so that the experimenter had knowledge of which cans contained water and which did not, then the experimenter himself might give some sign: a holding of his breath, a biting of his lips, or some other unconscious gesture, as the farmer moved his whalebone over the sections containing water.[4] The farmer's choice might be based on one of the unconscious gestures or reactions of the experimenter, without, of course, the experimenter, and perhaps even the farmer, being conscious of the fact that these cues were the basis of the farmer's choice. Fortunately, the experimenter, in this instance, anticipated this possibility and controlled for it by having his assistant prepare the cans.

[4] Perhaps the most famous case, described by Pfungst (1911), in which a subject was able to respond correctly on the basis of slight cues provided him by the presence of an experimenter who knew the correct response, was that of Clever Hans. Clever Hans was, in fact, a horse. By tapping his foot, he could give the correct answers to arithmetic problems involving addition, subtraction, multiplication, and division. Hans could also spell, read, and do a number of other things that horses ordinarily do not do. After a series of investigations, Pfungst found that Hans was responding on the basis of slight cues unknowingly provided by himself. A very slight and almost imperceptible forward movement of Pfungst's head would start Hans tapping and at another almost imperceptible upward movement of the head, or of the eyebrows, Hans would stop. Initially, Pfungst was completely unaware that he and others who questioned Hans were providing the horse with these cues. It was only after Pfungst happened to notice these almost imperceptible movements on the part of other questioners that he became aware that he too made these slight movements when he questioned Hans. Having discovered the principle by which Hans was obtaining his answers, Pfungst conducted a number of experiments in which another person played the role of questioner and he, Pfungst, played the role of the horse.

Pfungst had questioners concentrate on some number between one and ten or even some larger number. Then he would begin to tap out the answer using his right hand and would continue to tap until he believed he had perceived a final signal. He tested twenty-five different individuals of all ages and sex and differing in nationality and occupation. None of them was aware of the purpose of the experiment. Pfungst found that only in a few isolated instances were the questioners aware of any movements on their part. He also found that, with the exception of two individuals, they all made the same involuntary movements, the most important being a slight upward inclination of the head when Pfungst had tapped the correct number of times.

A book by Rosenthal (1966) provides an excellent survey of the various ways in which knowledge or expectancies of an experimenter *may* influence the outcome or results of an experiment. See also Rosenthal (1967).

In a well-designed experiment, the various factors that may influence the outcome of the experiment and that are not of themselves of interest must be controlled if sound conclusions are to be drawn concerning the results of the experiment. It is to be emphasized that these conclusions are derived from the structure of the experiment and the nature of the controls exercised. They do not come from the test of the null hypothesis. The statistical test of a null hypothesis indicates only the probability of a particular result given that the null hypothesis is true. If the experimenter rejects the null hypothesis, he must still examine the structure of his experiment and the nature of his experimental controls in making whatever explanation he does make as to *why* he obtained the particular outcome he did.

2.14 The Importance of Randomization

An essential notion in evaluating the outcome of the experiment described is that of randomness. We may recall that the experimenter mentioned that the selection of the five cans to be filled with water and the five to be left empty was determined on a random basis. The randomization, in this instance, served two functions. For one thing, because the randomization was done by the assistant, the experimenter (who made the observations of the farmer's choices) was ignorant as to which of the cans contained water and which were empty. Thus, we have some assurance that the experimenter himself could not provide cues that would assist the farmer in making his choices.

Randomization also ensures that the particular probability model used in evaluating the outcome of the experiment is applicable. Suppose, for example, that some slight but perceptible differences existed in the cans such that five of the cans had a slight dent. If the assistant had systematically, but not necessarily consciously, filled either the dented or the undented cans, the farmer may have reacted to this cue and used it as a basis for his choices. Randomization offers some assurance that a given characteristic of the cans will not, in turn, be associated with the presence or absence of water in the cans.

Similarly, in assigning the numbered sections of plywood to the cans, randomization is necessary. We would not want all of the even- or all of the odd-numbered sections of plywood to be assigned to the cans containing water. Randomization at this stage is necessary in order to ensure that there is no systematic association between the characteristics of the pieces of plywood and the numbers on them, on the one hand, and the presence and absence of water in the cans, on the other hand.

If the assistant had made a systematic division of the cans into two sets, he may have done so on the assumption that the manner in which the division was made could not in any way influence the selections made by the farmer in the course of the experiment. This assumption may, of course, be true, but it remains an assumption, and it may be difficult to convince others that it is true. The only convincing argument is that of appropriate randomization.

2.15 A Variation in Design

We consider a possible variation in the experimental procedure. Suppose that the ten cans are arranged at random into five pairs. One member of each pair is filled with water and the one to be filled is again determined at random. The farmer is told that he will be presented with five pairs of cans, one of which is filled with water and one of which is empty, and that he is to select the member of each pair that he believes contains water. What are the possible outcomes of this experiment?

There are two ways in which the farmer's first choice may be made and, independently of this choice, the second choice may be made in two ways; and, independently of this choice, there are two ways in which the third choice may be made; and so on for the five choices. Thus, there are a total of $2^5 = 32$ possible outcomes of this experiment.

If the farmer does not have the ability to detect the presence of water with his whalebone, then he should be just as likely to select the empty can as the water-filled can each time he has to make a choice. Let W be the selection of a water-filled can or a correct choice. Then $P(W) = 1/2$ for each of the five pairs. Let T be the number of W's in a sample of $n = 5$ choices and, obviously, T can take the possible values of 0, 1, 2, 3, 4, and 5.

There is only one way in which T can be equal to 5 and in this case the farmer would have to be correct on each of his five choices. Then the probability of $WWWWW$ will be equal to $(1/2)^5 = 1/32$. We also have as the probability of $WWWWD$, that is, the probability of the first four choices being correct and the last one wrong, $(1/2)^5 = 1/32$. But we note that, according to (2.4), we can permute the four W's and one D in $5!/4!1! = 5$ ways and the probability of each of these is $1/32$. Then $P(T = 4) = 5/32$. We also have $5!/3!2! = 10$ ways in which we can obtain $T = 3$ and $P(T = 3) = 10/32$. Similarly, we find that $P(T = 2) = 10/32$, $P(T = 1) = 5/32$, and $P(T = 0) = 1/32$. We note that the values of P assigned to the values of T satisfy the axioms stated previously for probabilities.

With this experimental design we find that $P(T = 5)$ is $1/32$ or approximately 0.031, whereas with the previously described design $P(T = 5)$ is approximately 0.004. Therefore, we should feel much more confident in rejecting the null hypothesis with the first experiment than with the second, if the outcomes of both experiments are $T = 5$.

Both experiments provide relatively little protection against a Type

II error and, as we have previously stated, one way in which we could increase the power of the tests of significance would be to increase the number of observations. The difficulty is that as we increase the number of observations, the methods described in this chapter for obtaining the probabilities associated with each possible outcome of the experiment become exceedingly laborious. In the next chapter, we shall describe methods that can be used to approximate the probabilities associated with the outcomes of the two types of experiments.

QUESTIONS AND PROBLEMS

2.1 A rat is placed upon a Lashley-type jumping stand and through a series of trials is trained to jump always to the smaller of two squares. The right and left positions of the smaller square are randomly alternated so that the experimenter has some confidence that the rat is not reacting to a position variable. The experimenter is interested in determining whether the established reaction pattern will be generalized to the extent that the rat will react similarly to the smaller of two circles. After the rat has learned to discriminate between the two squares, it is given a series of eight trials with two circles. The position of the smaller of the two circles is randomly alternated. We make the assumption that if generalization of the previous learning is not present, the rat will react to the two circles on the basis of chance. On the other hand, if the rat jumps to the smaller circle with a frequency greater than we are willing to attribute to chance, this hypothesis will be ruled out and we shall infer that generalization has taken place. (*a*) What is the probability of seven or more jumps to the smaller circle in eight trials if the null hypothesis is true? (*b*) If the number of trials is increased to twelve, what is the probability of ten or more jumps to the smaller circle if the null hypothesis is true?

2.2 It is claimed that infants stimulated by a loud sound show a response pattern that is differentiated from the pattern of response present when movements are restrained. Response to the loud sound is said to be that of "fear" and response to restraint is said to be that of "rage." An infant is stimulated four times by loud sound and four times by restraint of movement, and motion pictures are taken of the responses immediately after stimulation. Photographs are made from the film and printed in strips. There is a total of eight strips, four showing reaction to sound and the other four to restraint. This is explained to subjects who are to serve as judges. They are asked to select the set of four showing "fear." The questions that follow are related to the evaluation of the possible outcomes of the experiment, under the hypothesis that a correct selection is a matter of chance. (*a*) What is the probability of a single subject selecting the set of four correct photographs from the eight? (*b*) What is the probability of selecting a set of three correct and one wrong? (*c*) What is the probability of selecting a set of two correct and two wrong? (*d*) Suppose that the experiment had made use of twelve photographs, six of rage and six of fear. What are the possible outcomes of a subject's judgments and what is the probability of each?

2.3 An experimenter has a set of four cards, three of which are blank and one of which has an X printed on it. The cards are shuffled and placed face down in a row. The subject is to determine the position of the card with the X on it. (*a*) What is the probability that the subject will make a correct selection in a single trial, assuming that he is reacting by chance? (*b*) If the subject is given four trials, what is the probability that he will make precisely three correct choices by chance? (*c*) What is the probability that he will make precisely one right choice in three trials? (*d*) If there are 128 subjects who serve in the experiment and each subject is given three trials, then how many subjects would be expected to obtain perfect scores by chance alone? (*e*) How many of the subjects would be expected to obtain scores of two or more correct by chance?

2.4 An investigator has given a battery of fifteen tests to a group of students. If he should be interested in the relationship between each test with every other test, how many such relationships would he have to study?

2.5 A student claims that he can differentiate his own brand of cigarettes from three other popular brands. Outline an experiment which would provide evidence with respect to this claim. What kinds of experimental controls might be necessary in the experiment?

2.6 Outline an experiment in which one might test a student's claim that he can discriminate Beer A from Beer B. What experimental controls may be necessary? What role does randomization play in the experimental design? What outcomes of the experiment would be regarded as significant?

2.7 Suppose we wish to determine whether orange juice can be distinguished from onion juice and apple juice when visual and olfactory cues have been experimentally controlled. We block the nasal passages of a subject and blindfold him. He is then presented with a set of three test tubes. He is told that one of the test tubes contains onion juice, one orange juice, and one apple juice and that he is to pick out the one which he thinks contains the orange juice. Fifteen sets of three test tubes are presented to the subject. (*a*) What is the probability that he will make nine or more correct choices, if he is responding by chance? (*b*) What are some of the experimental controls that should be considered in planning this experiment? For example, what about the temperature of the juices?

2.8 In many experiments, the dependent variable is a rating assigned to a subject or an object by a judge. For example, judges may be asked to rate the improvement of patients in a mental hospital after they have been treated with a drug or after they have had several months of psychotherapy.

In taste laboratories, judges may be asked to rate the quality of foods that have been variously prepared. Discuss the nature of the experimental controls necessary in such studies. Discuss the role of randomization as a device for concealing from the judges which patients have been treated and which have not, and similarly which foods have been prepared in one way and which in another way.

2.9 We have eight disks, identified by the numbers 1, 2, 3, . . ., 8. (*a*) If a random sample of four disks is drawn without replacement, what is the probability that the sample will include Disk 8? Show how you arrive at your answer. (*b*) How many possible different samples of four disks can be drawn from the set of eight? (*c*) What is the probability of obtaining a sample that contains Disks 1, 8, 4, and 2? (*d*) If a sample of five is drawn without replacement, what is the probability of drawing Disks 8, 3, 1, 2, and 5, in that order? (*e*) What is the probability that the sample will contain Disks 8, 3, 1, 2, and 5, if the order is ignored?

2.10 We have eight disks. On four of the disks we have the letter A and on the other four the letter B. If a random sample of $n = 4$ is drawn without replacement, find the probability of obtaining: (*a*) 4 A's; (*b*) 3 A's and 1 B; (*c*) 2 A's and 2 B's; (*d*) 1 A and 3 B's; (*e*) 4 B's.

2.11 Under what conditions can we consider a sample of $n = 10$ observations drawn from some defined population to be a random selection from that population?

2.12 A sample of $n = 3$ is drawn from a set of ten objects, without replacement. The sample is drawn at random. How many different samples can be drawn if order does not count?

2.13 We have six males and four females. A random sample is drawn without replacement in such a way as to ensure that we have three males and two females in the sample. How many different samples can be drawn, if order does not count?

2.14 What is the difference between the number of permutations of N things taken n at a time and the number of combinations of N things taken n at a time?

2.15 What is the difference between a Type I error and a Type II error?

2.16 What is meant by the power of a statistical test?

2.17 If two events are described as being mutually exclusive, what does this mean?

2.18 What is meant by the significance level of a test?

2.19 What is the difference between a finite and an infinite population?

2.20 What is meant by a null hypothesis?

2.21 What is meant by a sample space for an experiment?

CHAPTER 3

APPROXIMATION OF THE PROBABILITIES ASSOCIATED WITH SAMPLING FROM BINOMIAL POPULATIONS

3.1 Introduction

Both of the experiments involving the farmer from Whidbey Island were limited, it was suggested, because if the farmer could do only somewhat better than chance, five observations would not be sufficient to result in a significant outcome. Thus, we might fail to reject the null hypothesis when it is in fact false, thereby making a Type II error. It was also suggested that, other things being equal, we could decrease the probability of a Type II error by increasing the number of observations made. But, as we increase the number of observations, we face the problem of evaluating a larger number of possible outcomes of the experiment, and the methods of the previous chapter (which give the exact probabilities associated with each possible outcome) become exceedingly laborious. It is fortunate that the probabilities associated with random sampling from a binomial population can be *approximated* quite satisfactorily by means of the standard normal distribution.

Because we have already obtained the exact probabilities for samples of $n = 5$ observations drawn from a finite population (the first experiment with the farmer) and from an infinite population (the second experiment with the farmer), we shall use these same two cases to illustrate the approximation method. With the exact probabilities available as a standard, we shall gain some notion of how well the approximation method works in these two specific cases. If they work fairly well for these two specific cases of $n = 5$ observations, we may feel even more confident in applying them to experiments involving a much larger number of observations.

3.2 Binomial Populations and Binomial Variables

Consider a variable X such that X can take only a value of 1 or 0. Assume that there exists a population of observations and that in this population there are F_1 observations with $X = 1$ and F_0 observations with $X = 0$. Then the total number of observations in the population will be $N = F_1 + F_0$. If an observation is randomly selected from this population, then $P(X = 1) = F_1/N$ and $P(X = 0) = F_0/N$ and

$$P(X = 1) + P(X = 0) = \frac{F_1}{N} + \frac{F_0}{N} = 1$$

It will be convenient to let P be the probability that $X = 1$ and $Q = 1 - P$ be the probability that $X = 0$.

A population in which a variable can take only one of two possible values is called a *binomial population* and the variable is called a *binomial variable.* Table 3.1 shows a binomial population in which $P(X = 1) = 0.5$ and $Q = P(X = 0)$ is also 0.5.

3.3 Mean of a Population

We define the mean of a population as

$$\mu = \sum P_i X_i \tag{3.1}$$

where the right-hand side means simply that we multiply each possible value of X by its associated probability and sum the respective products. For a binomial population where X can take only the values of 1 or 0 with corresponding probabilities of P and Q, we have

$$\begin{aligned}\mu &= P(1) + Q(0)\\ &= P\end{aligned} \tag{3.2}$$

and we see that for a binomial population, μ is just the probability associated with $X = 1$. As shown in Table 3.1, $\mu = \Sigma P_i X_i = 0.5$ and this is also the probability associated with $X = 1$.

3.4 Variance and Standard Deviation of a Population

We define the variance of a population as

$$\sigma^2 = \sum P_i(X_i - \mu)^2$$
$$= \sum P_iX_i^2 - 2\mu \sum P_iX_i + \mu^2 \sum P_i$$

Because $\Sigma P_iX_i = \mu$ and $\Sigma P_i = 1$, we have

$$\sigma^2 = \sum P_iX_i^2 - \mu^2 \qquad (3.3)$$

For a binomial population, we have

$$\sum P_iX_i^2 = P(1^2) + Q(0^2) = P$$

and because $\mu = P$, then

$$\sigma^2 = P - P^2$$

or

$$\sigma^2 = PQ \qquad (3.4)$$

The standard deviation is defined as the square root of the population variance and, for a binomial population, we have

$$\sigma = \sqrt{PQ} \qquad (3.5)$$

For the binomial population shown in Table 3.1, we have $P = 0.5$ and $Q = 0.5$ and for this population

$$\sigma^2 = (0.5)(0.5) = 0.25$$

and

$$\sigma = \sqrt{(0.5)(0.5)} = 0.5$$

Table 3.1
A BINOMIAL POPULATION WITH $P(X = 1) = 0.5$ AND $P(X = 0) = 0.5$

(1) X_i	(2) P_i	(3) P_iX_i	(4) $P_iX_i^2$
1	0.5	0.5	0.5
0	0.5	0.0	0.0
Σ	1.0	0.5	0.5

3.5 Expected Values[1]

We let $E(X) = \Sigma P_i X_i$ and $E(X)$ is called the expectation of X and is simply the long-run average value of X. Similarly, we let $E(X^2) = \Sigma P_i X_i^2$ and $E(X^2)$ is simply the long-run average value of X^2. Then

$$\mu = E(X) \tag{3.6}$$

and

$$\sigma^2 = E(X - \mu)^2$$

$$= E(X^2) - \mu^2 \tag{3.7}$$

3.6 Expected Value of a Sum and Variance of a Sum

Now suppose we draw a random sample of n observations from a given population with mean μ and we let T be the sum of the n values of X. Then

$$E(T) = E(X_1 + X_2 + \cdots + X_n)$$

$$= E(X_1) + E(X_2) + \cdots + E(X_n)$$

or

$$\mu_T = n\mu \tag{3.8}$$

For the special case of a binomial population with $\mu = P$, we have

$$E(T) = nP \tag{3.9}$$

If the n observations are independent, then the variance of the sum T will simply be n times the variance of X, and may be expressed as

$$\sigma_T^2 = E(T - \mu_T)^2$$

$$= n\sigma^2 \tag{3.10}$$

For the special case of a binomial population we have

$$\sigma_T^2 = nPQ \tag{3.11}$$

and

$$\sigma_T = \sqrt{nPQ} \tag{3.12}$$

[1] For a review of the algebra of mathematical expectation, see either Edwards (1971) or Edwards (1964).

3.7 *Expected Value of a Mean and Variance of a Mean*

We define the mean of a sample of n observations as

$$\bar{X} = \frac{X_1 + X_2 + \cdots + X_n}{n} = \frac{\sum X}{n}$$

and

$$E(\bar{X}) = \frac{1}{n} E(T)$$

or

$$E(\bar{X}) = \mu \tag{3.13}$$

The average long-run value of the mean of a random sample of n observations is, in other words, equal to the population mean μ. We note that for the special case of a random sample from a binomial population, we have

$$E(\bar{X}) = P \tag{3.14}$$

The variance of a mean of a random sample of n independent observations drawn from a given population will be

$$\sigma_{\bar{X}}^2 = \frac{\sigma^2}{n} \tag{3.15}$$

and for the special case of a binomial population, we have

$$\sigma_{\bar{X}}^2 = \frac{PQ}{n} \tag{3.16}$$

The square root of (3.15) is called the standard error of the mean. Thus, in general,

$$\sigma_{\bar{X}} = \frac{\sigma}{\sqrt{n}} \tag{3.17}$$

and for the special case of a binomial population, we have

$$\sigma_{\bar{X}} = \sqrt{\frac{PQ}{n}} \tag{3.18}$$

In the following sections we apply the formulas we have developed to the two experiments with the farmer described in the previous chapter. We shall see, in doing so, that it is possible to determine quickly and easily an approximation of the probabilities of the outcomes of these experiments.

3.8 *The Second Experiment with the Farmer: μ_T and σ_T*

In the second experiment the farmer was presented with a pair of cans and asked to choose which one contained water. Let X be the number of correct choices on a single trial or presentation of a pair of cans. Then, obviously, we have $X = 1$ if the farmer makes a correct choice, that is, if he selects the can with water, and $X = 0$ if he does not. The farmer was given $n = 5$ pairs and we let T be the sum of X in the $n = 5$ trials; T is just the number of correct choices in a set of $n = 5$.

We note that this experiment corresponds to drawing a random sample of $n = 5$ independent observations from a binomial population in which

Table 3.2

PROBABILITY DISTRIBUTION OF T, THE NUMBER OF CORRECT CHOICES, WHEN $n = 5$ INDEPENDENT OBSERVATIONS ARE DRAWN FROM A BINOMIAL POPULATION WITH $P(X = 1) = 0.5$

(1) T_i	(2) P_i	(3) P_iT_i	(4) $P_iT_i^2$
5	$\frac{1}{32}$	$\frac{5}{32}$	$\frac{25}{32}$
4	$\frac{5}{32}$	$\frac{20}{32}$	$\frac{80}{32}$
3	$\frac{10}{32}$	$\frac{30}{32}$	$\frac{90}{32}$
2	$\frac{10}{32}$	$\frac{20}{32}$	$\frac{40}{32}$
1	$\frac{5}{32}$	$\frac{5}{32}$	$\frac{5}{32}$
0	$\frac{1}{32}$	0	0
Σ	$\frac{32}{32}$	$\frac{80}{32}$	$\frac{240}{32}$

$P = 0.5$. Therefore, we should have

$$\mu_T = nP = 5(0.5) = 2.5$$

and

$$\sigma_T = \sqrt{nPQ} = \sqrt{5(0.5)(0.5)} = 1.118$$

Table 3.2 lists a sample space for the experiment. The possible values of T are given in column (1) and the probabilities P associated with each sample point or value of T are given in column (2). These probabilities are those that we obtained in the previous chapter. We see that

$$\mu_T = \sum P_i T_i = \frac{80}{32} = 2.5$$

and that

$$\begin{aligned} \sigma_T^2 &= \sum P_i T_i^2 - \mu_T^2 \\ &= \frac{240}{32} - (2.5)^2 \\ &= 7.50 - 6.25 \\ &= 1.25 \end{aligned}$$

and that

$$\sigma_T = \sqrt{1.25} = 1.118$$

The important point is that we were able to calculate $\mu_T = nP$ and $\sigma_T = \sqrt{nPQ}$ knowing only the values of n and P. In other words, it was not necessary for us to do all of the calculations involved in finding the probabilities associated with each possible outcome of the experiment and then to do the calculations shown in Table 3.2 in order to calculate μ_T and σ_T. We now show how μ_T and σ_T can be used in making a test of significance.

3.9 The Standard Normal Distribution

Figure 3.1 shows the probability distribution of T. We shall assume that this distribution can be reasonably approximated by means of a continuous normal distribution with the same mean $\mu = 2.5$ and the same standard deviation $\sigma = 1.118$. The normal distribution is a theoretical distribution and, because it refers to a continuous variable, can be represented by a curve. The equation of this curve for the *standard normal distribution* is

$$y = \frac{1}{\sqrt{2\pi}} e^{-(1/2)Z^2} \qquad (3.19)$$

where y = the height of the curve at any given point along the base line;

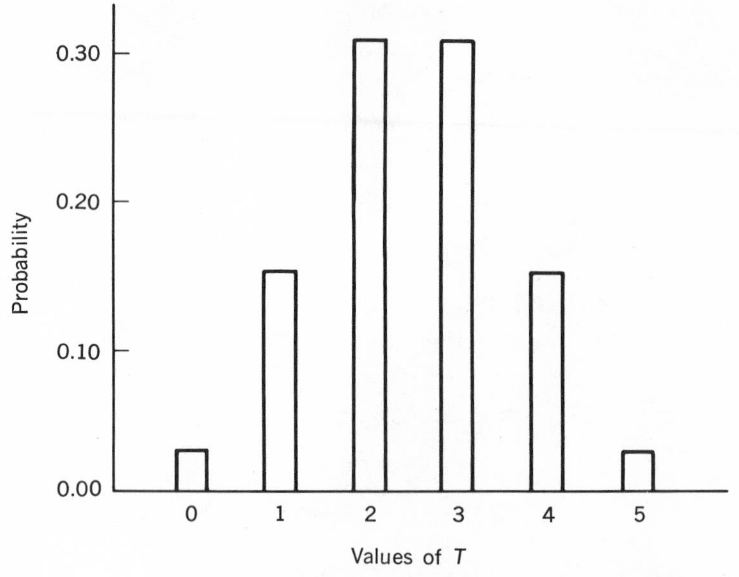

Figure 3.1
The probability distribution of T for random samples of $n = 5$ independent observations drawn from a binomial population in which $P(X = 1) = 0.5$.

$\pi = 3.1416$ (rounded), the ratio of the circumference of a circle to its diameter; $e = 2.7183$ (rounded), the base of the natural system of logarithms; and $Z = (X - \mu)/\sigma$. The area under the curve defined by (3.19) is equal to 1.00.

Table III in the Appendix is a table of the standard normal distribution. The first column gives values of $Z = (X - \mu)/\sigma$ in units of 0.01. The second column gives the proportion of the total area under the curve between the ordinates erected at $\mu_Z = 0$ and the value of Z given in the first column. The third column gives the proportion of the total area in the larger segment of the curve when an ordinate is erected at Z, and the fourth column gives the proportion of the total area in the smaller segment.

Any normally distributed variable can be transformed into a standard normal variable. A *standard normal variable* is defined by

$$Z = \frac{X - \mu}{\sigma} \tag{3.20}$$

Then

$$E(Z) = \frac{1}{\sigma} E(X - \mu)$$

or

$$\mu_Z = 0 \tag{3.21}$$

and because $\mu_Z = 0$, we have

$$\sigma_Z^2 = E(Z^2)$$

$$= \frac{1}{\sigma^2} E(X - \mu)^2$$

or

$$\sigma_Z^2 = 1 \tag{3.22}$$

A standard normal variable is a normally distributed variable with mean equal to 0 and variance and standard deviation equal to 1.

3.10 The Second Experiment with the Farmer: A Normal Curve Test

Assuming, for the moment, that T is a continuous and normally distributed variable with $\mu_T = 2.5$ and $\sigma_T = 1.118$, then

$$Z = \frac{T - \mu_T}{\sigma_T} \tag{3.23}$$

will be a standard normal variable. Suppose that the outcome of the experiment with the farmer is $T = 5$. We know from previous calculations that the probability associated with this outcome is approximately 0.031. We want to use the table of the standard normal distribution to approximate this probability. Because T is discrete and because the lower limit of $T = 5$ is 4.5, let us find the area falling to the right of 4.5. We have

$$Z = \frac{4.5 - 2.5}{1.118} = 1.79$$

and entering the table of the standard normal curve we find that 0.037 of the total area falls to the right of 1.79. In other words, $P(Z \geq 1.79) = 0.037$ and this is also the probability of $T \geq 4.5$ if T is a continuous and normally distributed variable with $\mu_T = 2.5$ and $\sigma_T = 1.118$.

We see that despite the fact that T in our experiment is discrete and is not actually normally distributed, the probability of 0.037 we have obtained approximates quite well the known probability of 0.031.

We have used the normal curve test with a relatively small value for n. It can be said that this approximate test improves as n increases. And obviously the methods of the last chapter become more laborious as n increases. We suggest the following rule for using the normal curve test: If both nP and nQ are at least equal to or greater than five, then, for all practical purposes, the normal curve test may be substituted for the exact test. This rule sets a *minimum* standard and, of course, we would be better off if both nP and nQ were considerably larger than five, not only because of the increased accuracy of the normal curve test, but also because of the increased power of the test with a larger n.

3.11 The First Experiment with the Farmer: A Normal Curve Test

We note that in the experiment where the farmer is presented with a pair of cans, the probability of a correct choice remained the same and equal to ½ for each successive pair. The model for this experiment is one that involves sampling from an infinite population or, what amounts to the same thing, sampling from a finite population with replacement after each draw. If we have a finite population and sample with replacement after each draw, the probabilities associated with the values of X will remain the same on each successive draw. But, in the experiment where the farmer was to select five cans from ten, the probabilities do not remain the same for each of his successive choices. Instead, the probability that the farmer's second choice was correct depended on whether his first choice was correct. For example, if his first choice was correct, then the probability of his second choice being correct was not ½ but 4⁄9. On the other hand, if his first choice was incorrect, then the probability that his second choice would be correct was 5⁄9. For all subsequent choices, the probability of a correct choice depends on the nature of the previous choices. In our model for this experiment, we said that we would regard the sampling procedure as one involving drawing a random sample of $n = 5$ from a finite population of $N = 10$, without replacement after each draw.

With this sampling procedure, we still have $E(T) = nP$ but we need to make a *finite population correction factor* in obtaining σ_T^2. The finite population correction factor is equal to $(N - n)/(N - 1)$ and the variance of T is then

$$\sigma_T^2 = \frac{N - n}{N - 1} nPQ \tag{3.24}$$

and

$$\sigma_T = \sqrt{\frac{N - n}{N - 1} nPQ} \tag{3.25}$$

With $N = 10$, $n = 5$, and with $P = 0.5$, we have

$$\sigma_T^2 = \left(\frac{10 - 5}{10 - 1}\right)(5)(0.5)(0.5) = 0.6944$$

and

$$\sigma_T = \sqrt{0.6944} = 0.833$$

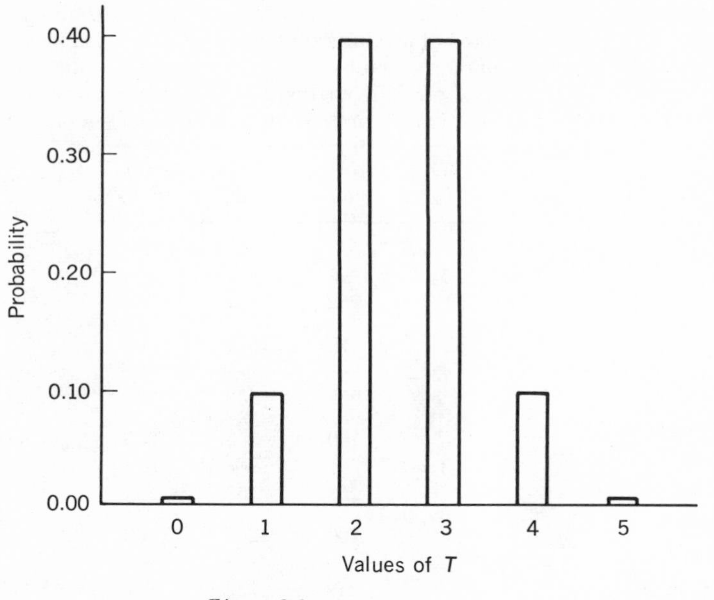

Figure 3.2
The probability distribution of T for random samples of $n = 5$ observations drawn without replacement from a binomial population of $N = 10$ in which $P(X = 1) = 0.5$.

Figure 3.2 shows the probability distribution of T for this experiment and Table 3.3 gives a sample space for the experiment. In column (1) we have the possible values of T and in column (2) we give the probabilities associated with each value of T. These probabilities are those we obtained in the previous chapter. Column (3) gives the values of P_iT_i and we note that $\mu_T = \Sigma P_iT_i = 630/252 = 2.5$. In column (4) we give the squared values of T multiplied by the corresponding probabilities. Then

$$\sigma_T{}^2 = \sum P_iT_i{}^2 - \mu_T{}^2$$
$$= \frac{1750}{252} - (2.5)^2$$
$$= 0.6944$$

which is equal to the value we obtained using (3.24). We also have

$$\sigma_T = \sqrt{0.6944} = 0.833$$

as before.

We have already found that for this experiment $P(T = 5)$ is approximately 0.004. Taking the lower limit of $T = 5$ as 4.5 and with $\sigma_T = 0.833$ and $\mu_T = 2.5$, we have, as a normal curve test,

$$Z = \frac{4.5 - 2.5}{0.833} = 2.40$$

Table 3.3

PROBABILITY DISTRIBUTION OF T, THE NUMBER OF CORRECT CHOICES, WHEN $n = 5$ OBSERVATIONS ARE DRAWN AT RANDOM AND WITHOUT REPLACEMENT FROM A BINOMIAL POPULATION OF $N = 10$ AND $P(X = 1) = 0.5$

(1) T_i	(2) P_i	(3) P_iT_i	(4) $P_iT_i^2$
5	$\frac{1}{252}$	$\frac{5}{252}$	$\frac{25}{252}$
4	$\frac{25}{252}$	$\frac{100}{252}$	$\frac{400}{252}$
3	$\frac{100}{252}$	$\frac{300}{252}$	$\frac{900}{252}$
2	$\frac{100}{252}$	$\frac{200}{252}$	$\frac{400}{252}$
1	$\frac{25}{252}$	$\frac{25}{252}$	$\frac{25}{252}$
0	$\frac{1}{252}$	0	0
Σ	$\frac{252}{252}$	$\frac{630}{252}$	$\frac{1750}{252}$

and from the table of the standard normal curve we find that $P(Z \geq 2.40)$ is approximately 0.008. This probability also is a not too inaccurate approximation of the probability of 0.004 that we obtained for the exact test.

The variance of T for the finite case differs from the variance of T for the infinite case only by the factor $(N - n)/(N - 1)$. It is clear that if n, the number of observations, is held constant, then, as N becomes indefinitely large, the correction factor approaches the limiting value of 1.00. In general, it can be said that if the ratio of n/N is less than $\frac{1}{5}$, then for all practical purposes we may neglect the correction factor in calculating σ_T.

3.12 Applications of the Models

The outcomes of many research problems can be evaluated in terms of a model that involves random sampling from a binomial population, either finite or infinite. In general, the experimental design is one in which there are only two possible responses that a subject may make in responding to some situation or experimental condition.

We might, for example, be interested in determining whether a subject, when presented with fresh and frozen orange juice, can correctly discriminate or select the fresh juice. Or can a subject correctly identify a given brand of cigarettes when presented with several brands? Can a subject distinguish between handwriting specimens of males and females, or correctly judge which of two tones is louder?

If a rat is placed in a maze and can turn either right or left at a choice point, is it plausible to assume that a right and left turn are equally probable? Or is it true that right turns occur with a frequency significantly greater than chance?

In a psychophysical experiment, we may be interested in determining by how much one weight must differ from a standard weight before a subject can detect significantly better than by chance that the weight is heavier than the standard. Or we may wish to find out how salty a solution must be before it can be discriminated significantly better than chance from a plain solution. Additional applications will occur to the reader.

It should be emphasized that the methods described in this chapter are perfectly general and do not require that $P(X = 1) = 1/2$, as in the two experiments with the farmer. We may have a binomial population in which $P(X = 1)$ is $1/3$, or $1/4$, or any other value. The first experiment with the farmer, for example, might have been modified by having four wet cans and six dry cans. The task set for the farmer would then be to select a set of four cans from the complete set of ten. In this case, P would be 0.4 rather than 0.5. Similarly, in the second experiment, we might have arranged the cans in triplets, rather than in pairs, so that each triplet contained one wet can and two dry cans. In this case, we would have $P = 1/3$ rather than $P = 1/2$.

Knowing that we have an experiment that involves random sampling from a binomial population with known or assumed value of P, then, as we have seen, we can quickly and easily determine μ_T and σ_T and evaluate, by means of a normal curve test, the probability of T equal to or greater than the value obtained in a given experiment.

It should also be emphasized that one of the reasons why the standard normal curve approximations are as good as they are in the two experiments to which they were applied is because in random sampling from a binomial population in which $P = Q$, the probability distribution of T is symmetrical. However, if $P \neq Q$, then the probability distribution of T will not be symmetrical but skewed. If P is larger than Q, the distribution of T will have a tail to the left; if P is smaller than Q, the distribution will have a tail to the right. As n, the number of observations, is increased the distribution of T becomes more symmetrical, even when P is not equal to Q. The rule suggested previously, that both nP and nQ be equal to or greater than five, provides for the possibility that P may be larger than or smaller than Q and that the distribution of T will be skewed when n is also small. For example, if $P = 0.8$, the rule suggests a *minimum* of twenty-five observations, if the normal curve test is to be applied in evaluating the outcome of the experiment.

3.13 Experiments Involving the Difference between Two Proportions

In many experiments our interest is in the difference between two proportions. For example, a group of rats may be divided at random into two groups of n_1 and n_2 rats. One group may then be given a series of training trials in a maze under one set of experimental conditions. The other group is also given a comparable training period but under a different set of experimental conditions. On the basis of a particular learning theory, the rats in the first group should, when placed in a new maze with two paths to the goal box, select one of the two paths more frequently than the rats in the second group. The outcome of the experiment is such that the proportion of rats selecting the critical path in the first group is larger than the proportion selecting the path in the second group, but the experimenter wants to know whether the difference between the two proportions is statistically significant.

To illustrate the methods involved in evaluating the outcomes of experiments such as the one described above, let us assume that we have randomly divided 80 rats into two groups with $n_1 = 50$ in one group and $n_2 = 30$ in the other.[2] Table 3.4 gives the number of rats in each group that use Path 1 and Path 2 on a test trial in going from the starting box to the goal box.

Let T_1 be the number of rats in Group 1 using Path 1, the critical path,

Table 3.4
NUMBER OF RATS IN EACH OF TWO GROUPS CHOOSING PATH 1 AND PATH 2 TO A GOAL BOX

	Path 2	Path 1	Total
Group 1	8	42	50
Group 2	12	18	30
Total	20	60	80

[2] In general, in comparing a difference between two groups, we are better off with an equal number of observations for each group. We have deliberately made the n's unequal in this example in order to illustrate a more general application of the methods of analysis.

and let T_2 be the corresponding number in Group 2. The difference between T_1 and T_2 is not meaningful because the two values of T are based on differing numbers of observations. We can, however, divide the two values of T by the corresponding n's to obtain the *mean* number of rats in each group using Path 1. These two means, as shown previously, are just the proportion of the number of rats in each group using Path 1. Then

$$p_1 = \bar{X}_1 = \frac{42}{50} = 0.84$$

and

$$p_2 = \bar{X}_2 = \frac{18}{30} = 0.60$$

and we wish to determine whether p_1 and p_2 differ significantly.

3.14 Standard Error of the Difference between Two Proportions

Let us assume, as a null hypothesis to be tested, that the two sample sets of observations have been drawn at random from a common binomial population in which the probability of Path 1 being used is $P = 60/80 = 0.75$. Then Q, the probability of Path 1 not being used, will be $1 - P = 0.25$.

If we have two independent and random samples of n_1 and n_2 observations, drawn from two populations with variances σ_1^2 and σ_2^2, then the standard error of the difference between the two sample means will be

$$\sigma_{\bar{X}_1 - \bar{X}_2} = \sqrt{\sigma_{\bar{X}_1}^2 + \sigma_{\bar{X}_2}^2} \tag{3.26}$$

We have previously stated that the variance of a sample mean is σ^2/n and, therefore,

$$\sigma_{\bar{X}_1 - \bar{X}_2} = \sqrt{\frac{\sigma_1^2}{n_1} + \frac{\sigma_2^2}{n_2}} \tag{3.27}$$

But $\bar{X}_1 = p_1$ and $\bar{X}_2 = p_2$, and for a binomial population, $\sigma^2 = PQ$. Then

$$\sigma_{p_1 - p_2} = \sqrt{\frac{P_1Q_1}{n_1} + \frac{P_2Q_2}{n_2}} \tag{3.28}$$

and because we have assumed that the two samples are from a common binomial population so that $P_1 = P_2$, the numerators of (3.28) are equal and we can drop the subscripts. Then

$$\sigma_{p_1 - p_2} = \sqrt{PQ\left(\frac{1}{n_1} + \frac{1}{n_2}\right)} \tag{3.29}$$

and (3.29) is the standard error of the difference between two proportions when we have two independent random samples drawn from a common binomial population.

3.15 The Normal Curve Test

Assuming that p_1 and p_2 are both normally distributed, then the difference, $p_1 - p_2$, will also be normally distributed. Then

$$Z = \frac{p_1 - p_2 - (P_1 - P_2)}{\sigma_{p_1 - p_2}} \tag{3.30}$$

will be a standard normal variable with mean equal to 0 and standard deviation equal to 1 and can be evaluated in terms of the table of the standard normal distribution.

For the data of Table 3.4, we have

$$\sigma_{p_1 - p_2} = \sqrt{(0.75)(0.25)\left(\frac{1}{50} + \frac{1}{30}\right)} = 0.10$$

Because the null hypothesis specifies that the two samples are from a common binomial population so that $P_1 = P_2$, then as a test of significance we have

$$Z = \frac{0.84 - 0.60}{0.10} = 2.4$$

By reference to the table of the standard normal distribution, we find that 0.0082 of the total area will fall to the right of an ordinate at $Z \geq 2.4$. Thus, if the null hypothesis is true, $P(Z \geq 2.4) = 0.0082$ and, therefore, $P(p_1 - p_2 \geq 0.24) = 0.0082$. Only about 82 times in 10,000 would we expect to obtain a difference between p_1 and p_2 as large as or larger than the one observed in the *same direction* ($p_1 > p_2$) as a result of random sampling when the null hypothesis we have tested is true.

3.16 Correction for Discreteness

In evaluating, by means of the standard normal distribution, the probability of T equal to or greater than some specified value, we made a correction that took into account the fact that T was a discrete variable. The correction was made by using the lower limit of T in the test of significance. In testing the significance of the difference between two proportions by means of the standard normal distribution we can also introduce a correction for the discreteness of T_1 and T_2. The correction is made in such a way as to reduce the difference between the two sample values of p; that is, for the sample with the larger value of p, we subtract 0.5 from T and for the sample with the smaller value of p, we add 0.5 to T. In the present example, making these corrections we have

$$p_1 = \frac{42.0 - 0.5}{50} = 0.8300 \quad \text{and} \quad p_2 = \frac{18.0 + 0.5}{30} = 0.6167$$

The difference between p_1 and p_2 will now be $0.8300 - 0.6167 = 0.2133$. Dividing this difference by $\sigma_{p_1-p_2} = 0.10$, we have $Z = 0.2133/0.10 = 2.133$ and from the table of the standard normal distribution we find $P(Z \geq 2.133)$ is approximately 0.017. Correcting the two values of T for discreteness will always result in a smaller value for Z than if the uncorrected values are used in finding p_1 and p_2.

To many individuals, the probability of $p_1 - p_2 \geq 0.24$, if it is in fact true that $P_1 = P_2$, is sufficiently small that they would reject the null hypothesis and, in doing so, would conclude that $P_1 > P_2$. In other words, the outcome of the experiment is such that it seems reasonable to believe that for rats trained under the same conditions as those involved in the experiment, the probability of a rat using Path 1 on the critical test trial is greater under Treatment 1 than under Treatment 2.

3.17 One- and Two-Sided Tests

If we have no prior hypothesis about the direction of the difference between p_1 and p_2, then our test of significance should take into account the probability of a positive difference ($p_1 > p_2$) and also the probability of a negative difference ($p_1 < p_2$). A *directional* test, such as the one we made, is often referred to as a *one-sided* or *one-tailed* test because only one tail or one side of the standard normal distribution is used in evaluating the outcome of the experiment. A *nondirectional* test is often referred to as a *two-tailed* or *two-sided* test. The nature of these two tests will be discussed in greater detail later. For the present, it is sufficient to point out that the probability for a two-sided test will be two times the probability for the one-sided test. In the present experiment, the probability associated with the two-sided test[3] will be (2) (0.0082) = 0.0164. The two-sided test takes into account the probability of obtaining either a positive or negative difference between p_1 and p_2.

[3] The probability cited is for the two-sided test without a correction for discreteness. With a correction for discreteness the probability for the two-sided test is (2)(0.017) = 0.034.

3.18 Suggested Rule for Using the Normal Curve Test

In using the standard normal distribution to obtain an approximation of the probabilities associated with the outcomes of an experiment involving a single random sample from a binomial population, we suggested that both nP and nQ should be at least equal to five. A similar rule may be stated for the use of the standard normal distribution in testing the difference between two proportions. The standard normal distribution will provide fairly good approximations to the exact probabilities if for both samples we have n_1P, n_2P, n_1Q, and n_2Q equal to or greater than five.

3.19 Testing the Same Subjects Twice

In some experiments we may have the same group of subjects tested twice. Suppose, for example, we have $n = 200$ subjects and they have been tested for the presence of a critical response before (Test 1) and after (Test 2) experiencing a set of experimental conditions. For each subject we will have a pair of observations. If the critical response is made by a subject, we shall call this a success (S), and if it is not, we shall call this a failure (F). On the first test we may have either S_1 or F_1 for each subject, and either of these two may be followed by S_2 or F_2 on the second test. We thus have $2 \times 2 = 4$ possible patterns: S_1F_2, S_1S_2, F_1F_2, and F_1S_2.

Let us assume that the outcome of the experiment is such that the frequencies with which each of these patterns occur are: 20, 80, 60, and 40, respectively. These frequencies are given in Table 3.5. Table 3.6 gives a schematic representation of the numbers in Table 3.5 and, in terms of the notation of Table 3.6, the proportion of successes on Test 1 is

$$p_1 = \frac{n_1}{n} = \frac{a + b}{n}$$

Table 3.5

FREQUENCY OF FAILURES AND SUCCESSES FOR 200 SUBJECTS TESTED BEFORE (TEST 1) AND AFTER (TEST 2) EXPERIENCING AN EXPERIMENTAL CONDITION

		Test 2		
		Failure	Success	Total
Test 1	Success	20	80	100
	Failure	60	40	100
	Total	80	120	200

Table 3.6

SCHEMATIC REPRESENTATION OF THE FREQUENCY OF FAILURES AND SUCCESSES WHEN n SUBJECTS ARE TESTED TWICE

		Test 2		
		Failure	Success	Total
Test 1	Success	a	b	n_1
	Failure	c	d	$n - n_1$
	Total	$n - n_2$	n_2	n

and the proportion of successes on Test 2 is

$$p_2 = \frac{n_2}{n} = \frac{b + d}{n}$$

Then the difference between p_1 and p_2 will be

$$p_1 - p_2 = \frac{a + b}{n} - \frac{b + d}{n} = \frac{a - d}{n}$$

Therefore, the test of significance is concerned only with the distribution of the frequencies for patterns S_1F_2 and F_1S_2 or the frequencies in cells a and d of Table 3.6.

3.20 The Test of Significance[4]

Under the null hypothesis that the probability of a success on Test 2 is equal to the probability of a success on Test 1,

$$E(p_1 - p_2) = \frac{1}{n} E(a - d) = 0$$

and the number of observations in cell a should be equal to the number in cell d. We may regard the set of $a + d$ observations as a random sample from a binomial population in which the probability of obtaining a is equal to the probability of obtaining d. Then the expected number of d's in a random sample of $a + d$ observations will be

$$E(d) = (a + d)P \qquad (3.31)$$

where $P = 1/2$. The standard deviation of d will be given by

$$\sigma = \sqrt{(a + d)PQ} \qquad (3.32)$$

[4] The test of significance presented in this section has been described by McNemar (1947) as a test for the difference between two correlated proportions. For the case of three or more correlated proportions, see Cochran (1950).

If d is approximately normally distributed, then

$$Z = \frac{d - (a + d)P}{\sqrt{(a + d)PQ}} \tag{3.33}$$

will be a standard normal variable and can be evaluated in terms of the standard normal distribution. Because $P = Q = \frac{1}{2}$, we have

$$Z = \frac{d - (a + d)\frac{1}{2}}{\frac{1}{2}\sqrt{a + d}}$$

$$= \frac{d - a}{\sqrt{d + a}} \tag{3.34}$$

With a correction for discontinuity, (3.34) becomes

$$Z = \frac{|d - a| - 1}{\sqrt{d + a}} \tag{3.35}$$

For the data of Table 3.5, we have $d = 40$ and $a = 20$. Then

$$Z = \frac{|40 - 20| - 1}{\sqrt{40 + 20}} = \frac{19}{7.746} = 2.45$$

By reference to the standard normal distribution we find that $P(Z \geq 2.45)$ is 0.0071. For a two-sided test, the probability is $(2)(0.0071) = 0.0142$.

QUESTIONS AND PROBLEMS

3.1 In a taste discrimination experiment a subject is presented with two brands of frozen orange juice and one of fresh orange juice. His task is to select the fresh orange juice in each presentation of the three samples. He is given fifteen trials and correctly selects the fresh orange juice in nine of the fifteen trials. Is the hypothesis tenable that he is responding by chance?

3.2 In a testing center it has been determined that the average test scorer is in error on 6 percent of the papers scored. A new employee scores 500 papers during a given day, and it is found that forty of his papers are scored incorrectly. We may regard the 500 papers that are scored as consisting of 500 trials of an event for which the probability of making an error in a single trial is 0.06. Has the new employee made a significantly large number of errors?

3.3 A subject is trained to push a key when the first of two tones that he hears is of greater intensity. The difference threshold for the subject is determined. In a new series of trials two tones are sounded that differ in intensity, but for which the difference is below the threshold for the subject. The louder tone is randomly alternated with the weaker tone so that it sometimes appears first and sometimes second. The subject claims that he is unable to distinguish between the two tones—that he would have to guess. The experimenter tells him to go ahead and guess, and the subject does so for a series of thirty trials. If his judgment is a guess, then on each trial we may regard the probability of a correct guess as $\frac{1}{2}$. What is the probability of twenty-one or more correct, if the null hypothesis is true?

3.4 A rat is trained to respond to the larger of two squares. The rat is now given a series of forty trials with two circles differing in size. If we assume that the rat will respond to the two circles by chance, that is, that no preference will be shown for the larger of the two circles, then what is the probability of twenty-six or more responses to the larger of the two circles? Assume that the null hypothesis is rejected. Can this finding be interpreted as evidence of generalization from the previous training? Could it also be interpreted as evidence of learning in the present situation? Describe modifications of the experiment such that it would be possible to distinguish between the effects of learning in the present situation and the effects of generalization from the previous situation.

3.5 A child is presented with three boxes, of which two are of the same color and one is of a different color. Without the child's knowledge, candy is placed under the box that is of the odd color. He is then allowed to lift the boxes until he discovers the one that has the candy under it. The situation is now changed by using boxes of the same color but with two of the same size and one that is of a different size. Let us assume that there is no transfer of training from his experience with the colored boxes to the boxes differing in size. The candy will always be placed under the box that differs in size from the other two and we shall assume that the probability of selecting this box is $\frac{1}{3}$. What is the probability, in a series of eighteen trials, that the child correctly selects the box with the candy under it ten or more times? Suppose a significant result is obtained. Would it be just as logical to attribute this finding to learning in the present situation as to transfer from the previous situation? Describe modifications of the experiment such that it is possible to differentiate between learning in the present situation and transfer of training from the previous situation.

3.6 In a large midwestern university it is known that 62 percent of the students are registered in the college of liberal arts. The campus daily draws a sample of 200 students for a public opinion poll and finds that in the sample there are 136 liberal arts students. If the sampling is random, how frequently would samples with 136 or more liberal arts students be expected by chance when the sample size is 200?

3.7 A child is seated at a table across from the experimenter. The child is shown a desired object and this is placed at the end of the table at the child's right. Directly in front of the child across the table is a spot marked by an X. The child is blindfolded and asked to move a disk toward the spot. Assume that the probability of moving the disk to the right of the spot is equal to the probability of an error to the left. In a series of twenty trials, the child makes fourteen right errors and six left errors. Does the child show a significant bias toward the position of the desired object? Would you wish to conclude, from the outcome alone, that the excess of right errors is the result of placing the desired object on the child's right? If so, then how would you answer the argument that the child might show a right-error bias if the desired object had been placed at the left? Describe modifications of this experiment such that the results can be interpreted as evidence of the desired object influencing the direction of the error.

3.8 Subjects are randomly assigned to two treatments A and B. After experiencing the treatments, both groups are given a critical test. A particular response is under investigation. In Group A, twenty-four out of sixty subjects make the response and in Group B, eight out of forty make the response. Can we conclude that these two proportions differ significantly?

3.9 Fifty-five subjects are given two shades of blue that differ only slightly with respect to saturation. They are asked to select the shade with the greater saturation and thirty-six of them make the correct selection. The same subjects are then retested with two shades of red that differ only slightly with respect to saturation. On this test it is found that thirty-one make the correct selection and that twenty-four of the thirty-one are also included in the thirty-six who made the correct selection in the test with blue. Can we conclude that the two proportions differ significantly?

3.10 The "Zeigarnik effect," which is concerned with the relative degree of recall and interrupted and completed tasks, has been studied by many psychologists. Tasks are presented to the subject and he is allowed to complete half of them and is interrupted on the other half. After the experimental session, the subject is asked to recall the tasks on which he has worked. A measure frequently used in such studies is the ratio (RI/RC) of the number of interrupted tasks recalled to the number of completed tasks recalled.

Lewis (1944) and Lewis and Franklin (1944) tested a group of twelve subjects under the usual conditions and found the median value of the RI/RC ratio to be 0.67. Thus, six subjects had values below 0.67 and six had values above this figure. A second group of fourteen subjects was tested under a cooperative work condition in which a co-worker was permitted to complete the tasks which had been interrupted for the subject. Under this test condition, twelve subjects had an RI/RC ratio greater than 0.67 and two had values less than 0.67. Can we conclude that the proportions exceeding 0.67 in the two groups differ significantly?

There is no evidence to indicate that the total of twenty-six subjects were *randomly* divided into two groups of fourteen and twelve subjects or randomly assigned to the two test conditions. Does this make any difference in the interpretation of the results of the experiment? Without randomization in the assignment of subjects to the two treatments, how can we know that the two groups of subjects do not differ systematically with respect to characteristics (organismic variables) that might account for the findings? Is there any reason to believe that if the second group was tested under the *same* conditions as the first group that the median RI/RC ratio would be 0.67 for this group?

Suppose the distribution of RI/RC ratios for all twenty-six subjects was examined and the median value for this complete distribution obtained. Then we would have thirteen subjects with values below and thirteen with values above, or the following 2 × 2 table:

	Below	Above	
Group 1			12
Group 2			14
	13	13	26

Assume random assignment was involved in the first study and also in the present one. Discuss the difference between the two designs.

What are some of the additional problems involved in the experiment? For example, how would one know that the nature of the tasks on which the subject is interrupted did not differ in some systematic way from those which he was allowed to complete? What could be done about this variable? Would randomization of the interrupted and completed tasks be of value?

3.11 Given an infinite binomial population with $P(X = 1) = 1/3$. Let T be the sum of $n = 15$ random observations drawn from this population. Find $E(T)$ and $E(T - \mu_T)^2$.

3.12 Given a finite binomial population with $N = 45$ and with $P(X = 1) = 1/3$. A random sample of $n = 15$ observations is drawn from this population without replacement. Let T be the sum of the n observations. Find $E(T)$ and $E(T - \mu_T)^2$.

3.13 Find the probability distribution of T for random samples of $n = 4$ drawn from an infinite binomial population in which $P(X = 1) = 1/3$. Use this distribution to calculate μ_T and σ_T^2. Show that these two values are equal to nP and nPQ, respectively.

3.14 We have a discrimination experiment in which the probability of a correct discrimination is $1/4$. If a subject is given $n = 48$ trials and if T is the number of correct discriminations, find $E(T)$ and $E(T - \mu_T)^2$.

3.15 One hundred subjects are divided at random into two groups of fifty subjects each. One group serves as a control group and the other as an experimental group. The proportion of subjects making a correct discrimination in the experimental group is 0.6 and the proportion making a correct discrimination in the control group is 0.4. Use a normal curve test to determine whether these two proportions differ significantly. Make a correction for discreteness.

3.16 What is the difference between a one-sided and a two-sided test of significance?

3.17 What are the essential properties of a standard normal variable?

3.18 If $P = 0.8$, what is the *minimum* number of observations we should have if we plan to make a normal curve test?

3.19 In a binomial experiment we have $P = \frac{1}{2}$. A subject is given $n = 100$ independent trials and makes $T = 65$ correct responses. Can we conclude that T is significantly greater than μ_T, if $\alpha = 0.01$? Make a correction for discreteness.

3.20 We have two groups of rats with $n = 50$ rats assigned to each group. The rats in Group 1 receive a treatment that is believed will increase the number of right turns in a maze. Group 2 is a control group and does not receive the treatment. On a critical test trial, the number of rats turning left and right in each group is as shown below:

	Left	Right
Group 1	20	30
Group 2	30	20

Under the assumption that the two samples have been drawn at random from a common binomial population, find the probability of $p_1 - p_2$ equal to or greater than the value obtained in the experiment. Make a correction for discreteness.

3.21 On an initial test, twenty-six out of fifty subjects make a given discrimination. The same subjects are then given a treatment and tested again. On the second test, thirty-nine of the fifty subjects make the discrimination. Of the thirty-nine subjects, twenty are those who made the discrimination on the first test. Test the null hypothesis that $E(p_1 - p_2) = 0$. Make a correction for discreteness.

3.22 If $Z = (X - \mu)/\sigma$, prove that $E(Z) = 0$ and $\sigma_Z^2 = 1$.

3.23 If X is a variable that can take only the values of $X = 0$ or $X = 3$ with corresponding probabilities of $\frac{2}{3}$ and $\frac{1}{3}$, find $\mu = E(X)$ and $\sigma^2 = E(X - \mu)^2$.

CHAPTER 4

SOME TESTS OF SIGNIFICANCE USING THE χ^2 DISTRIBUTION

4.1 Introduction

The methods of analysis described in the previous chapter can be used in evaluating the outcomes of experiments in which we have one or two sets of observations from a binomial population. We now consider techniques that can be used in evaluating the outcomes of experiments in which we have two or more classes of observations and/or in which we have two or more sets of observations.

For example, in a breeding experiment a cross between two plants results in 352 seedlings. According to genetic theory, the seedlings should segregate into four types in the ratio of 9:3:3:1. The observed frequencies for the four types are 200, 72, 60, and 20, respectively. Is this outcome of the experiment in accord with theory? Or do the frequencies deviate significantly from those expected on the basis of theory?

In a study of preferences, 120 subjects are presented with fresh, frozen, and canned orange juice. Each subject is asked to indicate the juice he prefers. The observed frequencies for the three juices are 60, 35, and 25, respectively. Can we conclude that these frequencies deviate significantly from a uniform chance distribution?

In evaluating outcomes of experiments of the kind described, we shall make use of the χ^2 distribution.[1] For these problems, we may define χ^2 as

$$\chi^2 = \sum_1^c \frac{(f_i - F_i)^2}{F_i} \tag{4.1}$$

[1] For a further discussion of the χ^2 test, see Cochran (1954).

where f_i is the *observed* frequency in the ith class, F_i is a corresponding *expected* frequency for that class, and the number of classes is equal to c. The expected frequencies are based on a null hypothesis of interest. If the probability associated with the obtained value of χ^2 is small, then the null hypothesis will be rejected.

4.2 One Sample with c Classes

For the genetic experiment, the observed frequencies for the four types of seedlings are

	Type 1	*Type* 2	*Type* 3	*Type* 4
f	200	72	60	20

According to theory, the seedlings should segregate in the ratio of 9:3:3:1. This is the null hypothesis we wish to test. If the theory is true, then $P_1 = 9/16$ should be the probability of Type 1, $P_2 = 3/16$ should be the probability of Type 2, $P_3 = 3/16$ should be the probability of Type 3, and $P_4 = 1/16$ should be the probability of Type 4. We have $n = 352$ observations and the corresponding expected frequencies will be

$$F_1 = 352\left(\frac{9}{16}\right) = 198$$

$$F_2 = 352\left(\frac{3}{16}\right) = 66$$

$$F_3 = 352\left(\frac{3}{16}\right) = 66$$

$$F_4 = 352\left(\frac{1}{16}\right) = 22$$

Then

$$\chi^2 = \frac{(200 - 198)^2}{198} + \frac{(72 - 66)^2}{66} + \frac{(60 - 66)^2}{66} + \frac{(20 - 22)^2}{22} = 1.293$$

To find $P(\chi^2 \geq 1.293)$, when the null hypothesis is true, we make use of the table of χ^2, Table IV in the Appendix. To use Table IV, we must enter the table with the number of degrees of freedom (d.f.) associated with the obtained value of χ^2. The number of degrees of freedom may be regarded as the number of deviations, $f_i - F_i$, that are free to vary. In the present problem, we note that

$$\sum_{1}^{c} (f_i - F_i) = 0$$

Therefore, only $c - 1$ of the deviations are free to vary and this is the number of degrees of freedom associated with the obtained value of χ^2 equal to 1.293. Entering the table of χ^2 with $c - 1 = 3$ d.f., we find that $P(\chi^2 \geq 1.293) > 0.70$. The outcome of this experiment does not appear to offer any significant evidence against the hypothesis that the seedlings should segregate in the ratio of 9:3:3:1.

Consider the preference study with respect to fresh, frozen, and canned orange juice. The outcome of the experiment is as given below:

	Fresh	*Frozen*	*Canned*
f	60	35	25

In this instance, the null hypothesis we wish to test is that each of the three juices has an equal probability of being chosen. If this hypothesis is true, then $P_1 = P_2 = P_3 = 1/3$, and the expected frequency for each type of juice, with $n = 120$ observations, will be $(120)(1/3) = 40$. Then

$$\chi^2 = \frac{(60 - 40)^2}{40} + \frac{(35 - 40)^2}{40} + \frac{(25 - 40)^2}{40} = 16.25$$

with $c - 1 = 2$ d.f. Entering the table of χ^2 with 2 d.f., we find that $P(\chi^2 \geq 9.21) = 0.01$. Then the probability of χ^2 equal to or greater than 16.25 must be considerably less than 0.01 and we may regard the outcome of this experiment as highly improbable, if the null hypothesis is true.

4.3 Two or More Samples with c Classes

Rosenzweig (1943) tested the recall of subjects for finished and unfinished tasks after they had worked on the tasks under differing sets of instructions. An "informal" group worked under the assumption that the experimenter was interested in studying work methods and that the ability of the subjects was not under investigation. A "formal" group worked on the same tasks under the impression that the problems were a kind of intelligence test. On some of the tasks both groups of subjects were interrupted and on other tasks they were allowed to work until the problem was completed. At the end of the experiment, subjects in each group were asked to recall the tasks on which they had worked. We shall assume that the $n = 60$ subjects were divided at random into two groups

Table 4.1

NUMBER OF SUBJECTS SHOWING A TENDENCY TO RECALL FINISHED AND UNFINISHED TASKS AND NUMBER OF SUBJECTS SHOWING NO TENDENCY OF DIFFERENTIAL RECALL IN TWO RANDOMIZED GROUPS[a]

Group	Finished	Unfinished	No Tendency	Total
Informal	7	19	4	30
Formal	17	8	5	30
Total	24	27	9	60

[a] Rosenzweig (1943).

of $n_1 = 30$ and $n_2 = 30$ subjects each. Table 4.1 gives the number of subjects in the formal and informal groups who recalled a larger number of finished tasks, a larger number of unfinished tasks, or who showed no tendency in differential recall of the finished and unfinished tasks.

If the difference in instructions to the two groups had no effect, we should expect the number of subjects in each of the classes of Table 4.1 to be similar for both groups. As a null hypothesis to be tested, we assume that both groups are from a common population in which the probabilities for each of the three classes of the table are $P_1 = 24/60$, $P_2 = 27/60$, and $P_3 = 9/60$, respectively. Then, because we have $n_1 = n_2 = 30$, the corresponding expected frequencies for each class for both groups will be

$$F_1 = 30\left(\frac{24}{60}\right) = 12.0$$

$$F_2 = 30\left(\frac{27}{60}\right) = 13.5$$

$$F_3 = 30\left(\frac{9}{60}\right) = 4.5$$

If we let $r =$ the number of rows or groups and $c =$ the number of classes as before, then, for problems of the kind described, we have

$$\chi^2 = \sum_1^r \sum_1^c \frac{(f_i - F_i)^2}{F_i} \qquad (4.2)$$

where the double summation sign means that we must sum over each of the c classes for each of the r groups or rows.

In the cells of Table 4.2 we have entered the terms, $f_i - F_i$, corresponding to each of the cell entries of Table 4.1. Then

$$\chi^2 = \frac{(7 - 12.0)^2}{12.0} + \frac{(19 - 13.5)^2}{13.5} + \cdots + \frac{(5 - 4.5)^2}{4.5} = 8.76$$

It may be observed in Table 4.2 that the deviations, $f_i - F_i$, sum to zero in each row and each column of the table. Therefore, only $(r - 1)(c - 1)$ of the deviations are free to vary. Accordingly, the χ^2 defined by (4.2) will have $(r - 1)(c - 1)$ d.f. For the present problem we have $\chi^2 = 8.76$ with 2 d.f. From the table of χ^2 we find that, with $\alpha = 0.05$, our obtained value is significant. We therefore reject the null hypothesis and conclude that the two groups are not random samples from a common population with probabilities as given for the various classes. Examination of Table 4.1 shows that for the informal group there is a tendency for more unfinished tasks to be recalled, whereas for the formal group there is a tendency for more finished tasks to be recalled.

4.4 *Two or More Samples with c = 2 Classes*

In a mental hospital, a new drug at a standard dosage was tested. All male first admissions between the ages of twenty and thirty-five were given the drug. The observation recorded was whether or not the patient showed a reaction to the drug. Records were kept separately for each of nine months. The number of patients showing a reaction and the number showing no reaction are given in Table 4.3 for each of the nine months. The null hypothesis to be tested is that the groups administered the drug each month are from a common population in which the probability of a reaction to the drug is $P = 198/360 = 0.55$ and the probability of a nonreaction is $Q = 1 - P = 0.45$.

When we have r samples and only $c = 2$ classes, there is a simplified method for calculating χ^2. We take the column of frequencies in Table 4.3

Table 4.2
THE $f_i - F_i$ TERMS FOR THE DATA OF TABLE 4.1

Group	Finished	Unfinished	No Difference
Informal	7 − 12.0	19 − 13.5	4 − 4.5
Formal	17 − 12.0	8 − 13.5	5 − 4.5

with the smaller total.[2] In the present instance this is column (2), headed f_1, which shows the frequency of nonreactors in each group. We now square each of the f_1 values to obtain the entries in column (5). In column (6) we have divided each f_1^2 value by n_i, the number of observations in the group. We then find the sums of the columns as shown at the bottom of the table. Then

$$\chi^2 = \frac{n^2}{\sum f_1 \sum f_2}\left[\sum \frac{f_1^2}{n_i} - \frac{(\sum f_1)^2}{n}\right] \tag{4.3}$$

where n = the total number of observations in all samples; Σf_1 = the total number of observations in one of the two classes; Σf_2 = the total number of observations in the other class; and n_i = the number of observations in the ith group or sample. Making the substitutions from Table 4.3 in (4.3), we obtain

$$\chi^2 = \frac{(360)^2}{(162)(198)}\left[74.562 - \frac{(162)^2}{360}\right] = 6.72$$

with degrees of freedom equal to $(r - 1)(c - 1) = (9 - 1)(2 - 1) = 8$. By referring to the table of χ^2, we find that $P(\chi^2 \geq 7.344)$ is 0.50 and that, consequently, $P(\chi^2 \geq 6.72)$ is slightly greater than 0.50. The outcome of the

Table 4.3

NUMBER OF NONREACTORS AND REACTORS TO A DRUG IN MONTHLY SAMPLES AT A MENTAL HOSPITAL

(1) Months	(2) Nonreactors f_1	(3) Reactors f_2	(4) Total n_i	(5) f_1^2	(6) f_1^2/n_i
January	18	32	50	324	6.480
February	20	25	45	400	8.889
March	22	20	42	484	11.524
April	19	19	38	361	9.500
May	14	22	36	196	5.444
June	21	19	40	441	11.025
July	22	21	43	484	11.256
August	16	20	36	256	7.111
September	10	20	30	100	3.333
Total	162	198	360		74.562

[2] It is not necessary to take the column of frequencies with the smaller total, but this simplifies the computations somewhat. If we choose the f_2 column rather than the f_1 column, we interchange the terms for f_1 and f_2 in (4.3).

experiment may be judged not to offer any significant evidence against the null hypothesis that the groups are from a common population in which the probability of a reaction is 0.55 and the probability of a nonreaction is 0.45.

It is of some interest to consider the nature of the test of significance in this example. Randomization was not involved in the assignment of patients to the various groups (months) and the treatment was the same for each group. If a significant value of χ^2 had been obtained, what would this mean? Statistically, it would mean that the proportions of reactors differ in the monthly samples. This, in turn, may indicate that the patients entering the hospital during certain months differed in some systematic way from the patients entering during other months or that the nature of the drug or its administration differed systematically between months. The test of significance, in the absence of randomization, may be interpreted as providing an indication of whether or not the row variable (months) and the column variable (reaction or no reaction) are independent or associated.[3] A nonsignificant value of χ^2 indicates that the two classifications are independent, whereas a significant value indicates that they are not independent. It is obvious that simply because two variables are not independent, this does not tell us which may be cause and which may be effect. We shall have more to say about this point in the next example to be considered.

4.5 Two Samples with c = 2 Classes

A reasoning problem that involved clamping together two sticks so that the length was just sufficient to wedge the joined sticks between the floor and ceiling of an experimental room was used in an investigation by Maier (1945). The subjects were instructed to construct a hat rack from the materials supplied, and the solution to the problem was as described, the projection of the clamp from the two sticks providing the necessary hook for hanging up a coat or hat. Men and women were used as subjects and they were tested under three different experimental conditions—the conditions involving different clues as to the solution of the problem. The data given in Table 4.4 are the totals for all three conditions.

Assume that the null hypothesis of interest is that the group of men and the group of women are from a common binomial population in which the probability of a solution is $P = 36/75 = 0.48$ and the probability of no solution is $Q = 1 - P = 0.52$. For the $r \times c = 2 \times 2$ table, that is, with

[3] For a discussion of various measures of association for cross classifications of the kind described in this chapter, see Goodman and Kruskal (1954; 1959).

Table 4.4

NUMBER OF MEN AND WOMEN REACHING NO SOLUTION OR A SOLUTION IN A REASONING PROBLEM[a]

	No Solution	Solution	Total
Men	13	26	39
Women	26	10	36
Total	39	36	75

[a] Maier (1945).

two rows or groups and two classes or columns, we have the schematic representation shown in Table 4.5. Then, in the notation of this table, we have

$$\chi^2 = \frac{n\left(|bc - ad| - \dfrac{n}{2}\right)^2}{(a + b)(c + d)(a + c)(b + d)} \tag{4.4}$$

where the factor $n/2$ is a correction for discreteness.

Substituting the data of Table 4.4 in (4.4), we have

$$\chi^2 = \frac{75\left(|676 - 130| - \dfrac{75}{2}\right)^2}{(39)(36)(39)(36)} = 9.84$$

with 1 d.f.[4] According to the table of χ^2, the obtained value is significant with $P(\chi^2 \geq 9.84) \leq 0.01$.

4.6 Problems in the Interpretation of Results

What may we conclude from the Maier study? For the particular sample involved, there is evidence that the row and column classifications are not independent. It is important to understand that the finding of an association between two variables cannot be interpreted in the same manner as when we have an experiment in which different treatments are involved and in which subjects are randomly

[4] If χ^2 has 1 d.f., then $\chi^2 = Z^2$ and it is possible to use the more complete table of the standard normal distribution to find the probability associated with χ^2. For example, if $\chi^2 = 4.0$ with 1 d.f., then $Z = 2.0$ or $Z = -2.0$. From the table of the standard normal distribution we find that $P(Z \geq 2.0) = 0.0228$ and $P(Z \leq -2.0) = 0.0228$. Then $P(Z^2 \geq 4.0) = (2)(0.0228) = 0.0456$, and this is also $P(\chi^2 \geq 4.0)$.

assigned to the treatments. With randomization and with a significant difference between two treatment groups, we have a basis for concluding that the observed difference between the two groups is the result of the difference in treatments. With randomization, we expect individual differences (organismic variables) to be randomized over the treatments. In the present example, we would not wish to attribute the difference in the proportion of females solving the problem and the proportion of males solving the problem to the one obvious way in which the two groups differ, that is, sex, because the two groups may also differ in many other respects.

Perhaps the point we wish to make can be emphasized by considering some fictitious but possible conditions that may be present in the study under consideration. Suppose, for example, that every subject classified as a male was also blue-eyed and that every subject classified as a female was brown-eyed. Then it would also be true that there is a significant association between eye color and failure and success in finding a solution to the problem.

If we wish to find out whether it is eye color or the sex classification that results in the greater probability of success for blue-eyed male subjects than for brown-eyed female subjects, then obviously the study should be repeated holding either the sex classification or eye color constant. For example, we might repeat the study with blue-eyed and brown-eyed females and also with blue-eyed and brown-eyed males. If it should be found that for both sex groups, the blue-eyed subjects have a higher probability of success than brown-eyed subjects, we would have some assurance that the outcome of the original experiment was not the result of a sex difference. On the other hand, if the outcome of the experiment is such that both brown-eyed and blue-eyed males have approximately the same probability of a success and that this probability is greater than the probability of a success for both blue-eyed and brown-eyed female subjects, we have some basis for believing it is the sex classification rather than eye color that is associated with success and failure. But, obviously, in order to rule out all other possible

Table 4.5

SCHEMATIC REPRESENTATION OF FREQUENCIES FOR $r = 2$ GROUPS AND $c = 2$ CLASSES

	Failure	Success	Total
Group 1	a	b	$n_1 = a + b$
Group 2	c	d	$n_2 = c + d$
Total	$a + c$	$b + d$	$n = n_1 + n_2$

variables that may be associated with sex differences, the study would have to be repeated holding each of these variables constant.

If we randomly assign a subject to two treatments and if we use a sufficiently large number of subjects, we hope that the randomization process will provide an adequate control with respect to the various organismic variables that might influence the outcome of the experiment, in that differences between the two treatment groups with respect to these variables should represent only chance or random differences and not systematic differences. We say that we hope randomization will provide an adequate control because, even though subjects have been randomly assigned to the two groups, they may still differ, as a result of chance, with respect to *some* organismic variables and the outcome of the experiment *may* be the result of these unknown systematic ways in which the two groups differ.

We can only reemphasize the principle stated by Fisher (1942) that we quoted earlier. This principle states that the outcome of any single experiment involving random variation is necessarily inconclusive. However, if we repeat the experiment, always using randomization in the assignment of subjects to treatments, it is relatively improbable that in each of the successive repetitions we will obtain the same systematic differences between the groups on the same organismic variable. If, in a series of repetitions of the experiment, the outcome is consistently in the same direction, this would appear to rule out the possibility that the outcome is the result of a difference between the two groups with respect to an organismic variable.

4.7 Test of Experimental Technique

In an experiment concerning the influence of a particular drug upon a physiological response, the drug was to be tested at two levels of concentration. The drug was to be administered by injection and the experimenter was not sure of his technique; that is, if comparable groups were tested a second time he was not sure whether he would obtain the same or comparable results. The experimental design provided for a test of the technique by repeating the complete experiment four times.

Subjects were divided at random into eight groups of twenty subjects each. Four of the groups were assigned at random to each level of the drug. The observation for each subject was the presence or absence of a specified reaction to the drug. The number of reactors and nonreactors in each group for each level of the drug is given in Table 4.6.

Consider only the four groups of subjects tested at the first level. If the experimenter's technique is under control, then we should expect to find the number of reactors and nonreactors in each of these four groups to

Table 4.6

NUMBER OF NONREACTORS AND REACTORS IN RANDOMIZED GROUPS WITH TWO LEVELS OF A DRUG

	(1) Groups	(2) Nonreactors f_1	(3) Reactors f_2	(4) Total n_i	(5) f_1^2	(6) f_1^2/n_i
First level	1	10	10	20	100	5.00
	2	12	8	20	144	7.20
	3	8	12	20	64	3.20
	4	15	5	20	225	11.25
	Total	45	35	80		26.65
Second level	1	6	14	20	36	1.80
	2	8	12	20	64	3.20
	3	5	15	20	25	1.25
	4	6	14	20	36	1.80
	Total	25	55	80		8.05

be comparable from group to group. The reason for this is that randomization was involved in the assignment of subjects to the groups and each group received the same treatment. Using (4.3), we have

$$\chi^2 = \frac{(80)^2}{(45)(35)}\left[26.65 - \frac{(45)^2}{80}\right] = 5.44$$

and this is a nonsignificant value for 3 d.f. The null hypothesis that these four groups are from a common population in which the probability of a reaction to the drug at the first level is $P = 35/80 = 0.4375$ is tenable. If a significant value of χ^2 had been obtained, in this instance, it would have indicated that something was apparently wrong with the experimental technique. Possible explanations for this might be found in systematic differences in the manner of injection, variations in the dosage injected, or in some other aspect of the experimental procedure. As it is, the nonsignificant value of χ^2 indicates that the proportions of reactors in the four groups are comparable.

Similarly, for the four groups tested with the second level of the drug, we have

$$\chi^2 = \frac{(80)^2}{(25)(55)}\left[8.05 - \frac{(25)^2}{80}\right] = 1.12$$

and this is also a nonsignificant value of χ^2 for 3 d.f. This test indicates that the outcome of the experiment is such that it provides no significant evidence against the null hypothesis that the four groups are random samples from a population in which the probability of a reaction to the drug at the second level is $P = 55/80 = 0.6875$.

Because the tests of technique indicate that the groups tested at the first level of the drug are homogeneous and that the groups tested at the second level are also homogeneous, we may pool the results for each level of the drug to obtain Table 4.7. We now wish to test the null hypothesis that the groups tested at the two different levels of the drug are from a common population in which the probability of a reaction is $P = 90/160 = 0.5625$ and the probability of a nonreaction is $Q = 1 - P$. Using (4.4), we have

$$\chi^2 = \frac{160\left(|\,875 - 2475\,| - \frac{160}{2}\right)^2}{(80)(80)(70)(90)} = 9.17$$

with 1 d.f. From the table of χ^2 we find that this is a significant value and we reject the null hypothesis. Examination of the data of Table 4.7 shows that a larger proportion of reactors is found with the second level of the drug than with the first level. Because randomization was involved in assigning subjects to the two levels of the drug, we have a basis for concluding that the observed difference between the two proportions is the result of the difference in the treatments, that is, in the level of the dosage.

4.8 χ^2 with More than 30 d.f.

The table of χ^2 provides entries for degrees of freedom equal to 30 or less. For a larger number of degrees of freedom, we may find

$$Z = \sqrt{2\chi^2} - \sqrt{2(\text{d.f.}) - 1} \tag{4.5}$$

The value of Z defined by (4.5) is approximately normally distributed with

Table 4.7
THE POOLED RESULTS FOR THE DATA OF TABLE 4.6

	Non-reactors	Reactors	Total
First level	45	35	80
Second level	25	55	80
Total	70	90	160

mean equal to 0 and standard deviation equal to 1 and may be considered a standard normal variable to be evaluated by means of the table of the standard normal distribution.

QUESTIONS AND PROBLEMS

4.1 Kuenne (1946) studied transposition behavior in two groups of children who differed with respect to age. Group 1 consisted of eighteen children ranging in age from approximately 34 to 46 months. Group 2 consisted of twenty-six children ranging in age from approximately 60 to 63 months. In the critical test trials, three of the children in Group 1 showed transposition behavior and fifteen did not. In Group 2, the number showing transposition behavior in the critical test trials was twenty, while six failed to meet the criterion. Can we conclude that the two proportions differ significantly?

Again, in this experiment, we must note that randomization was not involved in assigning subjects to the two groups. Age is an organismic variable and cannot therefore be randomly assigned to a subject. What bearing does this have upon the interpretation of the outcome of the study? Would you attribute the results to the age difference? If so, how would you answer the argument that the subjects may also differ with respect to important organismic variables other than age?

4.2 In a study by Hellman (1914) it is reported that of twenty breast-fed youngsters, four had normal teeth and sixteen showed malocclusion. Of twenty-two bottle-fed youngsters, one had normal teeth and the other twenty-one showed malocclusion. Can we conclude that the two proportions differ significantly?

Since randomization is not involved in this study, what bearing would this have upon the interpretation of the result of the experiment if it had been significant? Is it possible that mothers who breast-feed their youngsters may also differ in other respects from mothers who bottle-feed their youngsters? What other variables might be associated with breast-feeding and bottle-feeding which, in turn, might be associated with malocclusion and normal teeth?

4.3 Records were kept of the number of students who left a university auditorium through each of three main exits. For a sample of 795 students the counts were as follows: Exit 1: 245 students; Exit 2: 200 students; Exit 3: 350 students. Can we conclude that the exits are equally popular?

4.4 Kendall and Smith (1939) have described the tests they applied to their tables of random numbers. All the numbers in the published tables were run off by one operator using an electrical device constructed for the purpose. One of the tests applied to the numbers drawn was the frequency test which consisted of counting the frequencies of the digits from 0 to 9. Various sets of numbers were rejected, including this one:

Digit	f	Digit	f
0	1083	5	1007
1	865	6	1081
2	1053	7	997
3	884	8	1025
4	1057	9	948

Assuming randomness, the expected frequency for each digit is 1000. Can we conclude that the probability for each digit is the same?

4.5 Hartman (1939) tested men and women with various solutions of phenylthiocarbamide. The solutions were numbered in terms of strength from 0 to 10, and the threshold was recorded as the concentration below which they first tasted the presence of phenylthiocarbamide. Since some subjects tasted the weakest solution 0, the threshold for these subjects was recorded as below 0, giving rise to twelve classes. The frequency distributions of the thresholds for 290 men and 314 women were as below:

	Frequency		
Strength	Men	Women	Total
10	15	42	57
9	35	52	87
8	46	38	84
7	31	30	61
6	23	19	42
5	13	17	30
4	9	6	15
3	7	5	12
2	10	10	20
1	13	19	32
0	25	33	58
Below 0	63	43	106

Can we conclude that threshold and sex classification are independent?

4.6 Records were kept at a university medical clinic of students who had attacks of influenza. Some of these students had been given vaccinations against influenza and others had not. The students were also classified in terms of whether they had a severe attack or a minor attack. The data are as follows:

	Minor Attack	Severe Attack
Vaccinated	98	40
Not vaccinated	30	82

Can we conclude that these two variables are independent?

In the absence of randomization, would we want to attribute the severity of the attack to the presence or absence of vaccination? What are some of the possible systematic organismic differences that may exist between subjects who were vaccinated and those who were not?

Assume that a design could be worked out in which subjects would be randomly assigned to the vaccination and nonvaccination groups. What additional controls would be needed in this study? Would it make any difference if the physician who did the vaccinating also did the rating of the severity of the attack? How could this possible source of bias be controlled? A subject's knowledge of the fact that he has or has not been vaccinated might be of some importance. How could this be controlled? Should consideration be given to those subjects, vaccinated and nonvaccinated, who have no attacks?

4.7 Merritt and Fowler (1948) report a study in which the procedure was as follows: "...stamped, self-addressed, and sealed letters of two types were 'lost' by depositing them prominently but discreetly on sidewalks of various cities in the East and Midwest. Type A contained only a trivial message, while Type B contained, besides a message, a lead slug of the dimensions of a fifty-cent piece. The accompanying message indicated that the lead disk, as such, was of value to the addressee. Care was taken to drop the letters in locations sufficiently removed from one another to preclude the possibility of any one person finding more than one of the letters. All were put down in clear weather so that the envelopes would not become soiled and hence lose their appearance of value. Tests were made by night and day in both business and residential districts" (pp. 90–91).

Thirty-three letters of Type A were dropped and of these twenty-eight were returned by the person picking them up. Of Type B, 158 letters were dropped and eighty-six of these were returned. Can we conclude that the probability of a Type A letter being returned is the same as the probability of a Type B letter being returned?

4.8 Rats are placed in a starting box and at a choice point may take any one of four paths to a goal box. The number of rats using each path is given below:

Path 1	Path 2	Path 3	Path 4
10	20	30	40

(*a*) Test the null hypothesis that the probability of each path being used is ¼. (*b*) Assume that Paths 1 and 2 involve left turns and that Paths 3 and 4 involve right turns. Test the null hypothesis that right and left turns are equally probable.

4.9 We have an experiment involving a control group and two experimental groups. Each subject may make one of three mutually exclusive responses in the test situation. The number of subjects making each response in each of the groups is shown below:

	R_1	R_2	R_3
Control	50	40	10
Experimental 1	15	20	25
Experimental 2	15	10	15

Test the null hypothesis that the three groups are random samples from a common population.

4.10 We toss a coin $n = 10$ times. Assume that we obtain $T = 8$ heads. Test the null hypothesis that $P = \frac{1}{2}$ by finding Z and also χ^2. Do *not* make a correction for discreteness. You should find that $Z^2 = \chi^2$.

4.11 In a random sample of $n = 60$, we have the following number of observations in each of three categories:

	Category		
	1	2	3
f	15	10	35

Test the null hypothesis that $P_1 = P_2 = P_3$.

CHAPTER 5

THE *t* TEST FOR MEANS

5.1 Introduction

Assume that X is a normally distributed variable with population mean equal to μ and standard deviation equal to σ. If random samples of n observations each are drawn from this population, then the sum, T, of the values of n observations will also be normally distributed with mean equal to $\mu_T = n\mu$ and standard deviation equal to $\sigma_T = \sqrt{n\sigma^2}$. Then

$$Z = \frac{T - \mu_T}{\sigma_T} = \frac{T - n\mu}{\sigma\sqrt{n}} \tag{5.1}$$

will be a standard normal variable and can be evaluated in terms of the standard normal distribution.

If random samples of n observations each are drawn from a population in which X is not normally distributed, the distribution of T will not be normal in form if n is small. However, as n increases, the distribution of T will approach that of a normally distributed variable.[1]

5.2 The Distribution of *T* for Samples from Binomial Populations

We made use of the fact that the distribution of T approaches that of a normal distribution as n increases in our discussion of tests of significance involving random samples from a binomial population. If X is a binomial variable that can take only the values $X = 1$ and $X = 0$ with corresponding probabilities of P and $Q = 1 - P$, and if $P = Q$,

[1] This statement is based on the central limit theorem. A proof of this theorem can be found in Cramér (1946).

then the population distribution of X is rectangular in form. In this case the distribution of T is symmetrical and, with n as small as 10, a good approximation of the probabilities associated with the possible values of T was obtained by means of the standard normal distribution.

When $P \neq Q$, the population distribution of X is skewed and so also is the distribution of T when n is small. If $P = 1/3$, then the probabilities associated with the possible values of T can be approximated fairly well by means of the standard normal distribution, provided that n is at least equal to 15. The greater the skewness of the binomial population, the larger the number of observations needed in order to approximate the probabilities associated with the possible values of T by means of the standard normal distribution. This was the basis of the suggested rule that both nP and nQ should be *at least* equal to 5 in order to use the standard normal distribution in evaluating T. Obviously, the greater the difference between P and Q, the more skewed the distribution of X. Consequently, as the difference between P and Q increases, the suggested rule provides for an increase in n.

5.3 The Distribution of T: Three Examples in which X Is Not Normally Distributed

We have discussed the distribution of T for a random sample of n observations drawn from a binomial population in which X can take only the discrete values of $X = 1$ or $X = 0$, with corresponding probabilities of P and $Q = 1 - P$. But suppose, as is true of many variables in psychological research, that X can take possible values of

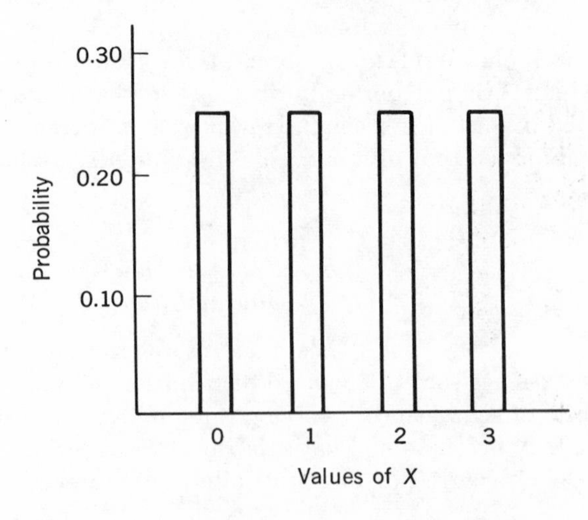

Figure 5.1

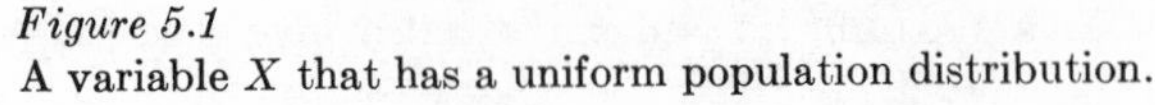

A variable X that has a uniform population distribution.

Table 5.1

PROBABILITY DISTRIBUTION OF THE SUM (T) OF n INDEPENDENT RANDOM VALUES OF X WHEN X HAS A UNIFORM DISTRIBUTION AND POSSIBLE VALUES OF 0, 1, 2, AND 3

	n					
T	1	2	3	4	5	6
0	0.250	0.062	0.016	0.004	0.001	0.000
1	0.250	0.125	0.047	0.016	0.005	0.002
2	0.250	0.188	0.094	0.039	0.015	0.005
3	0.250	0.250	0.156	0.078	0.034	0.014
4		0.188	0.188	0.121	0.063	0.029
5		0.125	0.188	0.156	0.099	0.053
6		0.062	0.156	0.172	0.132	0.082
7			0.094	0.156	0.151	0.111
8			0.047	0.121	0.151	0.133
9			0.016	0.078	0.132	0.142
10				0.039	0.099	0.133
11				0.016	0.063	0.111
12				0.004	0.034	0.082
13					0.015	0.053
14					0.005	0.029
15					0.001	0.014
16						0.005
17						0.002
18						0.000

not only 0 and 1, but also other possible values such as 2, 3, 4, 5, 6, and so on. The population distribution of X will be given by the corresponding probabilities associated with each possible value of X. We consider three examples: (1) a population distribution of X that is rectangular or uniform; (2) a population distribution of X that is U-shaped; and (3) a population distribution of X that is skewed. In all three examples, we shall assume that the possible values of X are 0, 1, 2, and 3.

We are interested in the distribution of T, the sum of the values of n random observations drawn from each of the three populations. We shall show that, even for a relatively small number of observations, the distributions of T for random samples drawn from these non-normally distributed populations begin to have a common form or shape and that this common form or shape of the distribution approaches that of a normal distribution as n increases. If this can be demonstrated for a relatively small number of observations, then we should have increased confidence that as n increases, all three distributions of T should, in the limit, tend to have distributions that can be described as approximately normal.

We shall not be concerned with possible differences in the means and standard deviations of the distributions of T, but only about the shape or

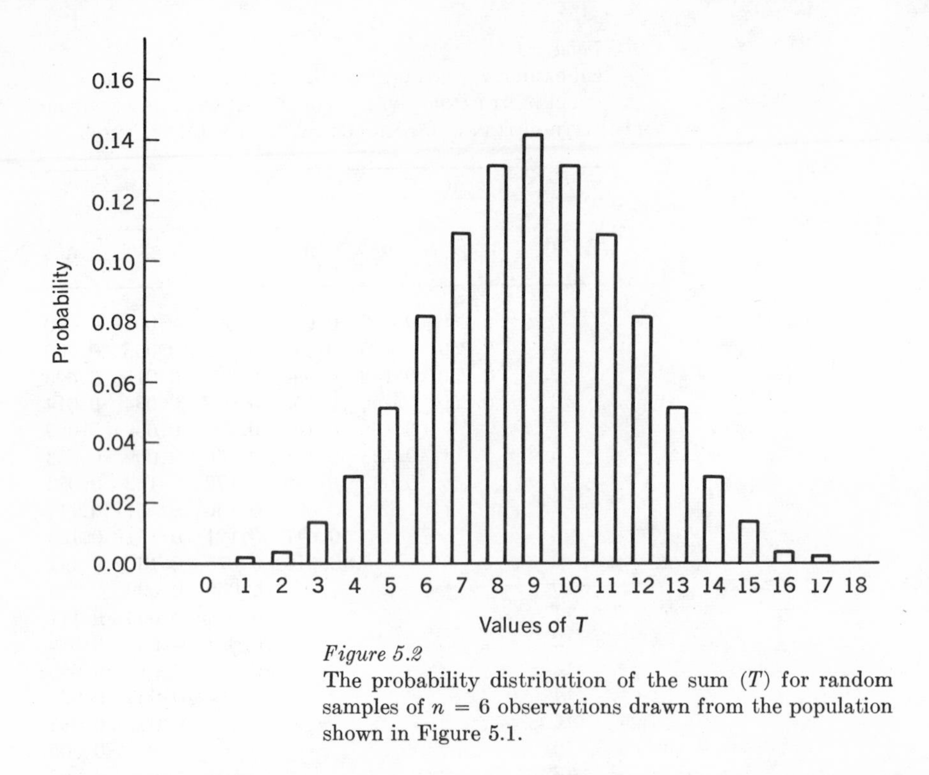

Figure 5.2
The probability distribution of the sum (T) for random samples of $n = 6$ observations drawn from the population shown in Figure 5.1.

form of the distributions. If the distributions of T all tend to approach a normal distribution, but with different means and standard deviations, we know that each can be transformed into standard normal variables by means of (5.1) and that each transformed variable will have a mean equal to 0 and a standard deviation equal to 1.

Figure 5.1 shows the population distribution of a variable X that can

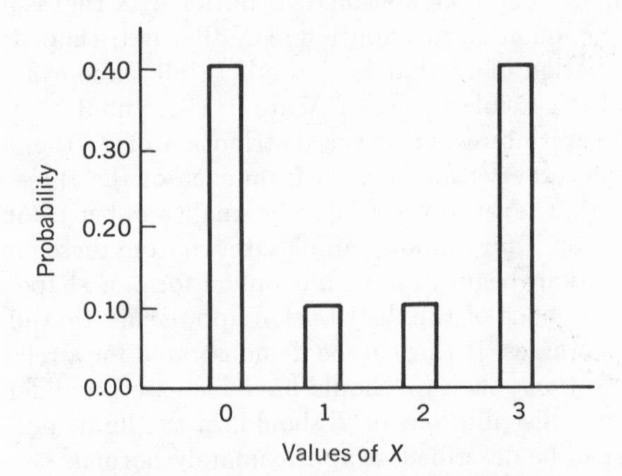

Figure 5.3
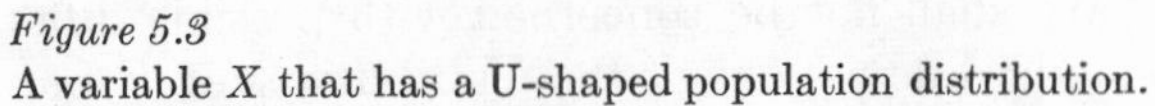
A variable X that has a U-shaped population distribution.

Table 5.2

PROBABILITY DISTRIBUTION OF THE SUM (T) OF n INDEPENDENT RANDOM VALUES OF X WHEN X HAS A U-SHAPED DISTRIBUTION AND POSSIBLE VALUES OF 0, 1, 2, AND 3, WITH CORRESPONDING PROBABILITIES OF 0.4, 0.1, 0.1, AND 0.4

	n					
T	1	2	3	4	5	6
0	0.400	0.160	0.064	0.026	0.010	0.004
1	0.100	0.080	0.048	0.026	0.013	0.006
2	0.100	0.090	0.060	0.035	0.019	0.010
3	0.400	0.340	0.217	0.123	0.066	0.034
4		0.090	0.111	0.091	0.062	0.038
5		0.080	0.111	0.101	0.076	0.051
6		0.160	0.217	0.196	0.147	0.099
7			0.060	0.101	0.107	0.090
8			0.048	0.091	0.107	0.099
9			0.064	0.123	0.147	0.139
10				0.035	0.076	0.099
11				0.026	0.062	0.090
12				0.026	0.066	0.099
13					0.019	0.051
14					0.013	0.038
15					0.010	0.034
16						0.010
17						0.006
18						0.004

take the possible values of 0, 1, 2, and 3, with corresponding probabilities of 0.25, 0.25, 0.25, and 0.25, respectively. The population distribution of X is rectangular or uniform. For the mean and variance of the population, we have

$$\mu = \sum P_iX_i = 1.5$$

and

$$\sigma^2 = \sum P_iX_i^2 - \mu^2 = 3.5 - (1.5)^2 = 1.25$$

Table 5.1 gives the probability distribution of T for random samples of $n = 1$ to $n = 6$ drawn from this population. Figure 5.2 shows the probability distribution of T for random samples when $n = 6$. It is obvious that the distribution of T, when $n = 6$, is no longer rectangular. Because the population distribution of X is symmetrical, the distribution of T is also symmetrical.

In Figure 5.3 we have the population distribution of a variable X that can take the possible values of 0, 1, 2, and 3, with corresponding probabilities of 0.4, 0.1, 0.1, and 0.4, respectively. In this instance the distribution of X is U-shaped. For the mean and variance of the population, we

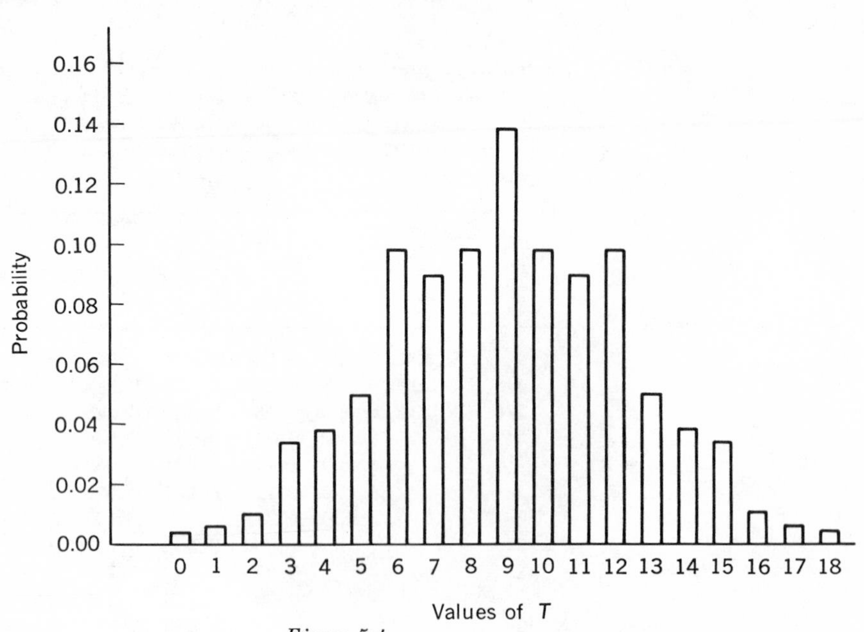

Figure 5.4
The probability distribution of the sum (T) for random samples of $n = 6$ observations drawn from the population shown in Figure 5.3.

have

$$\mu = \sum P_i X_i = 1.5$$

and

$$\sigma^2 = \sum P_i X_i^2 - \mu^2 = 4.1 - (1.5)^2 = 1.85$$

Table 5.2 gives the probability distribution of T for random samples of

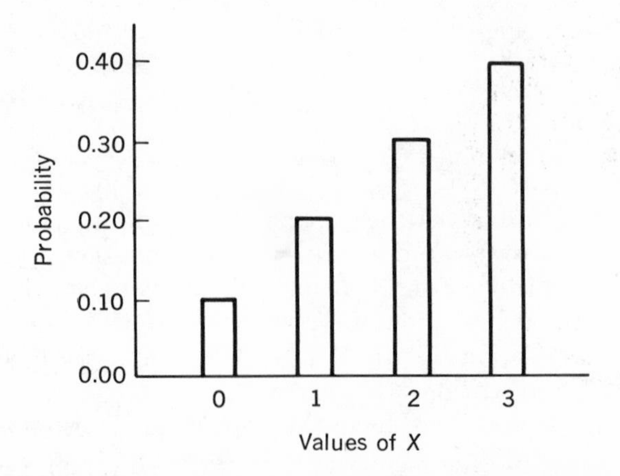

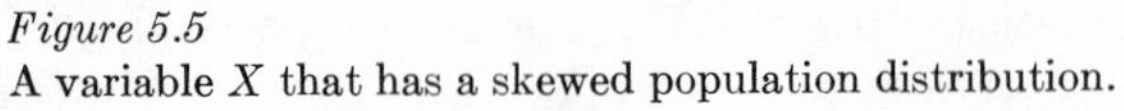
Figure 5.5
A variable X that has a skewed population distribution.

Table 5.3

PROBABILITY DISTRIBUTION OF THE SUM (T) OF n INDEPENDENT RANDOM VALUES OF X WHEN X HAS A SKEWED DISTRIBUTION AND POSSIBLE VALUES OF 0, 1, 2, AND 3, WITH CORRESPONDING PROBABILITIES OF 0.1, 0.2, 0.3, AND 0.4

	n					
T	1	2	3	4	5	6
0	0.100	0.010	0.001	0.000	0.000	0.000
1	0.200	0.040	0.006	0.001	0.000	0.000
2	0.300	0.100	0.021	0.004	0.001	0.000
3	0.400	0.200	0.056	0.012	0.002	0.000
4		0.250	0.111	0.031	0.007	0.001
5		0.240	0.174	0.065	0.018	0.004
6		0.160	0.219	0.112	0.038	0.010
7			0.204	0.161	0.070	0.023
8			0.144	0.190	0.111	0.044
9			0.064	0.184	0.150	0.073
10				0.138	0.172	0.110
11				0.077	0.167	0.140
12				0.026	0.133	0.158
13					0.083	0.154
14					0.038	0.127
15					0.010	0.087
16						0.047
17						0.018
18						0.004

$n = 1$ to $n = 6$ drawn from this population. In Figure 5.4 we show the probability distribution of T for random samples when $n = 6$ and it is obvious that the distribution of T, when $n = 6$, is no longer U-shaped. Because the distribution of X is symmetrical, the distribution of T is also symmetrical.

Figure 5.5 shows the population distribution of a variable X that can take the possible values of 0, 1, 2, and 3, with corresponding probabilities of 0.1, 0.2, 0.3, and 0.4, respectively. The distribution of X in this instance is skewed. For the mean and variance of the population, we have

$$\mu = \sum P_i X_i = 2.0$$

and

$$\sigma^2 = \sum P_i X_i^2 - \mu^2 = 5.0 - (2.0)^2 = 1.0$$

Table 5.3 gives the probability distribution of T for random samples of $n = 1$ to $n = 6$ drawn from this population and Figure 5.6 shows the probability distribution of T for random samples when $n = 6$. Again it is obvious that the distribution of T departs markedly from the population distribution of X. Because the distribution of X is skewed, the distribution

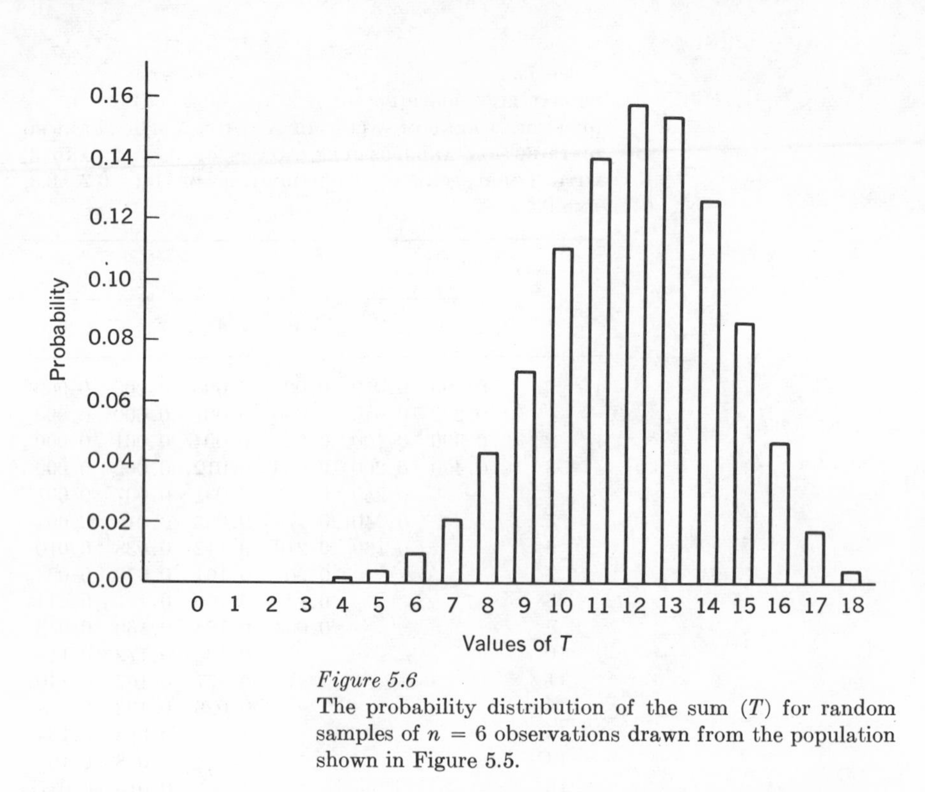

Figure 5.6
The probability distribution of the sum (T) for random samples of $n = 6$ observations drawn from the population shown in Figure 5.5.

of T is also skewed. We note, however, that the probabilities associated with the values of T in the left or skewed tail of the distribution are small and that the distribution is more symmetrical than is the population distribution of X.

It is apparent in all three of the examples that, even with a relatively small number of observations, the distributions of T are markedly different from the population distributions of X. In all three examples the distributions of T are beginning to approach a common form or shape. As n increases, then for all three of the populations the distributions of T will tend to approach, in form, a normal distribution and, in this case, the probabilities associated with the possible values of T for random samples of size n drawn from any of the three populations could be evaluated by means of (5.1) and the table of the standard normal distribution.

5.4 Normal Curve Approximation of the Probabilities of T

Although the three distributions of T are not normal for random samples as small as $n = 6$, let us see how well we can approximate the probabilities associated with extreme values of T by means of the standard normal distribution.

For the uniform or rectangular population we have $\mu = 1.5$ and $\sigma^2 = 1.25$. For the distribution of T based on random samples of $n = 6$ observations, we have

$$\mu_T = n\mu = (6)(1.5) = 9.0$$

and

$$\sigma_T = \sqrt{n\sigma^2} = \sqrt{(6)(1.25)} = 2.74$$

To estimate the probability of $T \geq 15$ we have, with a correction for discreteness or discontinuity,

$$Z = \frac{14.5 - 9.0}{2.74} = 2.01$$

and from the table of the standard normal distribution we find that $P(T \geq 15) = P(Z \geq 2.01) = 0.022$. With a correction for discontinuity, we also have $P(T \leq 3) = P(Z \leq -2.01) = 0.022$. The exact probability of $T \geq 15$, as shown in Table 5.1, is 0.021 and this is also the exact probability of $T \leq 3$.

For the U-shaped population we have $\mu = 1.5$ and $\sigma^2 = 1.85$. Then, for the distribution of T based on random samples of $n = 6$ observations, we have

$$\mu_T = n\mu = (6)(1.5) = 9.0$$

and

$$\sigma_T = \sqrt{n\sigma^2} = \sqrt{(6)(1.85)} = 3.33$$

To find $P(T \geq 15)$ using the normal curve approximation, we make a correction for discontinuity and obtain

$$Z = \frac{14.5 - 9.0}{3.33} = 1.65$$

From the table of the normal distribution we have $P(Z \geq 1.65) = 0.050$ and this is also the probability of $T \geq 15$. With a correction for discontinuity we also have $P(T \leq 3) = P(Z \leq -1.65) = 0.050$. The exact probability of $T \geq 15$, as shown in Table 5.2, is 0.054 and this is also the exact probability of $T \leq 3$.

For the skewed population we have $\mu = 2.0$ and $\sigma^2 = 1.0$. Then, for the distribution of T based on random samples of $n = 6$ observations, we have

$$\mu_T = n\mu = (6)(2.0) = 12.0$$

and

$$\sigma_T = \sqrt{n\sigma^2} = \sqrt{(6)(1.0)} = 2.45$$

In this case, let us find the probability of $T \geq 17$. With a correction for discontinuity, we have

$$Z = \frac{16.5 - 12.0}{2.45} = 1.84$$

From the table of the standard normal distribution we find $P(T \geq 17) = P(Z \geq 1.84) = 0.033$. The exact probability of $T \geq 17$, as shown in Table 5.3, is 0.022. With a correction for discontinuity, we also have $P(T \leq 7) = P(Z \leq -1.84) = 0.033$ and the exact probability, as given in Table 5.3, is 0.038.

We see that even with relatively small samples of $n = 6$ drawn from a uniform, a U-shaped, and a skewed population, the probabilities associated with the extreme values of T can be approximated fairly well by the standard normal distribution. For all three populations, the accuracy with which the normal distribution will approximate the probabilities associated with the values of T will increase as n increases.

5.5 Transforming a Sample Mean into a Standard Normal Variable

If Z, as defined by (5.1), is a standard normal variable that can be used to evaluate the probabilities associated with T, the sum of the values of n observations, and if we divide both numerator and denominator by n, then

$$Z = \frac{\bar{X} - \mu}{\sigma/\sqrt{n}} \tag{5.2}$$

and (5.2) is a standard normal variable that can be used to evaluate the probabilities associated with $\bar{X}$, the mean of a random sample of n observations. For the special case of samples from a binomial population, (5.2) becomes

$$Z = \frac{p - P}{\sqrt{PQ/n}} \tag{5.3}$$

because, for a binomial population, P is the population mean and $\sqrt{PQ}$ is the population standard deviation.

5.6 The Variance and Standard Error of a Mean When σ Is Known

In this and the following sections we shall be concerned with a variable X which we shall assume is normally distributed in the population. If we have a random sample of n observations from this population, the sample mean will be designated by $\bar{X}$ and will be equal to

$$\bar{X} = \frac{\sum X}{n} \tag{5.4}$$

and $\bar{X}$ will also be a normally distributed variable with expected variance given by

$$\sigma_{\bar{X}}^2 = \frac{\sigma^2}{n} \tag{5.5}$$

The square root of (5.5) is called the *standard error of the mean* and

$$\sigma_{\bar{X}} = \frac{\sigma}{\sqrt{n}} \tag{5.6}$$

It is obvious that the standard error of the mean is related to both the population standard deviation σ and the sample size n. If samples are drawn from a population with $\sigma = 10.0$, and if the sample size is $n = 25$, then $\sigma_{\bar{X}} = 2.0$. If $\sigma = 20.0$ and if $n = 25$, then $\sigma_{\bar{X}} = 4.0$. To reduce either of these standard errors by $\frac{1}{2}$, it would be necessary to quadruple the sample size. With $n = 100$ observations, and if $\sigma = 10.0$, then $\sigma_{\bar{X}} = 1.0$. If $\sigma = 20.0$ and if $n = 100$, then $\sigma_{\bar{X}} = 2.0$.

If the population standard deviation is known, then

$$Z = \frac{\bar{X} - \mu}{\sigma_{\bar{X}}} \tag{5.7}$$

would be a standard normal variable and could be used to test any null hypothesis regarding μ. In general, however, σ is not known.

5.7 The Variance and Standard Error of a Mean When σ Is Unknown

We define the variance of a sample of n observations as

$$s^2 = \frac{\sum (X - \bar{X})^2}{n - 1} \tag{5.8}$$

The variance defined by (5.8) is said to have $n - 1$ degrees of freedom and can be shown to be an unbiased estimate of the population variance σ^2 for any random sample of n independent observations. The sample standard deviation is defined as

$$s = \sqrt{\frac{\sum (X - \bar{X})^2}{n - 1}} \tag{5.9}$$

The variance of the mean of a random sample of n observations will be

$$s_{\bar{X}}^2 = \frac{s^2}{n} \tag{5.10}$$

and (5.10) can be shown to be an unbiased estimate of $\sigma_{\bar{X}}^2$. The standard error of a sample mean will then be

$$s_{\bar{X}} = \frac{s}{\sqrt{n}} \tag{5.11}$$

5.8 The t Distribution

We define the difference between a sample mean $\bar{X}$ and a population mean μ, divided by the standard error of the sample mean, as

$$t = \frac{\bar{X} - \mu}{s_{\bar{X}}} \tag{5.12}$$

with degrees of freedom associated with s^2 or $n - 1$.

The distribution of t depends on the number of degrees of freedom available in the set of n observations used in calculating s^2. Hence, the table of the t distribution is a two-dimensional table that must be entered with both the value of t and also the number of degrees of freedom. The distribution of t is not normal for small samples. Its distribution is symmetrical, as is the distribution of Z, but beyond a certain point (depending on the number of degrees of freedom available) the curve of t does not approach the base line as rapidly as does the curve of Z. This means that in order to cut off 5 percent of the total area in the right tail of the t distribution, we shall have to go out beyond the value of $Z = 1.65$ that cuts off 5 percent of the total area in the right tail of the Z distribution. Just how far out we shall have to go again depends on the number of degrees of freedom available.

The value of s^2 is itself subject to random variation. As n increases, the accuracy with which s^2 estimates σ^2 increases also. For very large values of n, the discrepancy between s^2 and σ^2 may be sufficiently small as to be negligible. In the limiting case, with n indefinitely large, the distribution of t is the same as the distribution of Z. In fact, with $n = 400$ observations, the distributions of t and Z are, for all practical purposes, the same.

If you look at the table of the standard normal distribution, Table III in the Appendix, you will find that $P(Z \geq 1.96) = 0.025$; that is, the ordinate at $Z = 1.96$ will cut off 0.025 of the total area in the right tail of this distribution. Now examine the table of the t distribution, Table V in the Appendix. Note that as the number of degrees of freedom increases, the value of t that cuts off 0.025 of the total area in the right tail of the t distribution also approaches 1.96. With 30 d.f., we have $P(t \geq 2.042) = 0.025$; with 100 d.f., $P(t \geq 1.984) = 0.025$; with 300 d.f., $P(t \geq 1.968) = 0.025$; with 500 d.f., $P(t \geq 1.965) = 0.025$; and with 1000 d.f., $P(t \geq 1.962) = 0.025$.

5.9 Confidence Interval for a Mean

Suppose we have a random sample of $n = 49$ observations, with $\bar{X} = 62.0$ and $s = 14.0$. Then $s_{\bar{X}} = 14.0/\sqrt{49} = 2.0$ and

$$t = \frac{62.0 - \mu}{2.0}$$

Now μ is unknown, but suppose we choose to regard any value of μ such that

$$P(\bar{X} \geq \mu) = 0.025 \qquad \text{and} \qquad P(\bar{X} \leq \mu) = 0.025$$

as improbable. With $49 - 1 = 48$ d.f., we find that

$$P(t \geq 2.01) = 0.025 \qquad \text{and} \qquad P(t \leq -2.01) = 0.025$$

We may set up the following inequality:

$$-t \leq \frac{\bar{X} - \mu}{s_{\bar{X}}} \leq t \tag{5.13}$$

Substituting in the above inequality with $t = 2.01$, $\bar{X} = 62.0$, $s_{\bar{X}} = 2.0$, and $-t = -2.01$, we have

$$-2.01 \leq \frac{62.0 - \mu}{2.0} \leq 2.01$$

or

$$(2.0)(-2.01) - 62.0 \leq -\mu \leq (2.0)(2.01) - 62.0$$

Multiplying by -1, remembering that the sense of an inequality is changed if the terms are multiplied by the same negative number, we obtain

$$62.0 + (2.0)(2.01) \geq \mu \geq 62.0 - (2.0)(2.01)$$

or

$$66.02 \geq \mu \geq 57.98$$

The interval 57.98 to 66.02 that we have just found is called a *confidence interval* and the limits of the interval are called *confidence limits*. The degree of confidence we have in the statement that μ falls within the confidence interval is called a *confidence coefficient*. In the illustrative example, we have determined a 95 percent confidence interval.

Confidence limits are statistics and like all statistics they are also subject to random variation. If we draw another sample of $n = 49$ observations from the same population as the first sample, both the sample mean and sample standard deviation may be expected to be different from the values we obtained for the first sample. Therefore, the 95 percent confidence interval established for the second sample would not necessarily be the same as the one established by the first sample. When we say we are 95 percent confident that μ falls within the 95 percent confidence interval, we are expressing our degree of confidence that, in repeated sampling, such an inference concerning μ will be correct 95 times in 100. For any particular sample, the inference will be right or wrong; that is, either μ falls within the interval or it does not.

We may note that when we establish a confidence interval the procedure implies a test of significance. In essence, with $\alpha = 0.05$ and a two-sided test of significance, we would reject, in the example being considered, any hypothesis that $\mu \leq 57.98$ or that $\mu \geq 66.02$.

With $n = 49$ and with $s = 14.0$, the 95 percent confidence limits are 57.98 and 66.02. Increasing n to 100 observations, that is, slightly more than doubling the sample size, will serve to reduce the confidence interval in two ways, assuming that s^2, the estimate of the population variance, remains the same. In the first place, the standard error of the mean will now be $s_{\bar{X}} = 14.0/\sqrt{100} = 1.4$, as compared with the value of 2.0 when the sample consisted of only 49 observations. In the second place, the values of t cutting off 0.025 of the total area in the two tails of the t distribution for 99 d.f. are -1.984 and 1.984 rather than the values of -2.01 and 2.01 for 48 d.f. We would then have as a 95 percent confidence interval

$$62.0 + (1.4)(1.984) \geq \mu \geq 62.0 - (1.4)(1.984)$$

or

$$64.78 \geq \mu \geq 59.22$$

This 95 percent confidence interval, based on $n = 100$ observations, has a range of $64.78 - 59.22 = 5.56$, whereas that based on $n = 49$ observations had a range of $66.02 - 57.98 = 8.04$. It should be clear that, if we wish a narrow confidence interval, we shall need to make a large number of observations when the estimated standard deviation of the population is as large as 14.0.

5.10 Standard Error of the Difference between Two Means

In an experiment on the influence of two treatments on retention, the treatments were assigned at random in such a way that twenty subjects received Treatment 1 (T_1) and twenty subjects

Table 5.4

RETENTION SCORES FOR TWENTY SUBJECTS ASSIGNED TO TREATMENT 1 AND TWENTY SUBJECTS ASSIGNED TO TREATMENT 2

Treat-ments	Scores				$\sum X$ and $\sum X^2$ for Each Treatment
	12	16	6	10	
	6	13	16	12	$\sum X_1 = 220$
T_1	7	14	13	11	
	12	9	10	9	$\sum X_1^2 = 2596$
	10	14	7	13	
	4	9	1	8	
	12	11	8	9	$\sum X_2 = 160$
T_2	9	0	10	9	
	9	9	8	10	$\sum X_2^2 = 1522$
	14	11	6	3	

received Treatment 2 (T_2). Subjects in both groups were presented with a series of paired words and were asked to guess which word in each pair was "correct." T_1 consisted of giving each subject a slight shock for each wrong response. In T_2 the subjects were not shocked; instead each wrong guess was followed by the flashing of a red light. Subjects in both groups were trained to a criterion set by the experimenter and then retested after a delay of 24 hours. The dependent variable X was the number of correct responses made on the delayed test. The "retention" scores for the subjects in the two groups are given in Table 5.4.

The means for the two treatments are

$$\bar{X}_1 = \frac{220}{20} = 11.0 \qquad \text{and} \qquad \bar{X}_2 = \frac{160}{20} = 8.0$$

The difference between these two means is $\bar{X}_1 - \bar{X}_2 = 11.0 - 8.0 = 3.0$. If the experiment were repeated under the same conditions an indefinitely large number of times, we would not expect to obtain the same values for $\bar{X}_1$ and $\bar{X}_2$ in these repetitions that we obtained in the particular experiment under consideration. The means of both samples are subject to random variation and this will also be true of the difference between the means. However, the distribution of $\bar{X}_1$ will be normally distributed about the population mean μ_1 and the distribution of $\bar{X}_2$ will be normally distributed about the population mean μ_2. The distribution of the difference, $\bar{X}_1 - \bar{X}_2$, will also be normally distributed about the population mean difference $\mu_1 - \mu_2$.

The standard error of the difference between the means of two independent random samples from populations with known standard deviations will be given by

$$\sigma_{\bar{X}_1-\bar{X}_2} = \sqrt{\sigma_{\bar{X}_1}^2 + \sigma_{\bar{X}_2}^2} \tag{5.14}$$

But $\sigma_{\bar{X}_1}^2 = \sigma_1^2/n_1$ and $\sigma_{\bar{X}_2}^2 = \sigma_2^2/n_2$ so that

$$\sigma_{\bar{X}_1-\bar{X}_2} = \sqrt{\frac{\sigma_1^2}{n_1} + \frac{\sigma_2^2}{n_2}} \tag{5.15}$$

In the present problem σ_1^2 and σ_2^2 are unknown, but each can be estimated by means of (5.8). Then the estimated standard error of the difference between the two means will be

$$s_{\bar{X}_1-\bar{X}_2} = \sqrt{\frac{s_1^2}{n_1} + \frac{s_2^2}{n_2}} \tag{5.16}$$

For the moment, let us assume that $\sigma_1^2 = \sigma_2^2$ so that s_1^2 and s_2^2 are both estimates of the same common population variance σ^2. If we have two or more estimates of a common parameter, these may be combined in such a way as to provide a single estimate. In the case of k sample variances, all of which are assumed to estimate the same common population variance, the single estimate is obtained by

$$s^2 = \frac{(n_1 - 1)s_1^2 + (n_2 - 1)s_2^2 + \cdots + (n_k - 1)s_k^2}{(n_1 - 1) + (n_2 - 1) + \cdots + (n_k - 1)} \tag{5.17}$$

with degrees of freedom equal to $\Sigma n_k - k$. If all the n's are equal, then the degrees of freedom will be equal to $k(n - 1)$, where n is the number of observations in each sample. We let

$$\sum x_k^2 = \sum (X - \bar{X})^2 \tag{5.18}$$

be the sum of squared deviations of the n observations in a given sample from the sample mean. Then we also have $\Sigma x_k^2 = (n_k - 1)s_k^2$ and

$$s^2 = \frac{\sum x_1^2 + \sum x_2^2 + \cdots + \sum x_k^2}{\sum n_k - k} \tag{5.19}$$

For the present problem, we have $k = 2$ and therefore

$$s^2 = \frac{\sum x_1^2 + \sum x_2^2}{n_1 + n_2 - 2} \tag{5.20}$$

with degrees of freedom equal to $n_1 + n_2 - 2$. Substituting with the single estimate s^2 for the separate estimates s_1^2 and s_2^2 in (5.16), we have

$$s_{\bar{X}_1-\bar{X}_2} = \sqrt{\frac{\left(\dfrac{\sum x_1^2 + \sum x_2^2}{n_1 + n_2 - 2}\right)}{n_1} + \frac{\left(\dfrac{\sum x_1^2 + \sum x_2^2}{n_1 + n_2 - 2}\right)}{n_2}} \tag{5.21}$$

and this may be written

$$s_{\bar{X}_1-\bar{X}_2} = \sqrt{\left(\frac{\sum x_1^2 + \sum x_2^2}{n_1 + n_2 - 2}\right)\left(\frac{1}{n_1} + \frac{1}{n_2}\right)} \tag{5.22}$$

We observe also that if $n_1 = n_2 = n$, then

$$s_{\bar{X}_1-\bar{X}_2} = \sqrt{\frac{2s^2}{n}} \tag{5.23}$$

where n is the number of observations in each group.

It should be emphasized that Σx_1^2 refers to the sum of squared deviations of the n_1 observations, obtained under Treatment 1, about the mean for Treatment 1, and similarly, Σx_2^2 refers to the sum of the squared deviations of the n_2 observations, obtained under Treatment 2, about the mean for Treatment 2. A convenient method for calculating these sums of squares is

$$\sum x_k^2 = \sum (X - \bar{X})^2 = \sum X^2 - \frac{(\sum X)^2}{n} \tag{5.24}$$

By (5.24) we find that the sum of squares for Treatment 1 is

$$\sum x_1^2 = 2596 - \frac{(220)^2}{20} = 176$$

and the sum of squares for Treatment 2 is

$$\sum x_2^2 = 1522 - \frac{(160)^2}{20} = 242$$

Substituting in (5.22) for the standard error of the difference between the means, we have

$$s_{\bar{X}_1-\bar{X}_2} = \sqrt{\left(\frac{176 + 242}{20 + 20 - 2}\right)\left(\frac{1}{20} + \frac{1}{20}\right)} = 1.049$$

5.11 Confidence Interval for a Difference between Two Means

We can, in the manner described previously, find 95 or 99 percent confidence limits for the population mean difference $\mu_1 - \mu_2$. With $n_1 + n_2 - 2 = 38$ d.f., we find that $t = -2.711$ will cut off 0.005 of the total area in the left tail and $t = 2.711$ will cut off 0.005 of the total area in the right tail of the t distribution. Then the 99 percent confidence limits will be given by

$$-2.711 \leq \frac{(11 - 8) - (\mu_1 - \mu_2)}{1.049} \leq 2.711$$

or

$$3 + (1.049)(2.711) \geq \mu_1 - \mu_2 \geq 3 - (1.049)(2.711)$$

$$5.84 \geq \mu_1 - \mu_2 \geq 0.16$$

and we can say that we are 99 percent confident that the population mean difference, $\mu_1 - \mu_2$, is within these limits.

5.12 Test of Significance for a Difference between Two Means

If our major interest is in determining whether a specified null hypothesis concerning $\mu_1 - \mu_2$ is to be rejected, then this hypothesis may be tested by finding

$$t = \frac{(\bar{X}_1 - \bar{X}_2) - (\mu_1 - \mu_2)}{s_{\bar{X}_1 - \bar{X}_2}} \qquad (5.25)$$

Specifically, if the null hypothesis is $\mu_1 = \mu_2$, so that $\mu_1 - \mu_2 = 0$, then

$$t = \frac{\bar{X}_1 - \bar{X}_2}{s_{\bar{X}_1 - \bar{X}_2}} \qquad (5.26)$$

and for the present problem we have

$$t = \frac{11 - 8}{1.049} = 2.86$$

with 38 d.f. With $\alpha = 0.01$ and a two-sided test, the null hypothesis would be rejected.

5.13 The Null Hypothesis and Alternatives

In general, we test a null hypothesis against a class of alternative hypotheses. If the null hypothesis is false, so that one of the alternative hypotheses is true, then we can define the power of a test of significance as

$$\text{Power} = 1 - P(\text{Type II error})$$

Because the probability of a Type II error is the probability of *not* rejecting the null hypothesis when it is false, the power of a test of significance can be said to be the probability of rejecting the null hypothesis when it *should* be rejected.

One way in which we can increase the power of a given test is to make α large. But we do not like to make α too large, because by doing so we

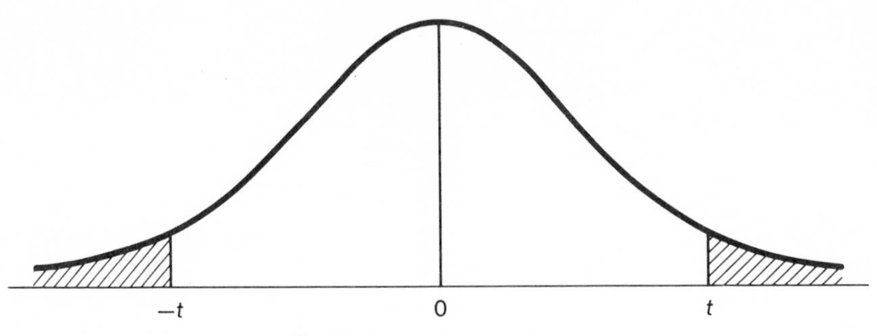

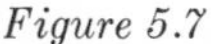

Figure 5.7
The two-sided test of significance of the null hypothesis $\mu_1 = \mu_2$ against the alternative $\mu_1 \neq \mu_2$. Each of the shaded areas in the two tails of the distribution is 0.025 of the total area. With $\alpha = 0.05$, the null hypothesis is rejected if the observed value of t falls in either of the two shaded areas.

increase the probability of a Type I error. If we hold α constant, then we can also increase the power of a given test by increasing the number of observations in the sample under consideration. If we hold both α and the number of observations in the sample constant, then we can increase the power of a test against a *selected* class of alternatives to the null hypothesis by the manner in which we choose the critical region of rejection in the t distribution. It is this latter method of increasing the power of a test that we now consider.

If we designate the null hypothesis as H_0 and the alternative to this hypothesis as H_1, then we may be interested in any one of the following three tests:

Test 1 $\quad H_0\colon \mu_1 = \mu_2$ with $H_1\colon \mu_1 \neq \mu_2$

Test 2 $\quad H_0\colon \mu_1 \leq \mu_2$ with $H_1\colon \mu_1 > \mu_2$

Test 3 $\quad H_0\colon \mu_1 \geq \mu_2$ with $H_1\colon \mu_1 < \mu_2$

Suppose we choose $\alpha = 0.05$. If we make Test 1, we shall reject the null hypothesis if the t we obtain falls in either of the two shaded areas of Figure 5.7. With 38 d.f., the critical values of t, those that would result in the rejection of the null hypothesis, are $t \leq -2.025$ and $t \geq 2.025$. These are the values of t cutting off 0.025 of the total area in each tail of the t distribution. Because the areas of rejection for Test 1 are in either one of the two tails of the t distribution, this test is called a *two-tailed* or *two-sided* test. Test 1 provides protection against the possibility that $\mu_1 > \mu_2$ and also the possibility that $\mu_1 < \mu_2$. In other words, it is sensitive to the absolute value of the difference between μ_1 and μ_2. Test 1 is the one we should use if we are interested in the absolute magnitude of the difference between μ_1 and μ_2 and not specifically in the direction of the difference.

If we make Test 2, then we shall reject the null hypothesis only if the

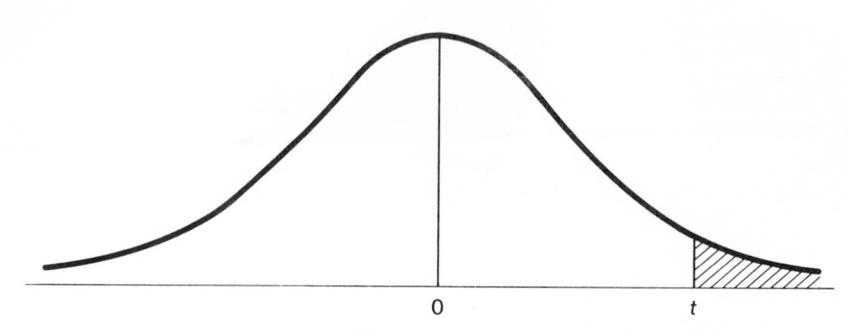

Figure 5.8
The one-sided test of significance of the null hypothesis $\mu_1 \leq \mu_2$ against the alternative $\mu_1 > \mu_2$. The shaded area in the right tail of the t distribution is 0.05 of the total area. With $\alpha = 0.05$, the null hypothesis is rejected if the observed value of t falls in the shaded area.

obtained value of t falls in the shaded area of Figure 5.8. If $\alpha = 0.05$, then we want the area in the right tail to correspond to 0.05 of the total area of the t distribution. With 38 d.f., the critical value of t, cutting off 0.05 of the total area in the right tail, is approximately 1.68. Because the area of rejection is the right tail of the t distribution, Test 2 is referred to as a *right-tailed*, a *one-tailed*, or a *one-sided* test. Test 2 provides protection against the class of alternatives $\mu_1 > \mu_2$ only. If it is true that $\mu_1 > \mu_2$, then Test 2 will be somewhat more powerful than Test 1 against this class of alternatives, but, unlike Test 1, Test 2 provides no protection against the possibility that $\mu_1 < \mu_2$. Test 2 should be used only if we have no interest whatsoever in the possibility that $\mu_1 < \mu_2$.

With Test 3, the region of rejection of the null hypothesis is the left tail of the t distribution, as shown in Figure 5.9. If $\alpha = 0.05$ and with 38 d.f.,

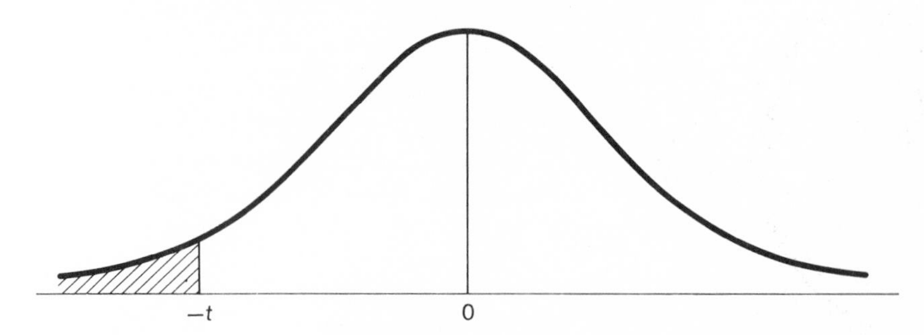

Figure 5.9
The one-sided test of significance of the null hypothesis $\mu_1 \geq \mu_2$ against the alternative $\mu_1 < \mu_2$. The shaded area in the left tail of the t distribution is 0.05 of the total area. With $\alpha = 0.05$, the null hypothesis is rejected if the observed value of t falls in the shaded area.

then for Test 3 the critical value of t is approximately -1.68. Test 3, like Test 2, is a one-tailed or one-sided test. Test 3 provides protection only against the class of alternatives $\mu_1 < \mu_2$. If it is true that $\mu_1 < \mu_2$, then Test 3 will be somewhat more powerful than Test 1 against this class of alternatives, but will not protect against the possibility that $\mu_1 > \mu_2$. Test 3 should be used, therefore, only if we have no interest whatsoever in the possibility that $\mu_1 > \mu_2$.

We have suggested elsewhere (Edwards, 1967b) that Test 1, almost without exception, is the preferred test in scientific research because it provides protection against both the possibility that $\mu_1 < \mu_2$ and that $\mu_1 > \mu_2$ and, in general, the scientist should be interested in both possibilities.

It should also be emphasized that if Test 2 or Test 3 is to be made in a given experiment, this decision *must* be made at the time the experiment is planned and should not be suggested by an examination of the outcome of the experiment. It may sometimes happen that the difference between two means will be declared significant with $\alpha = 0.05$ and if a one-sided test is made, but nonsignificant with $\alpha = 0.05$ if a two-sided test is made. To decide, after looking at the outcome of the experiment, that a one-sided test is to be made is not only unscientific, it is also dishonest.

We have previously discussed some experiments where a one-sided test is appropriate. In discrimination experiments, for example, we are ordinarily interested in alternatives to the null hypothesis that indicate a better than chance ability to make correct discriminations. As another example, consider the case of the farmer from Whidbey Island. The null hypothesis we tested was $P \leq \frac{1}{2}$ against the alternative $P > \frac{1}{2}$. In using the standard normal distribution to evaluate the outcome of this experiment, the region of rejection was the right tail of the standard normal distribution; that is, we made a one-sided test corresponding to Test 2. This particular test was made because we were only interested in the possibility that the farmer would do better than chance and we had no interest whatsoever in the possibility that his performance would be worse than chance.

5.14 Estimating the Number of Observations Needed in Comparing Two Treatment Means

Assume that on the basis of previous research or a pilot study that we have some knowledge as to the variability to be expected in a dependent variable X under a given set of treatments. It will, in fact, simplify the presentation if we can assume that the common population variance σ^2 is known and we shall make this assumption. Suppose also that we set $\alpha = 0.05$ and that we have decided upon a two-sided test of significance. For a two-sided test, with $\alpha = 0.05$, the critical values of Z are -1.96 and 1.96. In addition, we decide that the population mean difference, $\mu_1 - \mu_2$, must be equal to or greater than δ or equal to or less than $-\delta$, to be of either theoretical or practical interest. Furthermore, we want the probability of a Type II error to be no greater than 0.16 if the true difference between μ_1 and μ_2 is equal to or greater than δ or equal to or less than $-\delta$. We have previously defined the power of a test as $1 - P(\text{Type II error})$. Therefore, we desire the test to have a power of $1 - 0.16 = 0.84$, which is to say that we want the test to have a probability of at least 0.84 of rejecting the null hypothesis if it is true that

$$\delta \leq \mu_1 - \mu_2 \leq -\delta$$

Consider first only the possibility that $\mu_1 - \mu_2$ is at least δ. Figure 5.10 shows the distribution of $\bar{X}_1 - \bar{X}_2$, when $\mu_1 - \mu_2 = \delta$, at the right, and the distribution of $\bar{X}_1 - \bar{X}_2$, when $\mu_1 - \mu_2 = 0$, at the left. Let $Z_0 = 1.96$ be the critical value of Z resulting in the rejection of the null hypothesis when it is true, that is, when $\mu_1 - \mu_2 = 0$. If the null hypothesis is true, and if $n_1 = n_2 = n$, then we want the value of c in Figure 5.10, expressed as a standard normal variable, to be

$$Z_0 = \frac{c - 0}{\sqrt{\dfrac{2\sigma^2}{n}}} = 1.96$$

If the null hypothesis is false and if $\mu_1 - \mu_2 = \delta$, and if we obtain a difference that falls to the left of c, the null hypothesis will not be rejected and we shall make a Type II error. We want this probability to be no greater than 0.16. Thus, when $\mu_1 - \mu_2 = \delta$, we want 0.16 of the total area in the curve at the right in Figure 5.10 to fall to the left of c and 0.84 to the right of c. Let Z_1 be the Z value corresponding to c, in this instance, and from the table of the standard normal distribution we find that $Z_1 = -1.00$. Then we also

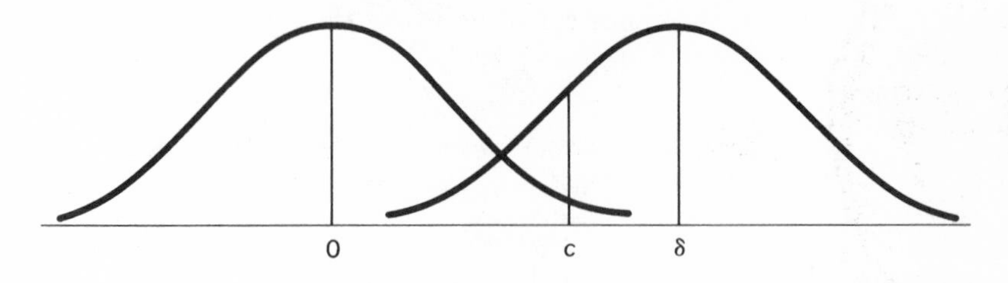

Figure 5.10
The distribution of $\bar{X}_1 - \bar{X}_2$ when $\mu_1 - \mu_2 = 0$ is shown at the left and the distribution of $\bar{X}_1 - \bar{X}_2$ when $\mu_1 - \mu_2 = \delta$ is shown at the right. The point c is located so as to cut off 0.025 of the total area in the right tail of the distribution at the left and 0.16 of the total area in the left tail of the distribution shown at the right.

have

$$Z_1 = \frac{c - \delta}{\sqrt{\dfrac{2\sigma^2}{n}}} = -1.00$$

Using the above two equations we have

$$c = 1.96\sqrt{\frac{2\sigma^2}{n}} \qquad \text{and} \qquad c = \delta - \sqrt{\frac{2\sigma^2}{n}}$$

Then

$$\delta = 1.96\sqrt{\frac{2\sigma^2}{n}} + \sqrt{\frac{2\sigma^2}{n}}$$

and

$$n = \frac{2\sigma^2}{\delta^2}(1.96 + 1.00)^2 \tag{5.27}$$

Assuming that $\sigma = 10.0$ and that $\delta = 5$, we have

$$n = \frac{2(10)^2}{(5)^2}(2.96)^2 = 70$$

and we would want to have at least $n = 70$ subjects in each treatment group if the test of significance is to have the properties described. We have considered only the possibility that $\delta = 5$. We would arrive at exactly the same value for n by considering the possibility that $\delta = -5$.

If the various assumptions we have made are correct, then for the two-sided test of significance the probability of a Type I error will be 0.05 and the probability of a Type II error will be 0.16, if $\delta = \mu_1 - \mu_2 = 5$ or if $\delta = \mu_1 - \mu_2 = -5$. If the absolute value of δ is greater than 5, then the probability of a Type II error will be less than 0.16.

If we express δ in terms of σ, so that $\Delta = \delta/\sigma$, then $\delta = \Delta\sigma$ and, substituting in (5.27), we have

$$n = \frac{2\sigma^2}{\Delta^2\sigma^2}(2.96)^2$$

$$= \frac{2}{\Delta^2}(2.96)^2 \tag{5.28}$$

Table 5.5 gives the number of observations needed in each of two groups for a two-sided test with the probability of a Type I error set at 0.05 and the probability of a Type II error set at 0.16 for various values of Δ. It is obvious that (5.28) could be modified to determine the number of

Table 5.5

NUMBER OF OBSERVATIONS NEEDED IN EACH OF TWO GROUPS FOR A TWO-SIDED TEST WITH P(TYPE I ERROR) = 0.05 AND P(TYPE II ERROR) = 0.16 FOR VARIOUS VALUES OF Δ

Δ	n	Δ	n
0.10	1752	0.60	49
0.15	779	0.65	42
0.20	438	0.70	36
0.25	281	0.75	31
0.30	195	0.80	28
0.35	143	0.85	25
0.40	110	0.90	22
0.45	87	0.95	20
0.50	70	1.00	18
0.55	58		

observations needed in each of two treatment groups, if we decide on some value other than 0.05 for the probability of a Type I error and also some value other than 0.16 for the probability of a Type II error.

5.15 *Optimum Allocation of the Total Number of Observations*

When we have an experiment in which two treatments are to be compared, the observations for Treatment 1 provide a variance estimate, s_1^2, and those for Treatment 2 also provide a variance estimate, s_2^2. If these two independent variances are both estimates of the same population variance, σ^2, then they may be combined, as shown previously, to obtain a single estimate of the population variance. When this is the case, then it can also be shown that the optimum allocation of the total number of observations is such that we should have the same number for each treatment, that is, so that $n_1 = n_2$. This is the optimum allocation, under the conditions described, in the sense that if $n_1 = n_2$, then the standard error of the difference between the two treatment means will be smaller than it would be if $n_1 \neq n_2$.

5.16 *Use of a Table of Random Numbers*

In behavioral science experiments a given observation is most often associated with a given subject or organism. To select at random two groups of subjects from a larger group or to divide at random a given number of subjects into two groups, we make use of a table of random numbers, Table I in the Appendix. The table consists of five blocks of 1000 random numbers each. For each block, the rows have been numbered 00 to 24 and the columns, reading downward, from 00 to 39. To find a point of entry into the table, some random procedure should be used. Many methods can be devised for determining a point of random entry into the table. For example, we might put the numbers 00, 01, 02, . . . , 39 on white disks—poker chips work fine. If we put the disks numbered 01 to 05 in a box, shake the box thoroughly, and draw one disk from the box, this number will give the block to be entered in the table. In the same manner, by putting the disks with numbers 00 to 24 in the box and selecting one at random, we can obtain a number corresponding to a row of the table. Then, with disks 00 to 39 in the box a random selection of one of the disks will provide a number corresponding to a column of the table. These three numbers will give a point of entry into the table. Once we have the point of entry, it makes no difference whether we read up, down, or across the table. Let us suppose the point of entry is block 02, row 01, and column 05.

Suppose we have one hundred subjects and we wish to select at random two groups of ten subjects each. We assign the numbers 00, 01, 02, . . . , 99 to the 100 subjects. It does not matter which subject receives which number; it is only necessary that each subject have a different number. Because the numbers assigned to the subjects consist of two-digit numbers, we shall make use of columns 05 and 06 and read downward. We read down these two columns at the point of entry, selecting the first twenty unlike numbers in the set 00 to 99. The first few numbers we encounter are 52, 98, 55, 94, 87, 42, and 30. We continue reading until we have twenty unlike numbers corresponding to twenty of the one hundred subjects. The first ten subjects selected this way may then be assigned to one of the two treatments, and the second ten subjects to the other treatment.

We can, obviously, use the same method to divide a fixed number of subjects at random into two groups of $n_1 = n_2$ subjects.

QUESTIONS AND PROBLEMS

5.1 Find a 95 percent confidence interval for a random sample of $n = 16$ observations with $\bar{X} = 22.4$ and $s = 4.3$.

5.2 The mean score on a standardized test for a random sample of 200 freshmen college students at University A is 133.8, with s equal to 14.7. For a random sample of 140 freshmen at University B, the mean score is 138.4, with s equal to 15.2. Determine whether the two means differ significantly.

5.3 Forty subjects are assigned at random to two treatments, with twenty subjects for each treatment. The measures on the dependent variable are given below:

Treatment 1				Treatment 2			
39	41	39	44	36	41	30	39
39	40	39	40	36	39	33	37
37	42	37	43	35	42	36	37
44	38	38	38	34	38	33	31
43	38	41	39	40	32	33	38

Determine whether the two treatment means differ significantly. If an automatic calculating machine is not available, the calculations may be somewhat easier if a constant, say thirty, is subtracted from each measure. If the same constant is subtracted from each measure, this will not influence the difference between the means, nor will it change the variance.

5.4 Morgan (1945) designed an experiment to test the hypothesis that failure to solve a problem tends to foster inductive reasoning more than immediate success. "*S*'s were confronted with the problem of discovering which of six cues to follow in order to make a bell ring. In one group (called the restricted hypothesis group) the cue which would make the bell ring was predetermined by the *E*. In another group (called the unrestricted hypothesis group) success followed the use of any cue by the *S*. Interspersed throughout the experiment were test series to determine how well the *S*'s in both groups could discover a predetermined cue" (p. 146). The question is whether the restricted hypothesis group profited by the mistakes made in searching for the correct cue and surpassed, on the test series, the performance of the subjects in the unrestricted group. The data are as follows:

Unrestricted Group					Restricted Group				
6	12	14	19	35	4	8	9	12	25
7	12	14	23		5	8	9	13	
8	12	15	24		6	8	10	13	
10	13	15	30		6	9	10	15	
10	14	16	34		7	9	11	15	

Determine whether the two treatment means differ significantly.

5.5 In an experiment the sum of squared deviations for one treatment group was 420 and for the other treatment group the sum of squared deviations was 482. Each group had $n = 25$ subjects. The difference between the treatment means was 3.03. Is this difference significant at the 5 percent level?

5.6 Measures obtained on a dependent variable for two treatment groups are given below:

Treatment 1				Treatment 2			
52	171	151	45	71	86	218	165
75	54	101		95	141	152	
170	104	74		151	52	120	
30	81	146		53	108	115	

Determine whether the difference between the treatment means is significant at the 5 percent level.

5.7 In an experiment, twenty rats were randomly assigned to each of two conditions. The experimental condition consisted of giving each rat a 12-hour period of exploration in a maze. The other group served as a control group and was not given a period of exploration. Both groups were deprived of food for the same length of time and tested in the maze. Records were kept of the number of trials required to learn the maze to a criterion of one run with no errors. Data for the two groups are given below:

Control				Experimental			
10	7	9	6	12	7	9	6
8	6	10	13	5	9	9	9
9	7	12	12	6	4	8	4
15	6	9	11	9	10	11	6
9	13	4	9	9	7	10	7

Determine whether the difference between the means for the control and experimental groups is significant at the 5 percent level.

5.8 In an experiment the standard error of the difference between two means was 1.42 with $n = 10$ subjects in each treatment group. A repetition of this experiment is planned and the experimenter wishes to be able to reject the null hypothesis if the absolute difference between the population means is 2.56 or greater. On the basis of the data available, it is possible to solve for s^2. Assume that s^2 is the population variance. (*a*) How many subjects should the experimenter have in each group, if $\alpha = 0.05$ and if the probability of a Type II error is to be no greater than 0.16? (*b*) How many subjects should the experimenter have in each group if $\alpha = 0.05$ and if the probability of a Type II error is to be no greater than 0.50?

5.9 Give a brief interpretation of the meaning of confidence limits.

5.10 Comment upon the following statement: Establishing confidence limits always implies a test of significance.

5.11 Discuss briefly the t test of a null hypothesis concerning a difference between two means in relation to the alternatives to the null hypothesis. (*a*) Under what conditions should we make a two-sided test? (*b*) Under what conditions should we make a right-tailed test? (c) Under what conditions should we make a left-tailed test? (d) Give an example where each test might be appropriate.

5.12 The null hypothesis to be tested in an experiment is $\mu_1 = \mu_2$. The experimenter sets $\alpha = 0.05$ for a two-sided test of the null hypothesis. Suppose that σ is known and that for the number of subjects in the two treatment groups, we have $\sigma_{\bar{X}_1-\bar{X}_2} = 1.5$. If the null hypothesis is false and if it is true that $\mu_1 - \mu_2 = 6.0$, what is the power of the test of significance?

5.13 In an experiment involving two treatments, we have the following results:

Group 1	Group 2
$\sum X_1 = 400$	$\sum X_2 = 520$
$\sum x_1^2 = 720$	$\sum x_2^2 = 800$
$n_1 = 20$	$n_2 = 20$

(a) Test the significance of the difference between the means. (b) Establish a 95 percent confidence interval for the difference between μ_1 and μ_2.

5.14 Ten rats are randomly divided into two groups of five rats each. One group is assigned to one treatment and the other to another treatment. Measures on a dependent variable for each rat are given below:

T_1	T_2
3	7
5	5
2	6
1	3
4	4

(a) If $\alpha = 0.05$, will the null hypothesis $\mu_1 = \mu_2$ be rejected? (b) Establish a 95 percent confidence interval for the difference between μ_1 and μ_2.

5.15 For a random sample of $n = 49$ observations, we have $s^2 = 196.00$ and $\bar{X} = 54.00$. Find a 95 percent confidence interval for μ.

5.16 We have two independent random samples of $n_1 = n_2 = 20$ observations each. For the two samples we have $\Sigma(X_1 - \bar{X}_1)^2 = 720.0$, $\Sigma(X_2 - \bar{X}_2)^2 = 800.0$, $\bar{X}_1 = 40.0$, and $\bar{X}_2 = 30.0$. (a) Use the t test to test the null hypothesis that the two samples have been drawn from the same normally distributed population. (b) Find a 95 percent confidence interval for $\mu_1 - \mu_2$.

5.17 What is meant by the power of a test of significance?

CHAPTER 6

HETEROGENEITY OF VARIANCE

6.1 Introduction

The methods used in determining the standard error of the difference between two means and the t test used in evaluating the difference between the means, as described in the last chapter, are based on the assumption that the separate variance estimates provided by the two samples are both estimates of the same population variance. In general, this is a reasonable assumption, but sometimes the effect of one treatment may be such as to either increase or decrease the variability of observations, whereas the effect of the other treatment may not.

In this chapter we describe a test of significance of the difference between two sample variances and a t test that may be used in evaluating the difference between two sample means, even though the sample variances do differ significantly. In addition, we describe some conditions that may result in a significant difference between two variances.

6.2 The F Distribution

Given two sample variances, s_1^2 and s_2^2, based on two independent random samples of n_1 and n_2 observations, we may test the null hypothesis $\sigma_1^2 = \sigma_2^2$ against the alternative hypothesis $\sigma_1^2 \neq \sigma_2^2$. The ratio of the two sample variances is distributed in a manner discovered by Fisher (1936) and the significant values of the ratio at the 0.05 and 0.01 levels of significance have been calculated by Snedecor (1956), who named the ratio F in Fisher's honor. The significant values of F at the 0.25, 0.10, 0.025, and 0.005 levels of significance have been calculated by Merrington and Thompson (1943). Now if F is defined as

$$F = \frac{s_1^2}{s_2^2} \qquad \text{or} \qquad F = \frac{s_2^2}{s_1^2} \tag{6.1}$$

or as the ratio of two variances, then whether F will be greater than 1.0 or smaller than 1.0 will depend merely on whether s_1^2 or s_2^2 is put in the numerator of the ratio. The tabled values of F, Table VIII and Table IX in the Appendix, are for a *one-sided* or *one-tailed* test and correspond to the probability of F greater than 1.0, when the null hypothesis is true.[1] Then to use the tables, we shall always find the value of F greater than 1.0 in (6.1) and this means that we shall always put the larger of the two sample variances in the numerator.

If the alternative to the null hypothesis is $\sigma_1^2 \neq \sigma_2^2$—and in experimental work it usually is—then to protect against this alternative we need to make a *two-sided* or *two-tailed* test; that is, we want to reject the null hypothesis if either $\sigma_1^2 > \sigma_2^2$ or if $\sigma_1^2 < \sigma_2^2$. For the two-sided test, with $\alpha = 0.05$, the critical value of F will be the tabled value with probability 0.025. Similarly, for a two-sided test with $\alpha = 0.01$, the critical value of F will be the tabled value with probability 0.005.

The F test, as defined by (6.1), is often referred to as a test of homogeneity of variance. If a nonsignificant value of F is obtained, the two sample variances are said to be *homogeneous*, that is, they are both assumed to be the estimates of the same population variance. With a significant value of F, the variances would be said to be *heterogeneous*.

6.3 Test of Homogeneity of Two Variances

In the experiment on retention described in the previous chapter, we said that we would assume that s_1^2 and s_2^2 did not differ significantly and that we would show later that this assumption was tenable. We had for Treatment 1, $\Sigma x_1^2 = 176$, and for Treatment 2, $\Sigma x_2^2 = 242$. Then

$$s_1^2 = \frac{176}{19} = 9.263 \qquad \text{and} \qquad s_2^2 = \frac{242}{19} = 12.737$$

Because s_2^2 is larger than s_1^2, we have

$$F = \frac{12.737}{9.263} = 1.375$$

[1] Table VIII is Snedecor's table and gives the 5 and 1 percent points for the distribution of F. Table IX is the Merrington and Thompson table and gives the 25, 10, 2.5, and 0.5 percent points for the distribution of F.

Table 6.1

MEANS AND VARIANCES FOR TWO TREATMENTS WITH UNEQUAL n'S

Treatment 1	Treatment 2
$\bar{X}_1 = 20.6$	$\bar{X}_2 = 16.0$
$s_1^2 = 28.42$	$s_2^2 = 6.72$
$n_1 = 10$	$n_2 = 20$

To determine whether $F = 1.375$ is significant, we enter the column of Table IX with the degrees of freedom corresponding to the numerator of the F ratio and the row with the degrees of freedom corresponding to the denominator. For the obtained $F = 1.375$, we have 19 d.f. for the numerator and 19 d.f. for the denominator. Table IX has no column corresponding to 19 d.f., but we find that the critical value of F, with $\alpha = 0.05$, for 20 and 19 d.f. is 2.51. The obtained value of $F = 1.375$ is less than this critical value and, with $\alpha = 0.05$, the null hypothesis would not be rejected.

6.4 The t Test with $\sigma_1^2 \neq \sigma_2^2$ and with $n_1 \neq n_2$

Let us suppose, in an experiment comparing two treatments, that we do not have $n_1 = n_2$ and that the two sample variances are heterogeneous. Consider, for example, the data of Table 6.1. Testing for homogeneity of variance we have, because s_1^2 is larger than s_2^2,

$$F = \frac{28.42}{6.72} = 4.23$$

with 9 and 19 d.f. With $\alpha = 0.05$, the critical value of F is 2.88. Because the obtained value of $F = 4.23$ exceeds the critical value, we reject the null hypothesis.

In the previous chapter we assumed that s_1^2 and s_2^2 were estimates of the same population variance and combined the separate estimates in such a way as to provide a single estimate of the common population variance. But in the present example we have rejected the null hypothesis $\sigma_1^2 = \sigma_2^2$. Therefore, s_1^2 and s_2^2 cannot be said to be estimates of the same population variance and because $n_1 \neq n_2$, then

$$s_{\bar{X}_1-\bar{X}_2} = \sqrt{\frac{\sum x_1^2 + \sum x_2^2}{n_1 + n_2 - 2}\left(\frac{1}{n_1} + \frac{1}{n_2}\right)} \qquad (6.2)$$

for the standard error of the difference between the two means is inappropriate. Instead of using a single estimate s^2, as in (6.2), to find the standard error of the difference between the two means, we shall use the separate estimates, s_1^2 and s_2^2. In this case we have

$$s_{\bar{X}_1-\bar{X}_2} = \sqrt{\frac{s_1^2}{n_1} + \frac{s_2^2}{n_2}} \tag{6.3}$$

as the standard error of the difference between two means. Then, using (6.3) for the data in Table 6.1, we have

$$s_{\bar{X}_1-\bar{X}_2} = \sqrt{\frac{28.42}{10} + \frac{6.72}{20}} = 1.783$$

and

$$t = \frac{20.6 - 16.0}{1.783} = 2.58$$

To determine whether $t = 2.58$ is significant we first find, from Table V, the critical values of t_1 for $n_1 - 1 = 9$ d.f. and t_2 for $n_2 - 1 = 19$ d.f. For a two-sided test, with $\alpha = 0.05$, these two values are $t_1 = 2.262$ and $t_2 = 2.093$. Then we find

$$t' = \frac{t_1 \dfrac{s_1^2}{n_1} + t_2 \dfrac{s_2^2}{n_2}}{\dfrac{s_1^2}{n_1} + \dfrac{s_2^2}{n_2}} \tag{6.4}$$

The value of t' obtained from (6.4) is the critical value in terms of which the obtained $t = 2.58$ will be evaluated. Substituting in (6.4), we have

$$t' = \frac{2.262 \dfrac{28.42}{10} + 2.093 \dfrac{6.72}{20}}{\dfrac{28.42}{10} + \dfrac{6.72}{20}} = 2.24$$

Because the obtained $t = 2.58$ exceeds $t' = 2.24$, the null hypothesis will be rejected. Because we have obtained both a significant value of F in testing the null hypothesis $\sigma_1^2 = \sigma_2^2$ and a significant value of t in testing the null hypothesis $\mu_1 = \mu_2$, we conclude that the two treatments have resulted in a significant difference in the treatment variances and also in the treatment means.

6.5 The t Test with $\sigma_1^2 \neq \sigma_2^2$ and with $n_1 = n_2$

For the variance of the difference between two means, assuming homogeneity of variance, we have the square of (6.2) or

$$s_{\bar{X}_1-\bar{X}_2}^2 = \frac{\sum x_1^2 + \sum x_2^2}{n_1 + n_2 - 2}\left(\frac{1}{n_1} + \frac{1}{n_2}\right) \tag{6.5}$$

If $n_1 = n_2 = n$, then (6.5) may be written as

$$s_{\bar{X}_1-\bar{X}_2}^2 = \frac{\sum x_1^2 + \sum x_2^2}{2n - 2}\left(\frac{2}{n}\right)$$

or

$$s_{\bar{X}_1-\bar{X}_2}^2 = \frac{\sum x_1^2 + \sum x_2^2}{n(n - 1)} \tag{6.6}$$

For the variance of the difference between two means, with heterogeneity of variance, we have the square of (6.3) or

$$s_{\bar{X}_1-\bar{X}_2}^2 = \frac{\dfrac{\sum x_1^2}{n_1 - 1}}{n_1} + \frac{\dfrac{\sum x_2^2}{n_2 - 1}}{n_2} \tag{6.7}$$

If $n_1 = n_2 = n$, then we can also write (6.7) as

$$s_{\bar{X}_1-\bar{X}_2}^2 = \frac{\sum x_1^2 + \sum x_2^2}{n(n - 1)} \tag{6.8}$$

We see that when $n_1 = n_2 = n$, then (6.2) and (6.3) are identical. Therefore, if $n_1 = n_2 = n$, we will obtain the same standard error of the difference between two means, regardless of whether we use (6.2) or (6.3).

The critical value for evaluating t, however, will depend on whether we have homogeneity of variance. With homogeneity of variance, the critical value of t, at some given significance level, will be the tabled value for $n_1 + n_2 - 2 = 2(n - 1)$ d.f. On the other hand, if we have heterogeneity of variance, then the obtained value of t will be evaluated in terms of t' as defined by (6.4). With $n_1 = n_2 = n$, we will have $t_1 = t_2 = t'$ and, if this is the case, then it is obvious that t' is simply the tabled value of t with $n_1 - 1 = n_2 - 1 = n - 1$ d.f. Thus, with equal n's and heterogeneity of variance, we may calculate

$$t = \frac{\bar{X}_1 - \bar{X}_2}{\sqrt{\dfrac{\sum x_1^2 + \sum x_2^2}{n(n - 1)}}} \tag{6.9}$$

but the t as defined by (6.9) would then be evaluated in terms of the tabled value for ½ the number of degrees of freedom that we would have with homogeneity of variance; that is, the critical value will be the tabled value with $n - 1$ d.f. and not the tabled value with $2(n - 1)$ d.f.

The t test described above, with $n_1 = n_2 = n$ and with heterogeneity of variance, is a conservative test, in the sense that the tabled value of t for $n - 1$ d.f. will be larger than the tabled value for $2(n - 1)$ d.f. for the same value of α. Other than the fact that it is a conservative test, there is no other basis for recommending it because, as will be pointed out later in the chapter, the t test for the difference between two means is quite insensitive to heterogeneity of variance when $n_1 = n_2$. Therefore, we suggest that one should calculate

$$t = \frac{\bar{X}_1 - \bar{X}_2}{\sqrt{\frac{s_1^2}{n_1} + \frac{s_2^2}{n_2}}} \tag{6.10}$$

and use t', as defined by (6.4), for evaluating the obtained value of t only when n_1 and n_2 differ considerably and if, in addition, the F test results in the rejection of the null hypothesis $\sigma_1^2 = \sigma_2^2$.

6.6 Nonrandom Assignment of Subjects

In this and the sections that follow we consider some conditions under which we may expect s_1^2 and s_2^2 to differ significantly. One of the most obvious conditions is when the subjects are not assigned at random to the two groups. If one of the groups initially included those subjects that are more homogeneous in their performance than the other, then we might also expect the two groups to differ in variability at the conclusion of the experiment. We hope to rule out this possible explanation by randomly assigning the subjects to the two groups. Then, if the two variances differ significantly, we might be justified in assuming that the difference is a result of the treatments. We say we *might* be justified in this assumption because, even with random assignment, we know that five times in one hundred a significant difference between the two variances will occur simply as a result of random variation.

6.7 Nonadditivity of a Treatment Effect

If the variances for two treatment groups differ significantly, this may be because the treatment effects are not additive. By "additive" we mean that if X_1 is the value of a given observation under a control condition, then under the treatment condition we would have $X_2 = X_1 + t_1$, where t_1 represents a constant treatment effect. If a treatment effect is additive, then it can easily be shown that $s_2^2 = s_1^2$ because the addition or subtraction of a constant has no influence on the variance.

On the other hand, suppose that the treatment, instead of acting in an additive fashion, acts in a multiplicative fashion. Then we would have $X_2 = X_1t_1$ and it can easily be shown that the variance of X_2 in relation to the variance of X_1 will be $s_2^2 = s_1^2t_1^2$, because multiplying each value of a variable by a constant serves to multiply the original variance by the square of the constant. Therefore, if a treatment effect is multiplicative, the variance for this treatment group may differ significantly from the variance of a control group or from the variance of another treatment group in which the treatment effect is additive.

6.8 Treatments that Operate Differentially on Organismic Variables

Let us consider another possible explanation for a significant difference between s_1^2 and s_2^2. Suppose we find that s_2^2 is significantly greater than s_1^2. A possible explanation of this result is that a treatment operates *differentially* with respect to an organismic variable. For example, if we have divided a group of subjects at random into two groups of n_1 and n_2 subjects each, we would expect that these two groups, if tested under identical conditions, would not differ significantly in their variances. Furthermore, if we were to obtain a measure of anxiety about competition by means of the Edwards Personality Inventory [see Edwards (1967a)], prior to the experiment itself, we would expect the two groups to show only a chance or random difference in their mean scores on this scale. This should be true, if we have randomly assigned subjects to the two groups.

Suppose, however, that one treatment operates differentially on those subjects with high anxiety scores and on those subjects with low anxiety scores. To be specific, let us assume that X is a measure of performance on a pursuit meter and that the treatment consists of telling the subjects that their performance is to be evaluated in comparison with the performance of other subjects. Now suppose that high- and low-anxiety subjects react differentially to these instructions. Assume, for example, that highly anxious subjects tend to be considerably disturbed by the instructions and, in turn, perform more poorly on the pursuit meter than they would if tested under a control or normal condition. For these subjects we would have $X_2 = X_1 - t_1$, where t_1 represents a constant-treatment effect subtracting from performance.

Let us also assume that subjects with a low degree of anxiety may be of such a nature that they do their best under conditions of competitiveness. For these subjects let t_2 be a constant-treatment effect that increases performance. Then for the low-anxiety subjects, we would have $X_2 = X_1 + t_2$; that is, their performance under the experimental condition would, in general, be better than under a control condition.

With the assumptions we have made and assuming also that degree of anxiety does not influence performance under a standard or control condition (that is, under a control condition the mean performance for high- and low-anxiety subjects is the same), then under the treatment condition we would have a distribution of measures that extends in both directions from the mean over the corresponding measures for the control condition. Because the X measures for the treatment group are being moved differentially in both directions from the control group mean, we would expect the treatment group to have a greater variance than the control group.

It is difficult to overemphasize the possible importance of organismic variables in accounting for differences in performance variability of subjects under different experimental conditions. With random assignment, and if the treatment effects are additive, it seems that one of the most probable explanations for a significant difference in variances is that of the differential operation of a given treatment on differences in an organismic variable. In many psychological experiments it may be of considerable value to obtain measures of one or more *relevant* organismic variables. To find that subjects with different values of an organismic variable will react differentially to a given treatment is of perhaps even greater psychological importance than to find that all subjects respond to the treatment in the same manner.

6.9 Two Samples from Different Poisson Distributions

When two treatment variances are heterogeneous, it is sometimes possible that a transformation of the original scale of measurement will result in a new scale such that the variances are homogeneous on the transformed scale. We now consider some transformations[2] that are appropriate under the conditions described.

For variables that have a Poisson distribution, the mean and variance are equal; that is, $\mu = \sigma^2$. Then, if we have two treatments such that the dependent variable for Treatment 1 corresponds to one Poisson distribution with $\mu_1 = \sigma_1^2$ and the dependent variable for the other treatment corresponds to another Poisson distribution with $\mu_2 = \sigma_2^2$, and if $\bar{X}_1$ and $\bar{X}_2$ differ significantly, then we should also expect s_1^2 and s_2^2 to differ significantly.

Poisson distributions are likely to be obtained when the observations consist of counts such as the number of responses of some kind made in a fixed period of time. If the treatment means tend to be proportional to the treatment variances, as occurs when we have two Poisson distributions with different means, then a transformation of the original observations to a new scale may stabilize the variances.

For the Poisson distribution, Bartlett (1936) recommends a square root transformation. For example, Bartlett suggests that we should transform each value of X by taking $\sqrt{X + 0.5}$. Freeman and Tukey (1950) have

[2] For a discussion of additional transformations, see Bartlett (1947), Mueller (1949), and Curtiss (1945).

Table 6.2

NUMBER OF ERRORS MADE BY TWO GROUPS OF SUBJECTS IN READING TWO DIFFERENT DIALS

	Round	Vertical
	2	6
	2	6
	0	10
	4	12
	3	6
$\sum X$	11	40
$\bar{X}$	2.2	8.0
s^2	2.2	8.0

suggested that for the Poisson distribution the variance stabilizing properties of the square root transformation[3] are improved by taking $\sqrt{X} + \sqrt{X+1}$.

To illustrate the square root transformation, we consider an experiment by Sleight (1948), who was interested in the legibility of readings of various dial types. Five different dial types were investigated: horizontal, open window, round, vertical, and semicircular types. In Table 6.2, we give only the data for the round and vertical types. We shall also assume that different subjects were assigned at random to each of the two treatments or dial types. For each treatment we have calculated the mean and variance, and these are also given in the table. We note that for each treatment the means are equal to the variances. This suggests that the square root transformation is appropriate.

The transformation $\sqrt{X + 0.5}$ is given in Table 6.3. The means and variances on the transformed scale are given at the bottom of the table. It is clear that the means and variances are no longer proportional and that the variances are more homogeneous on the transformed scale than on the original scale.

6.10 *The Case Where Treatment Standard Deviations and Means Are Proportional*

Another case in which heterogeneity of variance may be found is that in which the treatment standard deviations tend to be proportional to the treatment means. In this case a transformation to a logarithmic scale is recommended by Bartlett (1947). When values of X equal to zero are present, the transformation may take the form log $(1 + X)$.

In a study of the hoarding behavior of rats Morgan (1945), for example,

[3] Mosteller and Bush (1954) provide a table of the values of $\sqrt{X} + \sqrt{X+1}$.

Table 6.3
THE $\sqrt{X + 0.5}$ TRANSFORMATION FOR THE DATA OF TABLE 6.2

	Round	Vertical
	1.58	2.55
	1.58	2.55
	0.71	3.24
	2.12	3.54
	1.87	2.55
$\sum X$	7.86	14.43
$\bar{X}$	1.57	2.89
s^2	0.28	0.22

found that the logarithm of the number of pellets hoarded resulted in distributions that were approximately normal and with homogeneous variances. Similarly, Haggard (1945) found that a logarithmic transformation is suitable for measures of the galvanic skin response.

6.11 Two Samples from Different Binomial Populations

A transformation has also been suggested for counts based on samples drawn from different binomial populations. For example, for one treatment we may have a binomial population in which P_1, the probability of a successful response, is constant from trial to trial. We have a fixed number of n trials and the dependent variable is T_1, the number of successful responses. Then for subjects tested under this treatment, we have $E(T_1) = nP_1 = \mu_{T_1}$ and $E(T_1 - \mu_{T_1})^2 = nP_1Q_1 = \sigma_{T_1}^2$. If, for another treatment, P_2 (the probability of a successful response) is also constant from trial to trial and if we have the same number of trials as for the first treatment, then for subjects tested under the second treatment we have $E(T_2) = nP_2 = \mu_{T_2}$ and $E(T_2 - \mu_{T_2})^2 = nP_2Q_2 = \sigma_{T_2}^2$. Obviously, the only way in which $\sigma_{T_1}^2$ can be equal to $\sigma_{T_2}^2$ is if $P_1 = P_2$ or if $P_1 = Q_2$. If $P_1 = P_2$, then $\mu_{T_1} = \mu_{T_2}$ and $\bar{T}_1$ should not differ significantly from $\bar{T}_2$. If $P_1 = Q_2$, then we have $\sigma_{T_1}^2 = \sigma_{T_2}^2$ and heterogeneity of variance is not a problem. But if $P_1 = 0.5$ and $P_2 = 0.9$ and with $n = 100$ trials, then $\mu_{T_1} = 50$ and $\mu_{T_2} = 90$ with $\sigma_{T_1}^2 = 25$ and $\sigma_{T_2}^2 = 9$ and, in this case, obviously $\sigma_{T_1}^2 \neq \sigma_{T_2}^2$.

Instead of using T as the dependent variable, we could just as well use $p = T/n$, where n is the number of trials. Then, for a case such as the one described, the variance stabilizing transformation suggested is the inverse sine or angular transformation. Values of $\sin^{-1}\sqrt{p}$, where p is the percentage or proportion of correct responses in a fixed number of trials, have been tabled by Bliss (1937), but this reference is not readily available. Bliss' tables have been reproduced, however, by Snedecor (1956) and by Guilford (1954). Values of the transformation are also available in the Fisher and Yates (1948) tables.

6.12 Heterogeneity of Variance When Time Is a Dependent Variable

In studying the influence of varying amounts of incentive on speed of running in two groups of seven rats each, Crespi (1942) found a significant difference in the variances for a 1-unit and a 4-unit incentive group. $F = s_1^2/s_2^2$ was 72.4 when time was used as a measure of performance. For 6 and 6 d.f., $F = 72.4$ is highly significant. Because of this, and for other reasons which he discusses in some detail, Crespi transformed his unit of measurement to a new scale. Instead of using X, the time required to run the path, he transformed the measures to the scale $1/X$, the reciprocals of the original measures. With this transformation, the variances were stabilized, as indicated by the F ratio of 1.06 for the observations on the transformed scale.

A reciprocal transformation such as that used by Crespi may prove to be useful in other experiments where time is the dependent variable. For example, the transformation may be useful in word-association or reaction-time experiments or in studies of problem solving where the time taken to solve the problem is the dependent variable.

When time is a dependent variable, it may happen that for some of the subjects in one treatment the time measure is quite long. For example, if the dependent variable is the time required to solve a problem, then for one treatment a few subjects may take an excessively long period of time in arriving at a solution. The presence of a few extreme measurements will serve to increase the variance for this treatment, and the variance may be significantly greater than the variance for the other treatments. Transforming the time measures to $1/X$ may serve to make the variances more homogeneous on the transformed scale.

6.13 Transformation of Scale: General Considerations

We do not intend to give the impression that conclusions regarding means, based on the t test applied to the original data in which the treatment variances differ, will be changed if the data are transformed to another scale on which the variances are more homogeneous. There is considerable evidence to show that when we have an equal number of observations for each treatment group, that is, when $n_1 = n_2$, the t test is little influenced by heterogeneity of variance (Boneau, 1960). Thus, if $n_1 = n_2$ and if t is significant, this may, in general, be regarded as evidence that $\mu_1 \neq \mu_2$, even though it may also be true that $\sigma_1^2 \neq \sigma_2^2$.

6.14 Nonnormality: General Conclusions

Although we have assumed that X is a normally distributed variable, there is also considerable evidence to show that the t test for the difference between two means is relatively insensitive to departures from normality in the distribution of X. The important consideration is not the shape of the distribution of X, but rather the shape or form of the distribution of T, a sum of n values of X, or of $\bar{X}$, the mean of n values of X. In experimental work, our interest is primarily in the treatment means or sums. If X is normally distributed, then, for random samples, T and $\bar{X}$ will be normally distributed, as will also be the difference between the means or sums of two random samples.

It is the case, however, that for any variable X with mean μ and population variance σ^2, regardless of the shape or form of the distribution of X, the distribution of T or $\bar{X}$ for random samples approaches that of a normal distribution as n increases. How large n must be before the distribution of T or $\bar{X}$ approaches that of a normal distribution depends on the distribution of X.

In the previous chapter, we considered three different distributions of X. One was U-shaped, another was uniform or rectangular, and the third was skewed. In experimental work, if X is not normally distributed, the departures from normality will often resemble one of these three forms. But we found that even with a relatively small number of observations, the distribution of T, and consequently $\bar{X}$, for random samples from these three populations no longer resembled the population distribution. Judging from the shape of the distributions of T for a relatively small number of observations, it would seem reasonable to believe that if n were increased to fifteen or twenty observations, then, for all practical purposes, the distributions could be assumed to be approximately normal in form.

We wish to emphasize that the t test is a *robust* test,[4] primarily sensitive to

[4] A test of significance is commonly described as "robust" if it is insensitive to violations of assumptions underlying it. Under the conditions described above, there is considerable evidence to show that the t test is a robust test with respect to Type I errors. It is equally important that a test be robust with respect to Type II errors; that is, if the null hypothesis is false, then we would hope that the power of the test when assumptions are violated would be approximately the same as when the assumptions are met. Some limited investigations of the robustness of the t test with respect to Type II errors have been made under conditions of nonnormality and heterogeneity of variance. The evidence, although limited, indicates that the t test is also robust with respect to Type II errors. See, for example, Donaldson (1968).

differences in means and relatively insensitive to nonnormality of distribution and heterogeneity of variance, *provided* that $n_1 = n_2$ and that we also have a sufficient number of observations for each treatment. With $n_1 = n_2 \geq 25$, it seems reasonable to believe that a significant value of t obtained from the t test offers adequate assurance that $\mu_1 \neq \mu_2$, regardless of whether $\sigma_1^2 = \sigma_2^2$, and despite the fact that X may not be normally distributed, provided that the departures from normality are of the same kind for both treatment populations, as would ordinarily be the case in experimental work.

QUESTIONS AND PROBLEMS

6.1 We have the following measures obtained for a control and for an experimental group:

Control Group		Experimental Group			
11	15	4	15	10	10
11	10	4	3	7	12
10	8	8	13	6	14
12	10	9	9	1	8
8	8	12	9	5	5

(*a*) Can we assume that the two variances do not differ significantly? (*b*) Evaluate the significance of the difference between the means.

6.2 In another investigation, we have the following measures obtained for a control and an experimental group:

Control Group				Experimental Group			
12	10	16	11	15	4	10	15
8	13	10	10	12	4	7	3
12	11	11	9	10	8	6	13
11	9	10	9	5	9	1	9
11	11	9	7	8	12	5	9

(*a*) Can we assume that the two variances do not differ significantly? (*b*) Evaluate the significance of the difference between the means.

6.3 Twenty-five rats were deprived of food for a period of twenty-two hours before a test trial but were permitted to satisfy thirst immediately before the test trial. Another group of twenty-five rats was deprived of food for twenty-two hours and was permitted to satisfy thirst twelve hours before the test trial. The mean number of responses on the test trial for the first group was 21.36, and the variance was equal to 147.57. In the second group, the mean was 32.92, and the variance was equal to 489.91. The data are from Kendler (1945). (*a*) Can we assume that the two variances do not differ significantly? (*b*) Evaluate the significance of the difference between the means.

6.4 A group of forty-seven rats was tested in a maze placed in a room with temperature at 55 to 58 degrees Fahrenheit. Another group of forty-six rats was tested in a room with temperature at 75 to 79 degrees Fahrenheit. The mean number of trials required to learn a maze in the "cold" room was 19.8 with s equal to 7.36. The mean number of trials required for learning in the "normal" room was 25.9 with s equal to 13.3. The data are from Moore (1944). (*a*) Can we assume that the two variances do not differ significantly? (*b*) Evaluate the significance of the difference between the means.

6.5 A group of seventy-eight subjects was taught shorthand by the "word" method, and another group of 108 subjects was taught by the "sentence" method. At the end of the first semester, both groups were tested on a word test consisting of a list of words dictated slowly by the instructor and written in shorthand by the students. The mean score for the "word" group was 31.72 with s equal to 8.01. The mean score for the "sentence" group was 35.52 with s equal to 22.93. The data are from Clark and Worcester (1932). (*a*) Can we assume that the two variances do not differ significantly? (*b*) Evaluate the significance of the difference between the means. (*c*) If subjects were not randomly assigned to the two treatments, what bearing would this have on the interpretation of the results?

6.6 Scores on a visual-motor test were obtained for a group of seventy "control" psychiatric cases with diagnoses other than cerebral brain damage. Another group of seventy psychiatric cases with diagnoses of cerebral brain damage was also tested. The mean score for the "control" group was 3.5 with s equal to 4.8. The mean score for the "brain damage" group was 11.6 with s equal to 7.3. The data are from Graham and Kendall (1946). (*a*) Determine whether the variances and the means for the two groups differ significantly. (*b*) In the absence of randomization in the assignment of subjects to the two groups, how would you interpret the results of the tests of significance?

6.7 In a study by French and Thomas (1958) a group of ninety-two subjects was divided into two groups on the basis of their scores on an achievement test. The forty-seven subjects with scores of 8 or higher on the test were called the "high" group and the forty-five subjects with scores of 7 or lower were called the "low" group. Both groups were given a problem to solve and one of the variables measured was the time spent on the task. For the "high" group the mean was 27.14 minutes with a standard deviation of 10.02 minutes. For the "low" group the mean was 13.79 minutes with a standard deviation of 8.13 minutes. (*a*) Determine whether the variances and the means for the two groups differ significantly. (*b*) In the absence of randomization in the assignment of subjects to the two groups, how would you interpret the results of the tests of significance? (*c*) Is it possible to conclude that the "level" of achievement produced the difference in the mean times? (*d*) What are some other possible organismic differences between the two groups?

6.8 We have randomly assigned subjects to two groups so that we have $n = 20$ in each group. (*a*) What value of t will be required for significance at the 5 percent level if the variances differ significantly? (*b*) What value of t will be required for significance at the 5 percent level if the variances do not differ significantly?

6.9 For each of the above problems in which you found a significant difference between the two variances, discuss possible conditions that may account for the heterogeneity of variance.

6.10 Examine a recent issue of a journal which publishes the results of research. Try to find an article in which the investigator reports a significant difference in the variances of his two groups or in which you can demonstrate that the two variances differ significantly. Discuss possible conditions that may account for the heterogeneity of variance.

6.11 What is meant by the additivity of treatment effects?

6.12 If treatment means tend to be proportional to the treatment variances, what transformation is suggested?

6.13 If treatment means tend to be proportional to the treatment standard deviations, what transformation is suggested?

CHAPTER 7

THE ANALYSIS OF VARIANCE FOR A RANDOMIZED GROUP DESIGN

7.1 Introduction

In the last two chapters we have discussed the application of the t test to problems involving the significance of the difference between the means of two independent random samples. We are now ready to consider methods that can be used to test the significance of the differences between three or more means. The technique we shall use is known as the *analysis of variance.*

The early development of the analysis of variance as a powerful tool in experimental and research work was largely the accomplishment of Sir Ronald A. Fisher and his associates in England. In commenting on a paper presented by Wishart (1934) before the Royal Statistical Society, Fisher (1934, p. 52) had this to say concerning the analysis of variance:

> We were together learning how to use the analysis of variance, and perhaps it is worth while stating an impression that I have formed—that the analysis of variance, which may perhaps be called a statistical method, because that term is a very ambiguous one—is not a mathematical theorem, but rather a convenient method of arranging the arithmetic. Just as in arithmetical text-books—if we can recall their contents—we were given rules for arranging how to find the greatest common measure, and how to work out a sum in practice, and were drilled in the arrangement and order in which we were to put the figures down, so with the analysis of variance; its one claim to attention lies in its convenience. It is convenient in two ways: (1) because it brings to the eyes and to the mind a summary of a mass of statistical data in which the logical content of the whole is readily appreciated. Probably everyone who has used it has found that comparisons which they have not previously thought of may obtrude themselves, because

Table 7.1
NOTATION FOR A RANDOMIZED GROUP DESIGN WITH $k = 5$ TREATMENTS AND WITH $n = 8$ OBSERVATIONS FOR EACH TREATMENT

Treatments				
1	2	3	4	5
X_{11}	X_{21}	X_{31}	X_{41}	X_{51}
X_{12}	X_{22}	X_{32}	X_{42}	X_{52}
X_{13}	X_{23}	X_{33}	X_{43}	X_{53}
X_{14}	X_{24}	X_{34}	X_{44}	X_{54}
X_{15}	X_{25}	X_{35}	X_{45}	X_{55}
X_{16}	X_{26}	X_{36}	X_{46}	X_{56}
X_{17}	X_{27}	X_{37}	X_{47}	X_{57}
X_{18}	X_{28}	X_{38}	X_{48}	X_{58}
$\bar{X}_{1.}$	$\bar{X}_{2.}$	$\bar{X}_{3.}$	$\bar{X}_{4.}$	$\bar{X}_{5.}$

there they are, necessary items in the analysis. (2) Apart from aiding the logical process, it is convenient in facilitating and reducing to a common form all the tests of significance which we may want to apply. I do insist that its claim to attention rests essentially on its convenience. Nearly always we can, if we choose, put our data in other forms and other language. Naturally, like other logical arrangements, it is based on mathematical theorems previously proved, and in particular the tests of significance were based on problems of distribution the solution of which was published for the most part from 1921 to 1924.

We shall illustrate the necessary calculations in the analysis of variance for a *randomized group design* in which forty subjects have been assigned at random to one of $k = 5$ treatments with $n = 8$ subjects for each treatment.

7.2 Notation for a Randomized Group Design

In Table 7.1 we introduce a notation for the observations in a randomized group design with $k = 5$ treatments and with $n = 8$ observations for each treatment.[1] We shall find this notation very convenient in the analysis of variance. Each observation in the table is identified by two subscripts, the first corresponding to a particular treatment and the second to a particular observation for the treatment. For

[1] It is not necessary that we have equal n's for each treatment. However, there is no good reason for having unequal n's and there are many sound reasons why the n's should be equal. For one, the test of significance is relatively little influenced by heterogeneity of variance when the n's are equal.

example, X_{32} is the second observation for Treatment 3. We let X_{kn} be a general symbol for any observation, with the understanding that k and n when used as subscripts may correspond to variables. In Table 7.1, k can take any value from 1 to 5, because there are five treatments, and n can take any value from 1 to 8, because there are eight observations for each treatment. When k and n are used alone or as coefficients of other terms, they will always represent constants.

The sum of all kn observations will be represented by $\Sigma X_{..}$, where the dots that replace the subscripts kn indicate that we have summed all kn values of X_{kn}. Similarly, we let $\bar{X}_{..}$ be the mean of all kn observations. The various treatment sums can be represented by $\Sigma X_{1.}$, $\Sigma X_{2.}$, $\Sigma X_{3.}$, $\Sigma X_{4.}$, and $\Sigma X_{5.}$, where the dot that has replaced the subscript n means that we have summed over the n observations for a given treatment. Then $\Sigma X_{k.}$ will be a general symbol for any given treatment sum and $\bar{X}_{k.}$ will be a general symbol for any given treatment mean.

7.3 Sums of Squares for a Randomized Group Design

We may now write the following identity

$$X_{kn} - \bar{X}_{..} = (X_{kn} - \bar{X}_{k.}) + (\bar{X}_{k.} - \bar{X}_{..}) \tag{7.1}$$

If we square (7.1) and sum over the n observations for the kth treatment, we have

$$\sum_{1}^{n} (X_{kn} - \bar{X}_{..})^2 = \sum_{1}^{n} (X_{kn} - \bar{X}_{k.})^2 + 2(\bar{X}_{k.} - \bar{X}_{..}) \sum_{1}^{n} (X_{kn} - \bar{X}_{k.}) + n(\bar{X}_{k.} - \bar{X}_{..})^2$$

But

$$\sum_{1}^{n} (X_{kn} - \bar{X}_{k.}) = 0$$

because it is the sum of the deviations of the n observations for a given treatment from the treatment mean. Therefore,

$$\sum_{1}^{n} (X_{kn} - \bar{X}_{..})^2 = \sum_{1}^{n} (X_{kn} - \bar{X}_{k.})^2 + n(\bar{X}_{k.} - \bar{X}_{..})^2 \tag{7.2}$$

We can obtain an expression such as (7.2) for each of the k treatment groups and summing over the k groups, we have

$$\sum_{1}^{kn} (X_{kn} - \bar{X}_{..})^2 = \sum_{1}^{k} \sum_{1}^{n} (X_{kn} - \bar{X}_{k.})^2 + n \sum_{1}^{k} (\bar{X}_{k.} - \bar{X}_{..})^2 \tag{7.3}$$

The term on the left in (7.3) measures the variation of the kn observations about the mean of all kn observations and is called the *total sum of squares.* As (7.3) shows, the total sum of squares can be divided into two parts, corresponding to the two terms on the right of the expression. The first term on the right measures the variation of the n observations in each treatment group about the mean for the treatment group, and this sum of squares is called the *within treatment sum of squares.* The second term on the right measures the variation of the treatment means about the mean of all kn observations weighted by n, the number of observations in each treatment group, and is called the *treatment sum of squares.*

We have shown that whenever we have k groups of n observations each, it is always possible to analyze the total sum of squares into two parts: the within treatment sum of squares and the treatment sum of squares. The degrees of freedom associated with the total sum of squares will be $kn - 1$. The within treatment sum of squares will have $k(n - 1)$ d.f. and the treatment sum of squares will have $k - 1$ d.f.

7.4 Randomized Group Design: An Example

In Table 7.2 we give the values of a dependent variable of interest for a randomized group design with $k = 5$ treatments and with $n = 8$ subjects assigned at random to each treatment. For the

Table 7.2

RANDOMIZED GROUP DESIGN WITH FIVE TREATMENTS AND EIGHT SUBJECTS RANDOMLY ASSIGNED TO EACH TREATMENT

	Treatments					
	1	2	3	4	5	
	16	16	2	5	7	
	18	7	10	8	11	
	5	10	9	8	12	
	12	4	13	11	9	
	11	7	11	1	14	
	12	23	9	9	19	
	23	12	13	5	16	
	19	13	9	9	24	
$\sum X$	116	92	76	56	112	452
$\sum X^2$	1904	1312	806	462	1784	6268

total sum of squares, we have

$$\sum_{1}^{kn} (X_{kn} - \bar{X}_{..})^2 = \sum_{1}^{kn} X_{kn}^2 - \frac{(\sum X_{..})^2}{kn} \tag{7.4}$$

or, for the data of Table 7.2,

$$\sum_{1}^{kn} (X_{kn} - \bar{X}_{..})^2 = (16)^2 + (18)^2 + \cdots + (24)^2 - \frac{(452)^2}{40} = 1160.4$$

For the treatment sum of squares, we have

$$n \sum_{1}^{k} (\bar{X}_{k\cdot} - \bar{X}_{..})^2 = \sum_{1}^{k} \frac{(\sum X_{k\cdot})^2}{n} - \frac{(\sum X_{..})^2}{kn} \tag{7.5}$$

or, for the data of Table 7.2,

$$n \sum_{1}^{k} (\bar{X}_{k\cdot} - \bar{X}_{..})^2 = \frac{(116)^2}{8} + \frac{(92)^2}{8} + \cdots + \frac{(112)^2}{8} - \frac{(452)^2}{40} = 314.4$$

If we subtract the treatment sum of squares from the total sum of squares, the remainder will be the within treatment sum of squares. Thus,

$$\sum_{1}^{kn} (X_{kn} - \bar{X}_{..})^2 - n \sum_{1}^{k} (\bar{X}_{k\cdot} - \bar{X}_{..})^2 = \sum_{1}^{k} \sum_{1}^{n} (X_{kn} - \bar{X}_{k\cdot})^2 \tag{7.6}$$

and for the data of Table 7.2, we have $1160.4 - 314.4 = 846.0$ as the within treatment sum of squares.

The within treatment sum of squares is a pooled sum of squares based on the variation of the n measures within each treatment group about the mean of the treatment group. For example, if we consider each of the $k = 5$ treatment groups separately, we would have

$$\sum_{1}^{n} (X_{1n} - \bar{X}_{1\cdot})^2 = 1904 - \frac{(116)^2}{8} = 222.0$$

$$\sum_{1}^{n} (X_{2n} - \bar{X}_{2\cdot})^2 = 1312 - \frac{(92)^2}{8} = 254.0$$

$$\sum_{1}^{n} (X_{3n} - \bar{X}_{3\cdot})^2 = 806 - \frac{(76)^2}{8} = 84.0$$

$$\sum_{1}^{n} (X_{4n} - \bar{X}_{4\cdot})^2 = 462 - \frac{(56)^2}{8} = 70.0$$

$$\sum_{1}^{n} (X_{5n} - \bar{X}_{5\cdot})^2 = 1784 - \frac{(112)^2}{8} = 216.0$$

Table 7.3

ANALYSIS OF VARIANCE FOR THE RANDOMIZED GROUP DESIGN—ORIGINAL DATA IN TABLE 7.2

Source of Variation	Sum of Squares	d.f.	Mean Square	F
Treatments	314.4	4	78.60	3.25
Within treatments	846.0	35	24.17	
Total	1160.4	39		

and the sum of these sums of squares is equal to 846.0 and is the same as the value we obtained by subtraction.

The results of our calculations are summarized in Table 7.3. For the treatment sum of squares we have $k - 1 = 5 - 1 = 4$ d.f. and dividing the treatment sum of squares by its degrees of freedom, we obtain a variance estimate called the *treatment mean square*, or MS_T. Similarly, if we divide the within treatment sum of squares by its degrees of freedom, $k(n - 1)$, we obtain another variance estimate called the *within treatment mean square*, or MS_W. These two mean squares are shown in Table 7.3.

7.5 *Test of Significance for a Randomized Group Design*

For a randomized group design, we define F as

$$F = \frac{MS_T}{MS_W} \tag{7.7}$$

and this F will have $k - 1$ d.f. for the numerator and $k(n - 1)$ d.f. for the denominator. For the data given in Table 7.3, we have

$$F = \frac{78.60}{24.17} = 3.25$$

with 4 and 35 d.f.

The null hypothesis tested by (7.7) is that we have k independent random samples from the same normally distributed population so that

$$E(\bar{X}_{k\cdot}) = \mu \tag{7.8}$$

for all k treatment means and that

$$E(s_k^2) = \sigma^2 \tag{7.9}$$

for all k treatment variances.

If (7.9) is true, then MS_W is an unbiased estimate of the common population variance σ^2; that is, $E(MS_W) = \sigma^2$. If (7.8) and (7.9) are true, then it can be shown[2] that MS_T is also an unbiased estimate of σ^2, that is, $E(MS_T) = \sigma^2$. However, if (7.8) is not true, then, as we shall see, $E(MS_T) > E(MS_W)$. The tabled values of $F = MS_T/MS_W$ are those that would be expected to occur, at various levels of significance, when both (7.8) and (7.9) are true. Consequently, if the obtained value of F is greater than the tabled value for some defined level of significance, so that we may regard it as an improbable value when the null hypothesis is true, we may decide to reject the null hypothesis. In doing so, we may conclude that the treatment means are not all estimates of the same common population mean μ, under the additional assumption that the treatment variances are homogeneous; that is, the separate variance estimates for the k samples are all estimates of a common population variance.

We may now emphasize that the F test defined by (7.7) is relatively insensitive to heterogeneity of variance, provided that we have the same number of observations for each treatment. Consequently, if we have the same number of observations for each treatment and if we obtain a significant value of F, we shall ordinarily be safe in concluding that the population means for the treatments are not all equal to the same value, μ, even though it may also be true that the population variances for the treatments are not all equal to the same value, σ^2. As Box (1953) has pointed out, because the F test is very insensitive to nonnormality and because with equal n's it is also insensitive to variance inequalities, we can accept the fact that it can be safely used under most conditions as indicating that the treatment means differ significantly. In other words, the F test, like the t test, is a *robust* test in that it is relatively insensitive to violations of the assumptions of normality of distribution and homogeneity of variance.[3]

To evaluate the obtained value of $F = 3.25$, we enter the column of the F table with degrees of freedom corresponding to the numerator of the F ratio and find the row entry corresponding to the degrees of freedom in the denominator. We do not have a row entry corresponding to 35 d.f. in the table of F, but with $\alpha = 0.05$ we find that the critical value for 4 and 34 d.f. is $F = 2.65$ and for 4 and 36 d.f., the critical value is $F = 2.63$, and it is obvious that our obtained value of $F = 3.25$ is significant with $\alpha = 0.05$.

To conclude merely that the k treatment means are not all estimates of the same population value, μ, is alone not very satisfying. We would like to know something more specific about the nature of the differences in the

[2] The proof is given in the next section.

[3] There is a χ^2 test for the homogeneity of k variances but, as Box (1953) has emphasized, the test is as sensitive to nonnormality as it is to differences in the variances.

means. In the next chapter, we consider some tests that can be used in testing specific hypotheses about the various population means.

7.6 The Analysis of Variance Model When the Null Hypothesis Is True

Let us assume that the null hypothesis is true; that is, we assume that

$$E(\bar{X}_{1\cdot}) = E(\bar{X}_{2\cdot}) = \cdots = E(\bar{X}_{k\cdot}) = \mu$$

and that

$$E(s_1^2) = E(s_2^2) = \cdots = E(s_k^2) = \sigma^2$$

For the kth treatment, we have

$$\sigma^2 = E(X_{kn}^2) - \mu^2$$

so that

$$E(X_{kn}^2) = \sigma^2 + \mu^2$$

Then

$$E\left(\sum_1^n X_{kn}^2\right) = n\sigma^2 + n\mu^2 \qquad (7.10)$$

To find the expected value of $(\Sigma X_{k\cdot})^2/n$, we note that the square of a sum of n values of X_{kn} results in n^2 terms of which n are of the kind X_{kn}^2 and $n^2 - n = n(n-1)$ are of the kind $X_{ki}X_{kj}$. If the n values are drawn from a common population, then the expectation of each of the X_{kn}^2 terms is $\sigma^2 + \mu^2$. If, in addition, the n values of X_{kn} are independent, then the expectation of each of the $X_{ki}X_{kj}$ terms is μ^2. Consequently,

$$E\left[\frac{(\sum X_{k\cdot})^2}{n}\right] = \frac{1}{n}[n\sigma^2 + n\mu^2 + n(n-1)\mu^2]$$

which, when simplified, results in

$$E\left[\frac{(\sum X_{k\cdot})^2}{n}\right] = \sigma^2 + n\mu^2 \qquad (7.11)$$

Subtracting (7.11) from (7.10), we obtain

$$\begin{aligned} E\left[\sum_1^n (X_{kn} - \bar{X}_{k\cdot})^2\right] &= E\left[\sum_1^n X_{kn}^2 - \frac{(\sum X_{k\cdot})^2}{n}\right] \\ &= (n\sigma^2 + n\mu^2) - (\sigma^2 + n\mu^2) \\ &= (n-1)\sigma^2 \qquad (7.12) \end{aligned}$$

We note that the right side of (7.12) is the expected value of the sum of squares within the kth treatment group. The within treatment sum of squares is the sum of k such sums of squares and, under the null hypothesis, each will have the same expectation. Therefore,

$$E\left[\sum_1^k \sum_1^n (X_{kn} - \bar{X}_{k\cdot})^2\right] = k(n - 1)\sigma^2 \tag{7.13}$$

and dividing by the degrees of freedom, $k(n - 1)$, associated with the within treatment sum of squares, we have

$$E(MS_W) = \sigma^2 \tag{7.14}$$

The treatment sum of squares will be given by

$$n \sum_1^k (\bar{X}_{k\cdot} - \bar{X}_{\cdot\cdot})^2 = \sum_1^k \frac{(\sum X_{k\cdot})^2}{n} - \frac{(\sum X_{\cdot\cdot})^2}{kn} \tag{7.15}$$

If the null hypothesis is true, then each of the k values of $(\Sigma X_{k\cdot})^2/n$ will have the same expectation as that given by (7.11). Then

$$E\left[\sum_1^k \frac{(\sum X_{k\cdot})^2}{n}\right] = k\sigma^2 + kn\mu^2 \tag{7.16}$$

and we also have

$$E\left[\frac{(\sum X_{\cdot\cdot})^2}{kn}\right] = \sigma^2 + kn\mu^2 \tag{7.17}$$

Subtracting (7.17) from (7.16), we have the expected value of the treatment sum of squares, or

$$\begin{aligned} E\left[n \sum_1^k (\bar{X}_{k\cdot} - \bar{X}_{\cdot\cdot})^2\right] &= E\left[\sum_1^k \frac{(\sum X_{k\cdot})^2}{n} - \frac{(\sum X_{\cdot\cdot})^2}{kn}\right] \\ &= (k\sigma^2 + kn\mu^2) - (\sigma^2 + kn\mu^2) \\ &= (k - 1)\sigma^2 \end{aligned} \tag{7.18}$$

and, therefore, when the null hypothesis is true,

$$E(MS_T) = \sigma^2 \tag{7.19}$$

The model we have used for the case where the null hypothesis is true, that is, assuming both (7.8) and (7.9) to be true, is

$$X_{kn} = \mu_k + e_{kn} \tag{7.20}$$

or

$$X_{kn} = \mu + e_{kn} \tag{7.21}$$

because we assume that $\mu_k = \mu$ for all k treatments. We also assume that e_{kn} is an independent random error associated with each observation.[4] We note that $E(e_{kn}) = 0$, and, therefore, $E(e_{kn}^2) = \sigma^2$.

7.7 The Analysis of Variance Model When the Null Hypothesis Is False

We now consider the case where the null hypothesis is false in that (7.8) is not true, but (7.9) is true. What this means is that $E(\bar{X}_{k\cdot})$ is not equal to the same value, μ, for all k treatment means, but that $E(s_k^2)$ is equal to the same value, σ^2, for all k treatment variances.

As a model for the case where the null hypothesis is false, we assume that

$$X_{kn} = \mu + t_k + e_{kn} \tag{7.22}$$

which differs from (7.21) only in that X_{kn} involves a treatment effect t_k. The model assumes that each t_k is a constant so that

$$E(t_k) = t_k, \qquad E(t_k^2) = t_k^2, \qquad \text{and} \qquad E(t_i t_j) = t_i t_j$$

Without any loss in generality, we let

$$\mu = \frac{\mu_1 + \mu_2 + \cdots + \mu_k}{k} \tag{7.23}$$

and

$$t_k = \mu_k - \mu \tag{7.24}$$

Then

$$E\left(\sum_1^k t_k\right) = E(t_1 + t_2 + \cdots + t_k) = 0 \tag{7.25}$$

and

$$E\left(\sum_1^k t_k\right)^2 = E\,(t_1 + t_2 + \cdots + t_k)^2 = 0 \tag{7.26}$$

because the sum of the product terms, $t_i t_j$, obtained when we square the right side of (7.26) will be negative and equal to $\sum_1^k t_k^2$.

The model also assumes that the e_{kn}'s are random variables such that

$$E(e_{kn}) = 0 \tag{7.27}$$

and

$$E(e_{kn}^2) = \sigma^2 \tag{7.28}$$

Furthermore, the e_{kn}'s are assumed to be *independent* random variables so

[4] In addition, for tests of significance we assume that the e_{kn}'s are normally distributed.

that the expected value of the product between any two e_{kn} values which do not have the same subscripts is zero. Thus, even though

$$E\left(\sum_1^n e_{kn}\right) = E(e_{k1} + e_{k2} + \cdots + e_{kn}) = 0$$

we have

$$\begin{aligned} E\left(\sum_1^n e_{kr}\right)^2 &= E(e_{k1} + e_{k2} + \cdots + e_{kn})^2 \\ &= E(e_{k1}^2 + e_{k2}^2 + \cdots + e_{kn}^2) \\ &= E\left(\sum_1^n e_{kn}^2\right) \\ &= n\sigma^2 \end{aligned} \tag{7.29}$$

Using the model defined by (7.22), we now find the expected value of the within treatment sum of squares. For any one of the k treatment groups we have

$$E\left(\sum_1^n X_{kn}^2\right) = n\mu^2 + nt_k^2 + n\sigma^2 + 2n\mu t_k \tag{7.30}$$

For the same treatment group we also have

$$E\left[\frac{(\sum X_{k\cdot})^2}{n}\right] = \frac{1}{n} E\left(n\mu + nt_k + \sum_1^n e_{kn}\right)^2$$

We have already shown that $E\left(\sum_1^n e_{kn}\right)^2 = E\left(\sum_1^n e_{kn}^2\right) = n\sigma^2$, so that when we square and take the expectation of the right side of the above expression we obtain

$$\begin{aligned} E\left[\frac{(\sum X_{k\cdot})^2}{n}\right] &= \frac{1}{n}(n^2\mu^2 + n^2t_k^2 + n\sigma^2 + 2n^2\mu t_k) \\ &= n\mu^2 + nt_k^2 + \sigma^2 + 2n\mu t_k \end{aligned} \tag{7.31}$$

The expected value of the sum of squares within this treatment group will then be

$$E\left[\sum_1^n (X_{kn} - \bar{X}_{k\cdot})^2\right] = E\left[\sum_1^n X_{kn}^2 - \frac{(\sum X_{k\cdot})^2}{n}\right]$$

and subtracting (7.31) from (7.30), we obtain

$$E\left[\sum_1^n (X_{kn} - \bar{X}_{k\cdot})^2\right] = (n-1)\sigma^2 \tag{7.32}$$

Because the sum of squares within each of the k treatment groups will have the same expectation as (7.32), the expected value of the within treatment sum of squares will be

$$E\left[\sum_1^k \sum_1^n (X_{kn} - \bar{X}_{k\cdot})^2\right] = k(n-1)\sigma^2 \tag{7.33}$$

Dividing (7.33) by the degrees of freedom associated with the within treatment sum of squares, we have

$$E(MS_W) = \sigma^2 \tag{7.34}$$

and we see that MS_W is an unbiased estimate of the common population variance σ^2.

Using the model defined by (7.22), let us now find the expected value of the treatment mean square. From (7.31) we see that

$$E\left[\sum_1^k \frac{(\sum X_{k\cdot})^2}{n}\right] = kn\mu^2 + n\sum_1^k t_k^2 + k\sigma^2 + 2n\mu\sum_1^k t_k$$

But $\sum_1^k t_k = 0$ and therefore

$$E\left[\sum_1^k \frac{(\sum X_{k\cdot})^2}{n}\right] = kn\mu^2 + n\sum_1^k t_k^2 + k\sigma^2 \tag{7.35}$$

We also have

$$E\left[\frac{(\sum X_{\cdot\cdot})^2}{kn}\right] = \frac{1}{kn} E\left(kn\mu + n\sum_1^k t_k + \sum_1^{kn} e_{kn}\right)^2$$

But we have shown that $E\left(\sum_1^k t_k\right)^2 = 0$ and that $E\left(\sum_1^{kn} e_{kn}\right)^2 = E\left(\sum_1^{kn} e_{kn}^2\right) =$ $kn\sigma^2$ and therefore

$$E\left[\frac{(\sum X_{\cdot\cdot})^2}{kn}\right] = kn\mu^2 + \sigma^2 \tag{7.36}$$

Subtracting (7.36) from (7.35), we have the expectation of the treatment sum of squares or

$$E\left[\sum_1^k \frac{(\sum X_{k\cdot})^2}{n} - \frac{(\sum X_{\cdot\cdot})^2}{kn}\right] = (k-1)\sigma^2 + n\sum_1^k t_k^2 \tag{7.37}$$

and the expected value of the treatment mean square will be

$$E(MS_T) = \sigma^2 + \frac{n \sum_{1}^{k} t_k^2}{k - 1} \tag{7.38}$$

Then

$$\frac{E(MS_T)}{E(MS_W)} = \frac{\sigma^2 + \frac{n \sum_{1}^{k} t_k^2}{k - 1}}{\sigma^2} \tag{7.39}$$

Only if all μ_k's are equal to the same value, μ, will $\sum_{1}^{k} t_k^2 = 0$. If they are not, then the expected value of the numerator of the ratio defined by (7.39) will be larger than the expected value of the denominator. The null hypothesis tested by the sample value of $F = MS_T/MS_W$, in other words, is that $\sum_{1}^{k} t_k^2 = 0$ and this will be true only if the expected values of the k treatment means are all equal to the same population value. Sample values of $F > 1.0$ thus provide evidence that the treatment population means are not all equal to the same value μ.

7.8 The Analysis of Variance and Binomial Variables

Suppose we have a randomized group design and that the dependent variable is a binomial variable so that X can take only the values of $X = 1$ or $X = 0$. For example, the treatments might consist of various types of visual displays in which subjects are asked to read dials and their responses are classified as correct ($X = 1$) or incorrect ($X = 0$). If n subjects are assigned at random to each of the k treatments, then the treatment means will be $p_{1\cdot}, p_{2\cdot}, \ldots, p_{k\cdot}$. To test the null hypothesis that $P_{1\cdot} = P_{2\cdot} = \cdots = P_{k\cdot}$, we could use the χ^2 test described in a previous chapter. Alternatively, the robustness of the F test with respect to Type I errors suggests that the data might also be analyzed in terms of the analysis of variance under certain conditions, even though the values of X are obviously not continuous and normally distributed.

Hsu and Feldt (1969) have investigated the robustness of the F test for random samples from binomial populations. The populations investigated were those in which $P = 0.25$, $P = 0.40$, and $P = 0.50$. For each of these populations they generated by computer 10,000 random experiments with $k = 2$ or $k = 4$ treatments and with $n = 11$ or $n = 51$ subjects assigned to each treatment. They found that *approximately* 0.05 and 0.01 of these independently generated values of F equaled or exceeded the tabled values at the $\alpha = 0.05$ and $\alpha = 0.01$ levels of significance.

These results indicate that for binomial populations in which P ranges from 0.25 to 0.75, the probability of a Type I error given by the tabled values of F corresponds quite well with the empirically determined probability, provided that at least $n = 11$ subjects are assigned to each treatment. With $n > 11$ subjects assigned to each treatment, one might be even more confident that the probability of a Type I error is controlled at the significance level given by the F test.

The results of the Hsu and Feldt study apply to the case where the null hypothesis is true, that is, where $P_{1.} = P_{2.} = \cdots = P_{k.}$ so that $\sigma_1^2 = \sigma_2^2 = \cdots = \sigma_k^2$. It is obvious that if the k samples were drawn from different binomial populations, then the assumption of homogeneity of variance would be violated. But in this case the null hypothesis would also be false and the question of the probability of a Type I error is irrelevant. The relevant question, in this instance, is the probability of a Type II error or the power of the F test. We know, however, that by increasing the number of observations for each treatment we can also increase the power of the test of significance. Thus, with reasonably large n's for each treatment one might consider the F test as an alternative to the χ^2 test in analyzing the results of the experiment.

QUESTIONS AND PROBLEMS

7.1 Data for three treatment groups are given below:

Treatments

1	2	3
27	22	37
45	24	38
44	42	25
31	41	47
38	31	23

Find the value of $F = MS_T/MS_W$.

7.2 In the Morgan (1945) experiment, Problem 5.4, t was used to evaluate the difference between the means of two groups. For the same data, find the value of $F = MS_T/MS_W$. You should find that $F = t^2$.

7.3 We have below six samples that were drawn at random from a sampling box. Find the value of $F = MS_T/MS_W$.

Samples

1	2	3	4	5	6
9	9	6	6	12	10
12	10	12	9	6	6
7	8	9	12	8	8
14	3	9	7	7	12
5	7	8	5	3	9
8	8	10	5	13	4
7	5	2	8	7	8
7	9	10	9	13	7
8	3	9	6	6	6
3	8	12	3	6	2

7.4 Subjects were assigned at random to one of three treatment groups. Measures on the dependent variable are given below:

Treatments

1	2	3
22	21	32
35	44	23
45	35	22
24	40	41
43	35	44
38	22	32
23	50	18
30	28	22

Find the value of $F = MS_T/MS_W$.

7.5 In an experiment $n = 10$ subjects were assigned at random to each of $k = 5$ treatments. Find the value of $F = MS_T/MS_W$.

Treatments

1	2	3	4	5
13	7	12	10	13
9	4	11	12	6
8	4	4	9	14
7	1	9	7	12
8	10	5	15	13
6	7	10	14	10
6	5	2	10	8
7	9	8	17	4
6	5	3	14	9
10	8	6	12	11

7.6 The analysis of variance should prove to be useful in problems involving a "test of technique," that is, where the experimenter is not sure that he can reproduce his results. Such failures may be the result of inability to standardize and thus control the conditions of the experiment. They may also be due to unreliable observers or unreliable measuring devices or other factors. The problem cited here happens to involve the observers, and was carried out under the direction of Professor Roger B. Loucks at the University of Washington. Subjects were assigned at random to one of four graduate assistants—here referred to as Operators A, B, C, and D. The operators did not test an equal number of subjects, but each operator observed his subjects under supposedly the same set of conditions. Records were kept of a number of different variables. The one reported here concerns but one phase of the research, the errors made in making turns in an airplane trainer. The table below gives the records obtained by each operator for his particular group of subjects. Find the value of $F = MS_T/MS_W$.

Operator A	Operator B	Operator C	Operator D
6	5	4	8
4	3	7	5
3	4	3	6
7	3	7	7
13	3	4	7
9	4	8	9
4	0	7	7
10	3	4	10
8	4	8	4
9	4	4	3
8	3	11	3
5	3	13	
5	4	9	
10	2		
9	5		
15	3		
10	3		
6	1		
4	2		
5			
7			

7.7 In the above example, there is a tendency for the means of the various groups to be proportional to the standard deviations. (*a*) Find the means and standard deviations to determine this for yourself. (*b*) Transform the data to the logarithmic scale, $\log(X + 1)$. Find the means and standard deviations on the transformed scale. Are the means and standard deviations still proportional? (*c*) Find the value of $F = MS_T/MS_W$ for the transformed data.

7.8 Subjects were assigned at random to one of three treatments. The outcome of the experiment is given below:

Treatment 1	Treatment 2	Treatment 3
12	0	4
12	0	4
19	2	0
24	4	8
12	4	6
11	5	4
19	5	5
22	0	0
11	1	9
11	3	7

(*a*) Find the mean and variance for each treatment. Note that they tend to be proportional. (*b*) Find the value of $F = MS_T/MS_W$. (*c*) Transform the data to the scale $\sqrt{X + 0.5}$. Find the mean and variance for each treatment on the transformed scale. Has the variance been stabilized by the transformation? (*d*) Find the value of $F = MS_T/MS_W$ for the transformed data.

7.9 Why is it possible to regard the treatment mean square, when the null hypothesis is true, as an estimate of the common population variance?

7.10 We have $k = 3$ treatments with $n = 3$ subjects assigned to each treatment. Find the value of $F = MS_T/MS_W$.

Treatments		
1	2	3
4	8	2
0	5	4
2	2	3

7.11 We have two treatments with $n_1 = n_2 = 5$ subjects assigned to each treatment. The outcome of the experiment is given below:

Treatments	
1	2
3	7
5	5
2	6
1	3
4	4

(*a*) Test the null hypothesis that $\mu_1 = \mu_2$ by means of the t test. (*b*) For the same data find $F = MS_T/MS_W$. You should find that $t^2 = F$.

7.12 We have n independent random values of X drawn from a common population with mean μ and standard deviation σ. Give the expected value of each of the following. The summation is, in all cases, over the n values of X.

(*a*) $E(X)$
(*b*) $E(\bar{X})$
(*c*) $E(X - \mu)^2$
(*d*) $E(\sum X)$
(*e*) $E(\sum X^2)$
(*f*) $E[\sum (X - \bar{X})^2]$
(*g*) $E(s^2)$
(*h*) $E(X_iX_j)$
(*i*) $E(n\bar{X}^2)$
(*j*) $E(\bar{X}^2)$
(*k*) $E(X^2) - \mu^2$
(*l*) $E(\bar{X} - \mu)^2$
(*m*) $E(X^2)$
(*n*) $E(\sum X)^2$
(*o*) $E(X_iX_j) - \mu^2$
(*p*) $E(\sum X^2/n)$

CHAPTER 8

MULTIPLE COMPARISONS ON TREATMENT MEANS AND SUMS

8.1 Introduction

Suppose we have tested a set of k means by the analysis of variance and have concluded that the means differ significantly. This, alone, as we pointed out in the last chapter, is not very satisfactory. What we would usually like to know is how the means differ. Is every mean significantly different from every other? Are there significant differences between some of the means and not between others?

A variety of methods have been proposed for investigating the differences existing among a set of k means. These test procedures are useful whenever we are concerned with *multiple comparisons* among the means. In this chapter, we shall consider some *selected* methods for making and testing the significance of multiple comparisons.[1]

In making multiple comparisons among the treatment means, it is not necessary that the treatment mean square of the analysis of variance be significant.[2] In other words, we may have a nonsignificant treatment mean square and still use the methods to be described for making multiple comparisons and testing the significance of them. However, this does not

[1] The problem of multiple comparisons is not a simple one and statisticians who have worked on the problem are not themselves in complete agreement as to procedures. Further discussions can be found in the references cited in this chapter and in Federer (1955). See also Ryan (1959).

[2] This is true of all of the tests described except Scheffé's. If the analysis of variance results in a nonsignificant treatment mean square with $\alpha = 0.05$, then no comparison using Scheffé's test will be significant with $\alpha = 0.05$.

mean that we should indiscriminately apply the methods, one after another, with the anticipation that one or another may result in some finding that meets the requirements of statistical significance. Our choice of method should, instead, be guided by questions of experimental interest.

8.2 Duncan's Multiple Range Test

The first case we shall consider is a comparative experiment in which $k = 8$ treatments are tested with $n = 4$ observations for each treatment.[3] The summary of the analysis of variance for the experiment is given in Table 8.1. The F of 30.96 with 7 and 24 d.f. is highly significant.

The observed means, each based on $n = 4$ observations, are given in Table 8.2. We shall assume that we had no *a priori* hypotheses as to the differences to be found between the eight means, and in the table they are simply arranged in order of magnitude and identified by the letters A to H. If we wish to determine which of the differences among these means are significant and which are not, the test procedure we suggest is Duncan's (1955) *multiple range test.* We shall illustrate the multiple range test only for the case where we have the same number of observations in each treatment group or for each mean.

The first step in applying the multiple range test is to arrange the means in order of magnitude, as in Table 8.2. We then find the standard error of a single mean as given by

$$s_{\bar{X}} = \frac{s}{\sqrt{n}} \tag{8.1}$$

where s is the square root of the within treatment mean square, MS_W, of

Table 8.1

ANALYSIS OF VARIANCE FOR $k = 8$ TREATMENTS WITH $n = 4$ OBSERVATIONS FOR EACH TREATMENT

Source of Variation	Sum of Squares	d.f.	Mean Square	F
Treatments	7803.16	7	1114.74	30.96
Within treatments	864.00	24	36.00	
Total	8667.16	31		

[3] An extension of Duncan's multiple range test for the case of unequal n's is given by Kramer (1956). See also Duncan (1957).

Table 8.2

DUNCAN'S MULTIPLE RANGE TEST APPLIED TO THE DIFFERENCES BETWEEN $k = 8$ TREATMENT MEANS—THE ANALYSIS OF VARIANCE FOR THE SAME EXPERIMENT IS GIVEN IN TABLE 8.1

Any two treatment means not underscored by the same line are significantly different. Any two treatment means underscored by the same line are not significantly different.

		(1) *A*	(2) *B*	(3) *C*	(4) *D*	(5) *E*	(6) *F*	(7) *G*	(8) *H*	(9) Shortest Significant Ranges
Means		24.7	41.7	55.6	56.4	60.1	66.3	70.3	77.0	
A	24.7		17.0	30.9	31.7	35.4	41.6	45.6	52.3	$R_2 = 11.88$
B	41.7			13.9	14.7	18.4	24.6	28.6	35.3	$R_3 = 12.39$
C	55.6				0.8	4.5	10.7	14.7	21.4	$R_4 = 12.72$
D	56.4					3.7	9.9	13.9	20.6	$R_5 = 12.96$
E	60.1						6.2	10.2	16.9	$R_6 = 13.17$
F	66.3							4.0	10.7	$R_7 = 13.32$
G	70.3								6.7	$R_8 = 13.44$

A	*B*	*C*	*D*	*E*	*F*	*G*	*H*
					———	———	———
				———	———	———	
		———	———	———	———		

the analysis of variance, and n is the number of observations on which each of the means is based. In the present problem, $s = \sqrt{36} = 6$ with $8(4 - 1) = 24$ d.f. Then

$$s_{\bar{X}} = \frac{6}{\sqrt{4}} = 3$$

8.3 Shortest Significant Ranges

Table X in the Appendix gives the *significant studentized ranges* for Duncan's multiple range test with α equal to 0.10, 0.05, 0.01, 0.005, and 0.001. Let us choose $\alpha = 0.01$. Then we enter the row of Table Xc with the number of degrees of freedom for the within treatment mean square or s^2. We have 24 d.f. in the present problem and $k = 8$ means. From the table we find the significant studentized ranges for $k = 2$, 3, 4, 5, 6, 7, and 8 means. These values, as given in Table Xc for 24 d.f., are 3.96, 4.13, 4.24, 4.32, 4.39, 4.44, and 4.48, respectively. We now multiply each significant studentized range by the standard error of the mean. In the present instance, the standard error is 3. The resulting values are called the *shortest significant ranges*. The shortest significant ranges, $R_2, R_3, \ldots, R_k$, are, for the present problem, as follows: $R_2 = 11.88$, $R_3 = 12.39$, $R_4 =$

12.72, $R_5 = 12.96$, $R_6 = 13.17$, $R_7 = 13.32$, and $R_8 = 13.44$. These values of R are recorded in column (9) at the right of Table 8.2 for convenient reference.

8.4 Tests of Significance

The differences between pairs of means are tested in the following order: the largest minus the smallest, the largest minus the second smallest, and so on, up to the largest minus the second largest. Then we test the difference between the second largest minus the smallest, the second largest minus the second smallest, and so on, until we have tested the second smallest minus the smallest. Because we have ranked the means in order of magnitude, with A being the smallest and H the largest, the order of testing will involve first finding the differences in column (8) of Table 8.2, then those in column (7), and so on, moving from right to left.

Each difference in Table 8.2 is significant if it exceeds the corresponding shortest significant range. If it does not, then it is not significant. The only exception to this rule is that no difference between two means can be significant if the two means are both contained in a subset of means that has a nonsignificant range. The term "subset" may refer to the complete set of k means. Because of the exception noted, it is convenient to group these two means and all of the intervening means by underscoring, as shown at the bottom of Table 8.2. No additional tests are made between the means of a subset underscored in the manner described.

Because $H - A$ is the range of eight means, the difference must exceed $R_8 = 13.44$, the shortest significant range for eight means. Because $H - B$ is the range of seven means, it must exceed $R_7 = 13.32$, the shortest significant range for seven means, and so on. When we come to $H - F$, we find that the difference, 10.7, does not exceed $R_3 = 12.39$. Therefore, this difference is not significant and no further tests are made between the means of the subset F, G, and H. That F, G, and H do *not* differ significantly is shown by the underscoring of these three treatment means at the bottom of Table 8.2.

In column (7) of Table 8.2, the difference $G - E = 10.2$ is a range of three means and does not exceed $R_3 = 12.39$. The underscoring of E, F, and G at the bottom of the table shows that these means do *not* differ significantly. In column (6) we find that $F - A = 41.6$, the range of six means, exceeds $R_6 = 13.17$, and $F - B = 24.6$, the range of five means, exceeds $R_5 = 12.96$. However, $F - C = 10.7$, the range of four means, does not exceed $R_4 = 12.72$. Therefore, no further tests are made between the means in the subset C, D, E, and F, and the fact that they form a subset is shown by the underscoring at the bottom of the table.

The final test we make is $B - A = 17.0$, and because this difference exceeds $R_2 = 11.88$, it is significant. The complete results of the various tests are summarized by the underscoring at the bottom of Table 8.2. Any two means *underscored* by the same line do not differ significantly. Any two means *not underscored* by the same line do differ significantly.

It should be obvious that it is *not* necessary to record the complete table of differences, as we have done in Table 8.2. For example, once we find that $H - F$ is not significant, we know that $H - G$ and $G - F$ are not to be tested. Similarly, having found that $F - C$ is not significant, no tests are to be made between any of the differences in the subset C, D, E, and F, and there would be no need to find any of these differences.

8.5 Protection Levels

Duncan's multiple range test is based on the concept of *protection levels.* A two-mean protection level is given by $1 - \alpha$. For example, if $\alpha = 0.01$, then the two-mean protection level is $1 - 0.01 =$ 99 percent. If the two population means are in fact equal, then $\alpha = 0.01$ is the probability that we will wrongly declare them to be significantly different. The two-mean protection level against this erroneous conclusion is $1 - 0.01 = 99$ percent. The k-mean protection level of the multiple range test is given by $(1 - \alpha)^{k-1}$. In the present example, with $k = 8$ means and $\alpha = 0.01$, the protection level is $(1 - 0.01)^{8-1} = 93$ percent, which is the minimum probability of making no erroneous statements about significant differences when the multiple range test is applied to $k = 8$ treatment means, and when the two-mean protection level is $1 - 0.01 = 99$ percent. We thus have somewhat less protection against erroneous conclusions about significance in comparing the differences among eight means than we would have if we were testing the significance of the difference between only two means.

The exponent, $k - 1$, for the protection level is given by the number of orthogonal comparisons that can be made on a set of k means and is equal to the number of degrees of freedom associated with the treatment mean square in the analysis of variance. If we had chosen $\alpha = 0.05$, then the protection level, based on degrees of freedom would be, for a set of eight means, $(1 - 0.05)^{8-1} = 70$ percent.

If $k > 2$ means are being compared, it seems reasonable to expect that we are more likely to have some real differences among the means than would be the case if only $k = 2$ means are compared. Duncan has argued, therefore, that in testing the differences between $k > 2$ means, the test of significance should be more powerful, that is, more likely to detect real differences, than when testing the difference between $k = 2$ means. With the multiple range test, the increased power is obtained by risking a lowered protection level as k increases. In essence, the experimenter who uses Duncan's test in evaluating the differences among $k > 2$ means is likely to make fewer Type II errors and somewhat more Type I errors than he would if the protection level for $k > 2$ was the same as that for $k = 2$ means.

8.6 Comparisons on Treatment Means

If we have an experiment in which k treatment means are involved and the results are analyzed by the analysis of variance, then we shall have a treatment sum of squares with $k - 1$ d.f. It is always possible, as we shall see later, to analyze the treatment sum of squares into $k - 1$ component parts, each with 1 d.f., and such that the sum of these component parts is equal to the treatment sum of squares. Each of the component parts may involve a comparison between two or more of the treatments. We shall consider only the case where we have k means with an equal number of observations for each.

Table 8.3 gives the summary of the analysis of variance for an experiment in which $k = 4$ treatments were tested with $n = 10$ observations for each treatment. We have $F = 9.09$, with 3 and 36 d.f., a significant value.

Table 8.4 shows three comparisons that might be made on the set of $k = 4$ treatment means. The numbers given in each row of the table are called coefficients of the treatment means and we shall use a with appropriate subscripts, as shown at the right of the table, to represent these coefficients. The first subscript refers to a particular treatment mean which is to be multiplied by the coefficient and the second subscript corresponds to a particular comparison.

Multiplying the treatment means by the coefficients in the first row, we obtain the comparison

$$d_1 = \bar{X}_{1\cdot} - \bar{X}_{2\cdot}$$

Table 8.3

ANALYSIS OF VARIANCE FOR $k = 4$ TREATMENTS WITH $n = 10$ OBSERVATIONS FOR EACH TREATMENT

Source of Variation	Sum of Squares	d.f.	Mean Square	F
Treatments	83.50	3	27.83	9.09
Within treatments	110.16	36	3.06	
Total	193.66	39		

Table 8.4

THREE ORTHOGONAL COMPARISONS ON $k = 4$ TREATMENT MEANS WITH VALUES OF THE COEFFICIENTS FOR EACH COMPARISON AT THE LEFT AND NOTATION FOR THE COEFFICIENTS AT THE RIGHT

	$\bar{X}_{1.}$	$\bar{X}_{2.}$	$\bar{X}_{3.}$	$\bar{X}_{4.}$	$\bar{X}_{1.}$	$\bar{X}_{2.}$	$\bar{X}_{3.}$	$\bar{X}_{4.}$	
Comparison	Values of Coefficients				Notation for Coefficients				$\sum a_{.i}^2$
d_1	1	-1	0	0	a_{11}	a_{21}	a_{31}	a_{41}	2
d_2	0	0	-1	1	a_{12}	a_{22}	a_{32}	a_{42}	2
d_3	$\frac{1}{2}$	$\frac{1}{2}$	$-\frac{1}{2}$	$-\frac{1}{2}$	a_{13}	a_{23}	a_{33}	a_{43}	1

Multiplying the treatment means by the coefficients in the second row, we obtain the comparison

$$d_2 = \bar{X}_{4.} - \bar{X}_{3.}$$

If we multiply the treatment means by the coefficients in the last row, we find that this gives the comparison

$$d_3 = \tfrac{1}{2}(\bar{X}_{1.} + \bar{X}_{2.}) - \tfrac{1}{2}(\bar{X}_{3.} + \bar{X}_{4.})$$

or the difference between the average of the means for Treatments 1 and 2 and the average of the means for Treatments 3 and 4.

Comparisons of the kind shown in Table 8.4 are linear functions of the treatment means. Any linear function of the treatment means

$$d_i = a_{1i}\bar{X}_{1.} + a_{2i}\bar{X}_{2.} + \cdots + a_{ki}\bar{X}_{k.} \tag{8.2}$$

is called a comparison, if the sum of the coefficients is equal to zero, that is, if $\Sigma a_{.i} = 0$.

8.7 Orthogonal Comparisons

If we make two comparisons, d_i and d_j, on the same set of k treatment means, then d_i and d_j are said to be *orthogonal* or *independent*, if the sum of the products of the corresponding coefficients for the two comparisons is equal to zero, that is, if

$$\sum a_{ki}a_{kj} = a_{1i}a_{1j} + a_{2i}a_{2j} + \cdots + a_{ki}a_{kj} = 0 \tag{8.3}$$

We note that the comparisons shown in Table 8.4 are *mutually orthogonal*, because the sums of the products of coefficients for all possible pairs of comparisons are equal to zero. For example, for comparisons d_1 and d_2, we have

$$(1)(0) + (-1)(0) + (0)(-1) + (0)(1) = 0$$

The sum of the products of the coefficients for comparisons d_1 and d_3 and for comparisons d_2 and d_3 also sum to zero.

8.8 Standard Error of a Comparison

The standard error of any comparison d_i, that is, the standard error of the corresponding weighted difference obtained by multiplying the means by the coefficients for the comparison, will be

$$s_{d_i} = \sqrt{s^2\left(\frac{a_{1i}^2}{n_1} + \frac{a_{2i}^2}{n_2} + \cdots + \frac{a_{ki}^2}{n_k}\right)} \tag{8.4}$$

where s^2 is the within treatment mean square of the analysis of variance. If the number of observations is the same for each mean, then (8.4) may be written

$$s_{d_i} = \sqrt{\frac{s^2}{n}\sum a_{.i}^2} \tag{8.5}$$

where n is the number of observations for a single mean. We note that for any comparison between two means, say $\bar{X}_{1.}$ and $\bar{X}_{2.}$, the corresponding coefficients will be 1 and -1 and $\Sigma a_{.i}^2 = 2$. Then (8.5) is simply $\sqrt{2s^2/n}$, the standard error of the difference between two means when $n_1 = n_2 = n$.

8.9 Test of Significance of a Comparison

Under the general null hypothesis, all of the k treatment means have the same expected value μ. Then, because $\Sigma a_{.i} = 0$ for any comparison d_i, the expected value of d_i will also be equal to zero. For example, if

$$d_i = 3\bar{X}_{1.} - (\bar{X}_{2.} + \bar{X}_{3.} + \bar{X}_{4.})$$

then

$$E(d_i) = 3\mu - (\mu + \mu + \mu) = 0$$

The significance of the difference of the comparison represented by d_i can then be evaluated by finding

$$t = \frac{d_i}{s_{d_i}} \tag{8.6}$$

with degrees of freedom equal to the number of degrees of freedom associated with the error (within treatment) mean square of the analysis of variance. Confidence limits for d_i may be established in the usual way.

In Table 8.5 we give the means for the treatments of the analysis of

Table 8.5

APPLICATION OF THE COMPARISONS OF TABLE 8.4

The means are those obtained in the experiment summarized in Table 8.3.

	$\bar{X}_{1.}$	$\bar{X}_{2.}$	$\bar{X}_{3.}$	$\bar{X}_{4.}$	
Comparison	17.2	19.4	15.8	19.0	Value of d_i
d_1	1	-1	0	0	-2.2
d_2	0	0	-1	1	3.2
d_3	$\frac{1}{2}$	$\frac{1}{2}$	$-\frac{1}{2}$	$-\frac{1}{2}$	0.9

variance reported in Table 8.3 and the coefficients for the three comparisons of Table 8.4. Multiplying the means by the corresponding coefficients for each comparison, we obtain $d_1 = -2.2$, $d_2 = 3.2$, and $d_3 = 0.9$.

Summing the squares of the coefficients for each comparison, we have

$$\sum a_{\cdot 1}^2 = 2, \qquad \sum a_{\cdot 2}^2 = 2, \qquad \text{and} \qquad \sum a_{\cdot 3}^2 = 1$$

From the analysis of variance of Table 8.3, we have $s^2 = 3.06$. Then the standard errors, given by (8.3), for each of the three comparisons will be

$$s_{d_1} = \sqrt{\frac{3.06}{10}(2)} = 0.782$$

$$s_{d_2} = \sqrt{\frac{3.06}{10}(2)} = 0.782$$

$$s_{d_3} = \sqrt{\frac{3.06}{10}(1)} = 0.553$$

Dividing each d_i by its standard error, as in (8.4), we obtain

$$t_1 = \frac{-2.2}{0.782} = -2.813$$

$$t_2 = \frac{3.2}{0.782} = 4.092$$

$$t_3 = \frac{0.9}{0.553} = 1.627$$

Each of these t's has 36 d.f., the number of degrees of freedom associated with s^2 of the analysis of variance. If $\alpha = 0.01$, and with a two-sided test, the first two comparisons, d_1 and d_2, are significant, whereas the third, d_3, is not.

8.10 Comparisons on Treatment Sums

Instead of making the comparisons on the treatment means, we may choose to use the treatment sums. If n is the same for each treatment, and if d_1 is a comparison on the treatment means, then

$$d_1 = \frac{1}{n}\left(a_{11} \sum X_{1\cdot} + a_{21} \sum X_{2\cdot} + \cdots + a_{k1} \sum X_{k\cdot}\right) \tag{8.7}$$

Let the difference obtained by multiplying each of the treatment sums by the coefficients be D_1; that is, let

$$D_1 = a_{11} \sum X_{1\cdot} + a_{21} \sum X_{2\cdot} + \cdots + a_{k1} \sum X_{k\cdot} \tag{8.8}$$

We shall refer to D_1 as a comparison on the *treatment sums*. Then it can also be shown that

$$\frac{D_1^2}{n \sum a_{\cdot 1}^2}$$

will be a component of the treatment sum of squares with 1 d.f. Because $D_1^2/n\Sigma a_{\cdot 1}^2$ is a sum of squares with 1 d.f., it follows that it is also a mean square with 1 d.f. and, therefore,

$$MS_{D_1} = \frac{D_1^2}{n \sum a_{\cdot 1}^2} \tag{8.9}$$

If a second comparison D_2 is made on the treatment sums, and if D_1 and D_2 are orthogonal, that is, if the sum of the products of the corresponding coefficients is zero, then

$$MS_{D_2} = \frac{D_2^2}{n \sum a_{\cdot 2}^2} \tag{8.10}$$

will be a component of the *residual treatment* sum of squares or a component of

$$\text{Residual} = \text{Treatment sum of squares} - MS_{D_1}$$

with 1 d.f.

Similarly, after partitioning the treatment sum of squares into two orthogonal components, MS_{D_1} and MS_{D_2}, each with 1 d.f., we may then choose a comparison D_3 that is orthogonal to both D_1 and D_2. If all of the comparisons are mutually orthogonal, that is, if every pair is orthogonal,

Table 8.6

APPLICATION OF THE COMPARISONS OF TABLE 8.4 TO TREATMENT SUMS

The treatment sums are those obtained in the experiment summarized in Table 8.3.

Comparison	$\sum X_{1\cdot}$ 172	$\sum X_{2\cdot}$ 194	$\sum X_{3\cdot}$ 158	$\sum X_{4\cdot}$ 190	Value of D_i
D_1	1	-1	0	0	-22
D_2	0	0	-1	1	32
D_3	$\frac{1}{2}$	$\frac{1}{2}$	$-\frac{1}{2}$	$-\frac{1}{2}$	9

then the sum of $MS_{D_1}, MS_{D_2}, \ldots, MS_{D_{k-1}}$ will be equal to the treatment sum of squares in the analysis of variance, or

$$\text{Treatment sum of squares} = MS_{D_1} + MS_{D_2} + \cdots + MS_{D_{k-1}}$$

Each MS_{D_i} is not only a mean square but also a sum of squares because each has 1 d.f.

8.11 Test of Significance of a Comparison on the Treatment Sums

Each of the mean squares, MS_{D_i}, may be tested for significance by finding

$$F = \frac{MS_{D_i}}{MS_W} \tag{8.11}$$

The F defined by (8.11) will have 1 d.f. for the numerator. The degrees of freedom for the denominator will be equal to those associated with MS_W, the within treatment mean square.

In Table 8.6 we give the treatment *sums* corresponding to the treatment means of Table 8.5. The coefficients given in the table are the same as those given in Table 8.4. The values of D_i are found by multiplying the treatment sums by the corresponding coefficients for each comparison. Squaring each of the values of D_i and dividing by the corresponding values of $n\Sigma a_{\cdot i}^2$, we obtain the mean squares MS_{D_i} for the comparisons. We note that

$$MS_{D_1} + MS_{D_2} + MS_{D_3} = 24.2 + 51.2 + 8.1 = 83.5$$

and that 83.5 is the treatment sum of squares in the analysis of variance shown in Table 8.3.

Dividing each MS_{D_i} by $MS_W = 3.06$, we have

$$F_1 = \frac{24.2}{3.06} = 7.91$$

$$F_2 = \frac{51.2}{3.06} = 16.73$$

$$F_3 = \frac{8.1}{3.06} = 2.65$$

With $\alpha = 0.01$, $F = 7.39$ will be significant for 1 and 36 d.f. Thus, we would conclude that D_1 and D_2 are significant, whereas D_3 is not.

8.12 Proof that $t^2 = F$

The conclusions concerning significance reached by means of the F test are exactly the same as those we arrived at by means of the t test for the same data. If we square the t's obtained previously for the same data, we may note that each t^2 for a given comparison on the means, d_i, is equal, within rounding errors, to the corresponding value of F for the same comparison, D_i, on the treatment sums. For example, we have $t_1^2 = (-2.813)^2 = 7.91$, $t_2^2 = (4.092)^2 = 16.74$, and $t_3^2 = (1.627)^2 = 2.65$ and these values are, within rounding errors, equal to the corresponding values of F.

That F as defined by (8.11) is exactly equal to the square of t as defined by (8.6) can easily be shown. For a given comparison, d_i, we have

$$t^2 = \frac{d_i^2}{\frac{s^2}{n}\sum a._i^2} = \frac{\frac{1}{n^2}D_i^2}{\frac{s^2}{n}\sum a._i^2} = \frac{\left(\frac{n}{\sum a._1^2}\right)\left(\frac{1}{n^2}\right)D_i^2}{s^2} = \frac{MS_{D_i}}{MS_W} = F$$

8.13 Multiplying the Coefficients of a Comparison by a Constant

If we have a comparison D_i so that $\Sigma a._i = 0$, then we can multiply the coefficients for the comparison by any constant without changing the nature of the comparison. For example, in Table 8.6, if we multiply the coefficients for D_3 by 2, then we would have 1, 1, -1, and -1. Multiplying each of the treatment sums by these new coefficients, we have

$$D_3 = (172 + 194) - (158 + 190) = 18$$

with $\Sigma a_{.3}^2 = 4$, and

$$MS_{D_3} = \frac{D_3^2}{n \sum a_{.3}^2} = \frac{(18)^2}{(10)(4)} = 8.1$$

as before. We would obtain exactly the same value for MS_{D_3}, regardless of the value of the constant used in multiplying the coefficients. The value of D_3 would, of course, be changed, but so would the value of $\Sigma a_{.3}^2$, with the result that we would obtain exactly the same value for MS_{D_3}. The fact that we can multiply the coefficients for a given comparison by a constant, without changing the nature of the comparison or the value of MS_{D_i}, is sometimes useful in simplifying the computations of the mean square.

8.14 Choosing a Set of Orthogonal Comparisons

It is possible to analyze the treatment sum of squares into more than one *set* of orthogonal comparisons. For example, with $k = 4$ means, the two *sets* of orthogonal comparisons given in Table 8.7 differ from each other and also from the *set* of orthogonal comparisons given in Table 8.4. That each of the two sets of comparisons given in Table 8.7 is orthogonal can easily be determined by showing that the sum of the products of the corresponding coefficients for each possible pair of comparisons within each set is equal to zero.

Because more than one set of orthogonal comparisons is possible for a given group of $k > 2$ means, which particular set of comparisons is to be made should be determined by experimental interests and planned at the same time the experiment is planned. The particular comparisons shown in Table 8.4, for example, might have been of experimental interest and planned in advance, if the dependent variable of interest were a measure of maze performance and if four groups of rats had been tested under the following conditions:

Group 1: a group tested after 12 hours of water deprivation
Group 2: a group tested after 24 hours of water deprivation
Group 3: a group tested after 12 hours of food deprivation
Group 4: a group tested after 24 hours of food deprivation.

The first comparison of Table 8.4 would test for the difference between the 12- and 24-hour water-deprived groups; the second comparison would test for the difference between the 12- and 24-hour food-deprived groups; and the third comparison would test for the difference between the average performance of the water-deprived and the food-deprived groups.

Again, we emphasize that the methods described for testing orthogonal comparisons for significance should be used only if the comparisons have been planned in advance. It may also be emphasized that we do not need to

Table 8.7

TWO DIFFERENT SETS OF ORTHOGONAL COMPARISONS ON $k = 4$ TREATMENT SUMS

	$\sum X_{1.}$	$\sum X_{2.}$	$\sum X_{3.}$	$\sum X_{4.}$
Set 1	Coefficients			
D_1	-1	$\frac{1}{3}$	$\frac{1}{3}$	$\frac{1}{3}$
D_2	0	-1	$\frac{1}{2}$	$\frac{1}{2}$
D_3	0	0	-1	1
Set 2	Coefficients			
D_1	$\frac{1}{2}$	$\frac{1}{2}$	$-\frac{1}{2}$	$-\frac{1}{2}$
D_2	$-\frac{1}{2}$	$\frac{1}{2}$	$-\frac{1}{2}$	$\frac{1}{2}$
D_3	$-\frac{1}{2}$	$\frac{1}{2}$	$\frac{1}{2}$	$-\frac{1}{2}$

make all of the possible $k - 1$ orthogonal comparisons in a given set. In some cases the experimenter may only be interested in several of the possible comparisons. The methods described may be used to test the significance of any planned orthogonal comparisons equal to or less than $k - 1$, where k is the number of treatment groups.

8.15 Treatments as Values of an Ordered Variable

In some experiments the treatments may consist of an ordered variable. For example, we might test different groups of rats after 0, 6, 12, 18, and 24 hours of food or water deprivation. In other cases, the treatments may consist of increasing intensities of shock, of increasing amounts of reward, or of increasing numbers of reinforcements.

If the treatments consist of an ordered variable and if we can assume that the differences between the treatments are uniform (that is, equal), then we may be interested in determining whether the treatment means (or sums) are functionally related to the different values of the treatment variable. We may, for example, be interested in finding out whether the

Table 8.8

ANALYSIS OF VARIANCE FOR AN EXPERIMENT IN WHICH THE TREATMENTS CONSIST OF FIVE EQUALLY INCREASING LEVELS OF REINFORCEMENT

Source of Variation	Sum of Squares	d.f.	Mean Square	F
Treatments	2665.48	4	666.37	31.51
Within treatments	951.73	45	21.15	
Total	3617.21	49		

treatment means are linearly related to the values of the treatment variable or whether they deviate significantly from a linear relation. If the deviations from linearity are significant, then we may wish to determine whether the trend of the means can be adequately described by a quadratic or second-degree equation.

Suppose, for example, that we have an experiment in which the treatments consist of five equally increasing levels of reinforcement, which we designate by 1, 2, 3, 4, and 5. With $n = 10$ rats assigned to each treatment, assume that the analysis of variance for the experiment is as given in Table 8.8.

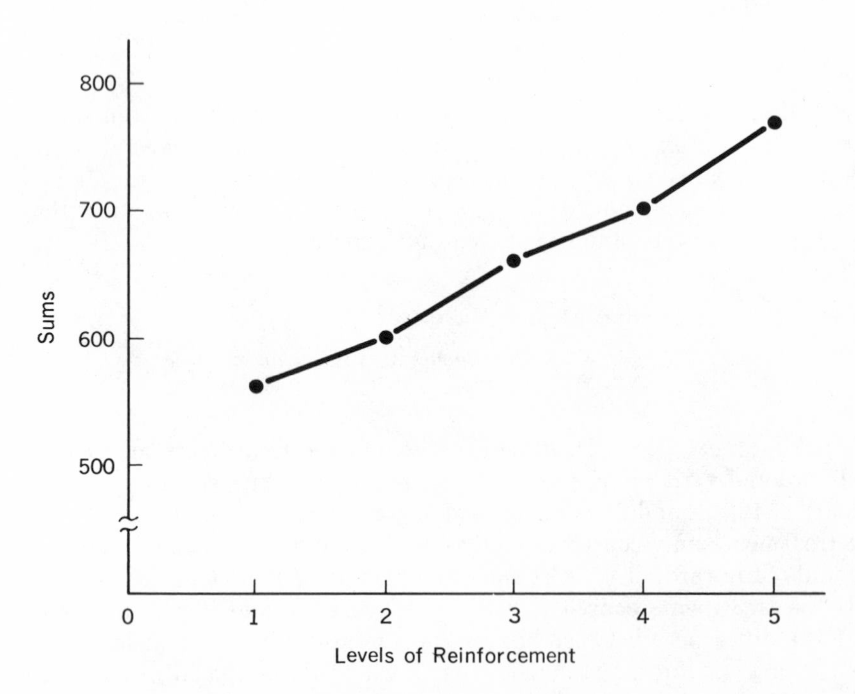

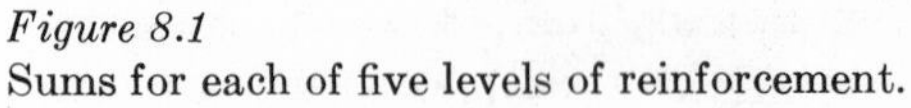
Figure 8.1
Sums for each of five levels of reinforcement.

Table 8.9
LINEAR AND QUADRATIC COMPARISONS ON THE TREATMENT SUMS WITH $k = 5$ TREATMENTS

	Levels of Reinforcement					
	1	2	3	4	5	
Comparison	564	601	663	703	770	D_i
Linear	−2	−1	0	1	2	514
Quadratic	2	−1	−2	−1	2	38

Let us assume that the treatment sums, given in Table 8.9, represent the number of bar presses made by the rats at each reinforcement level. In Figure 8.1, the treatment sums are plotted against the levels of reinforcement.

8.16 Orthogonal Polynomials

To determine whether the linear component of the trend of the sums is statistically significant and also whether the treatment sums deviate significantly from linearity, we make use of a table of coefficients for orthogonal polynomials, Table XI in the Appendix.[4] This table gives the coefficients to be used in finding the linear and quadratic components of the treatment sums of squares. It may be noted that the coefficients in each row sum to zero and that, for any fixed value of k, the sum of the products of the coefficients for the linear and quadratic components is also zero. The linear and quadratic comparisons, therefore, meet the requirement for orthogonality discussed earlier.

For any given value of k, orthogonal coefficients corresponding to $k - 1$ components can be written. For example, if $k = 5$, the successive sets of coefficients would correspond to the linear, quadratic, cubic, and quartic components of the treatment sum of squares. Successive application of these coefficients would enable one to determine how well the trend of the treatment sums (or means) is represented by a polynomial of the first, second, third, and fourth degree, respectively. Orthogonal coefficients for the higher degree polynomials can be found in the Fisher and Yates (1948) tables.

The coefficients for the linear and quadratic components for $k = 5$ treatments, as obtained from Table XI, are shown in Table 8.9.

[4] The coefficients for orthogonal polynomials given in Table XI are for the case of equal intervals between treatments. If the intervals are unequal, the coefficients given in Table XI should not be used. For procedures to be used with unequal intervals, see Grandage (1958) or Gaito (1965).

8.17 *Test of Significance of Linearity and Deviations from Linearity*

Using the treatment sums rather than the means, and multiplying these sums by the coefficients for the linear comparison, as given in Table 8.9, we have

$$D_1 = (-2)(564) + (-1)(601) + (0)(663) + (1)(703) + (2)(770)$$
$$= 514$$

Then $\Sigma a_{.1}^2 = 10$ and, because we have $n = 10$ observations for each sum, we find

$$MS_{D_1} = \frac{(514)^2}{(10)(10)} = 2641.96$$

as the mean square for the linear component with 1 d.f.

The sum of squares for deviations from linear regression will be equal to the treatment sum of squares minus the sum of squares for the linear component. Thus,

$$\text{Deviations from linearity} = \text{Treatment} - \text{linear} \qquad (8.12)$$

or, for the example under discussion,

$$\text{Deviations from linearity} = 2665.48 - 2641.96 = 23.52$$

and the sum of squares defined by (8.12) will have $k - 2$ d.f.

Table 8.10 summarizes the analysis. Testing the linear comparison for significance, we have

$$F = \frac{MS_{D_1}}{MS_W} = \frac{2641.96}{21.15} = 124.92$$

with 1 and 45 d.f. The value $F = 124.92$ is highly significant and we conclude that there is a significant linear trend in the treatment sums (or means).

Table 8.10

TEST OF SIGNIFICANCE FOR LINEAR REGRESSION AND DEVIATIONS FROM LINEAR REGRESSION

Source of Variation	Sum of Squares	d.f.	Mean Square	F
Linear regression	2641.96	1	2641.96	124.92
Deviations	23.52	3	7.84	
Within treatments	951.73	45	21.15	
Total	3617.21	49		

It is obvious that the mean square for deviations from linearity is not significant because this mean square is 7.84 and the error mean square is 21.15 and, consequently, for this test $F < 1.0$. If the mean square for deviations from linearity is larger than the error mean square, we could test it for significance by finding

$$F = \frac{\text{Mean square for deviations from linearity}}{\text{Error mean square}} \tag{8.13}$$

with $k - 2$ d.f. for the numerator of the F ratio.

8.18 Test of Significance of Curvature

If F as defined by (8.13) is significant, then we might also determine whether there is a significant curvature in the trend of the means by finding the quadratic component of the treatment sum of squares.[5] Merely to illustrate the calculations, because we already know that the quadratic component cannot be significant, we multiply each of the treatment sums by the coefficients for the quadratic comparison, as given in Table 8.9, to obtain

$$D_2 = (2)(564) + (-1)(601) + (-2)(663) + (-1)(703) + (2)(770)$$
$$= 38$$

Then $\Sigma a_{.2}^2 = 14$ and

$$MS_{D_2} = \frac{(38)^2}{(10)(14)} = 10.31$$

with 1 d.f. If MS_{D_2} is larger than the error mean square, we would test it for significance by

$$F = \frac{MS_{D_2}}{MS_W} \tag{8.14}$$

If MS_{D_2} is significant, this would mean that there is a significant curvature in the trend of the treatment sums (or means).

8.19 Comparing Each Treatment with a Control

In some experiments the major objective is to compare each of a number of different treatments with a standard or control. Under these circumstances, the test procedure suggested is one

[5] It is possible, but not in the present example, that the mean square for deviations from linearity may not be significant, yet one of the mean squares for a higher order polynomial is significant.

Table 8.11

VALUES OF A DEPENDENT VARIABLE FOR A CONTROL GROUP AND $k = 4$ TREATMENT GROUPS WITH $n = 8$ OBSERVATIONS FOR EACH GROUP

	Control	T_1	T_2	T_3	T_4
	5	16	16	2	7
	8	18	7	10	11
	8	5	10	9	12
	11	12	4	13	9
	1	11	7	11	14
	9	12	23	9	16
	5	23	12	13	24
	9	19	13	9	19
Σ	56	116	92	76	112
Mean	7.0	14.5	11.5	9.5	14.0

developed by Dunnett (1955). For example, in an experiment on the influence of incentives on learning, one group of subjects may be tested under a standard set of instructions and in the absence of any added incentives. This group may be designated the *control* group. The different treatments may then consist of varying kinds of incentives, introduced in an effort to improve performance over that of the control group. Our interest is in finding out which, if any, of the incentives result in performance significantly *better* than that of the control group.

Suppose, for example, that we have one control group and $k = 4$ treatment groups with $n = 8$ observations for each group.[6] We designate the mean of the control group by $\bar{X}_{0\cdot}$ and the means of the treatment groups by $\bar{X}_{1\cdot}, \bar{X}_{2\cdot}, \ldots, \bar{X}_{k\cdot}$. Each of the k treatment means is to be tested for significance by comparison with $\bar{X}_{0\cdot}$. As we have stated above, our concern with the treatment means involves only the question of whether they are significantly *greater* than the control mean. We are not concerned with guarding against the alternative that a treatment mean may be significantly smaller than the control mean. Table 8.11 gives the values of the dependent variable for the control group and for the $k = 4$ treatment groups.

Assuming that the variances of the five groups are all estimates of a common population variance, the estimate based on the combined variances of the control and the k treatment groups will be the mean square within groups with $(n - 1) + k(n - 1) = 35$ d.f. The mean square within

[6] Dunnett (1955) indicates that the optimum allocation of subjects to the control and to each of the k treatment groups is approximately $n_0/n_1 = \sqrt{k}$, where n_0 is the number of observations for the control and n_1 is the number for each of the k treatment groups. For example, with $k = 4$ treatments we should have approximately twice as many subjects in the control group as in each of the treatment groups.

groups, obtained in the usual way, is 24.17 for the observations in Table 8.11. Then the standard error of the difference between two means will be

$$s_{\bar{X}_1-\bar{X}_2} = \sqrt{\frac{2s^2}{n}}$$

or, for the data of Table 8.11, with $s^2 = 24.17$ and $n = 8$,

$$s_{\bar{X}_1-\bar{X}_2} = \sqrt{\frac{(2)(24.17)}{8}} = 2.46$$

8.20 Significance of the Difference between a Control Mean and Each Treatment Mean

Because we decided in advance that we were interested only in finding out whether the k treatment means exceed significantly the control mean, the tests of significance will be one-sided and we shall make k of them. Table XIIa in the Appendix gives the values of t for a one-sided test with probability 0.95 that all k statements concerning the difference between a treatment mean and a control mean are correct. Table XIIc gives the corresponding values of t for the two-sided test, also with probability of 0.95 that all k statements concerning the differences are correct.[7] For the one-sided test, we enter Table XIIa with $k = 4$ and d.f. $= 35$, the degrees of freedom associated with the error mean square. We have no entry for 35 d.f., but by interpolation between 30 and 40 we find $t = 2.24$.

Instead of making successive t tests to determine whether the k differences between the treatment means and the control mean result in $t \geq 2.24$, we solve for the magnitude of the difference itself that will be significant. In order for a difference to be declared significant, we must have

$$t = \frac{(\bar{X}_{k\cdot} - \bar{X}_{0\cdot}) - (\mu_k - \mu_0)}{2.46} \geq 2.24$$

or, since the null hypothesis specifies that $\mu_k - \mu_0 = 0$,

$$\bar{X}_{k\cdot} - \bar{X}_{0\cdot} \geq (2.46)(2.24)$$

or

$$\bar{X}_{k\cdot} - \bar{X}_{0\cdot} \geq 5.51$$

Then any observed difference between a treatment mean and the control mean will be judged significantly greater than zero, if $\bar{X}_{k\cdot} - \bar{X}_{0\cdot} \geq 5.51$.

[7] Tables XIIb and XIId give, respectively, the values of t for a one-sided test and a two-sided test with probability 0.99 that all k statements concerning the difference between a treatment mean and the control mean are correct.

The observed differences are

$$\bar{X}_{1\cdot} - \bar{X}_{0\cdot} = 14.5 - 7.0 = 7.5 > 5.51$$
$$\bar{X}_{2\cdot} - \bar{X}_{0\cdot} = 11.5 - 7.0 = 4.5 < 5.51$$
$$\bar{X}_{3\cdot} - \bar{X}_{0\cdot} = 9.5 - 7.0 = 2.5 < 5.51$$
$$\bar{X}_{4\cdot} - \bar{X}_{0\cdot} = 14.0 - 7.0 = 7.0 > 5.51$$

and we conclude that the means for Treatments 1 and 4 are significantly greater than the mean of the control group and the means for Treatments 2 and 3 are not, with probability of 0.95 that these statements are all correct.

8.21 Scheffé's Test for Multiple Comparisons

Scheffé (1953) has suggested a test that is appropriate for making *any* and *all* comparisons of interest on a set of k means, including those comparisons that may be suggested by the values of the means themselves. In other words, to use Scheffé's test *we do not need to plan the comparisons in advance.*[8] Table 8.12 shows the various comparisons that *might* be made with respect to a set of $k = 4$ means, with $n = 10$ observations for each mean. Each row is a comparison on the *treatment sums* and the sums are those given previously in Table 8.6, where MS_W was equal to 3.06 with 36 d.f. It is not necessary that all of the comparisons shown in the table be made, but we may make any that are of interest or all of them, using Scheffé's test, with probability equal to or greater than $1 - \alpha$ that all statements concerning the significance of the comparisons, including those comparisons we do not make, are true. Then, if $\alpha = 0.05$, the probability that all statements regarding the comparisons in Table 8.12 are true will be equal to or greater than 0.95.

Let any given comparison of the treatment sums be represented by D_i. The values of D_i, given in column (7) of Table 8.12, are obtained by multiplying each of the treatment sums, given at the top of the table, by the corresponding coefficients in each row. For a given D_i to be a comparison, we require only that $\Sigma a_{\cdot i} = 0$ for the comparison, that is, for the sum of the coefficients for the comparison to be equal to zero. It is not necessary that the various comparisons be orthogonal.

Then, with each sum based on the same number of observations, we have

$$MS_{D_i} = \frac{D_i^2}{n \sum a_{\cdot i}^2} \qquad (8.15)$$

[8] Scheffé's test can, of course, be used in testing planned orthogonal comparisons of the kind described earlier. The test will be more conservative than the procedures described for testing planned orthogonal comparisons; that is, larger differences will be required for significance. Scheffé suggests that with his test one might consider taking $\alpha = 0.10$ rather than $\alpha = 0.05$.

Table 8.12
POSSIBLE COMPARISONS ON $k = 4$ TREATMENT SUMS

(1)	(2)	(3)	(4)	(5)	(6)	(7)	(8)	(9)
	$\sum X_{1\cdot}$	$\sum X_{2\cdot}$	$\sum X_{3\cdot}$	$\sum X_{4\cdot}$				
Comparison	172	194	158	190	$\sum a_{\cdot i}^2$	D_i	D_i^2	MS_{D_i}
1 vs. 2	1	−1	0	0	2	−22	484	24.20
1 vs. 3	1	0	−1	0	2	14	196	9.80
1 vs. 4	1	0	0	−1	2	−18	324	16.20
2 vs. 3	0	1	−1	0	2	36	1296	64.80
2 vs. 4	0	1	0	−1	2	4	16	0.80
3 vs. 4	0	0	1	−1	2	−32	1024	51.20
1 vs. 2 + 3	2	−1	−1	0	6	−8	64	1.07
1 vs. 2 + 4	2	−1	0	−1	6	−40	1600	26.67
1 vs. 3 + 4	2	0	−1	−1	6	−4	16	0.27
2 vs. 1 + 3	−1	2	−1	0	6	58	3364	56.07
2 vs. 1 + 4	−1	2	0	−1	6	26	676	11.27
2 vs. 3 + 4	0	2	−1	−1	6	40	1600	26.67
3 vs. 1 + 2	−1	−1	2	0	6	−50	2500	41.67
3 vs. 1 + 4	−1	0	2	−1	6	−46	2116	35.27
3 vs. 2 + 4	0	−1	2	−1	6	−68	4624	77.07
4 vs. 1 + 2	−1	−1	0	2	6	14	196	3.27
4 vs. 1 + 3	−1	0	−1	2	6	50	2500	41.67
4 vs. 2 + 3	0	−1	−1	2	6	28	784	13.07
1 + 2 vs. 3 + 4	1	1	−1	−1	4	18	324	8.10
1 + 3 vs. 2 + 4	1	−1	1	−1	4	−54	2916	72.90
1 + 4 vs. 2 + 3	1	−1	−1	1	4	10	100	2.50
1 vs. 2 + 3 + 4	3	−1	−1	−1	12	−26	676	5.63
2 vs. 1 + 3 + 4	−1	3	−1	−1	12	62	3844	32.03
3 vs. 1 + 2 + 4	−1	−1	3	−1	12	−82	6724	56.03
4 vs. 1 + 2 + 3	−1	−1	−1	3	12	46	2116	17.63

as the mean square for the comparison of interest. The mean squares for each row comparison are given in column (9) of Table 8.12. Then, as a test of significance, we may find

$$F = \frac{MS_{D_i}}{s^2} \tag{8.16}$$

where s^2 is the error mean square of the analysis of variance.

Instead of evaluating F as defined by (8.16) in the usual way, by finding the tabled value for the 1 d.f. corresponding to the numerator and the degrees of freedom associated with s^2 in the denominator, we compare the obtained value of F with

$$F' = (k - 1)F \tag{8.17}$$

where F' is $(k - 1)$ times the *tabled value* of F for $k - 1$ and $k(n - 1)$ d.f. *F' is the standard in terms of which the F of* (8.16) *is to be evaluated.*

In our example, we have s^2, the within treatment mean square, equal to 3.06 with 36 d.f. We have $k = 4$ means and the tabled value of F for 3 and 36 d.f. is 2.86, with $\alpha = 0.05$. Then

$$F' = (4 - 1)(2.86) = 8.58$$

For any mean square in Table 8.12, we know that $MS_{D_i}/3.06$ must be equal to or greater than $F' = 8.58$ to be judged significant. Solving for the *smallest significant value* of MS_{D_i}, we have

$$(F')(s^2) = (8.58)(3.06) = 26.25$$

and any MS_{D_i} in Table 8.12 that equals or exceeds 26.25 will be judged significant.

Scheffé presents his test for comparisons on the treatment means rather than the treatment sums; that is, a comparison is given by d_i rather than by D_i. If the comparisons are made on the treatment means, then the standard error of comparison will be

$$s_{d_i} = \sqrt{\frac{s^2}{n} \sum a_{.i}^2} \qquad (8.18)$$

and the test of significance will be given by

$$t = \frac{d_i}{s_{d_i}} \qquad (8.19)$$

The t defined by (8.19) can then be evaluated for significance by comparing it with $\sqrt{F'}$, that is, the square root of (8.17). Confidence limits for the d_i's can also be established in the manner described previously.

Scheffé's test has much in its favor because it permits the investigator to examine his data and to make any and all comparisons he wishes to make. It should be pointed out, however, that it is useless to apply Scheffé's test to the results of any experiment in which the treatment mean square is not significantly larger than the within treatment mean square. If the treatment mean square is not significant at some significance level α, then no comparison will be judged significant by Scheffé's test at the same significance level. If the treatment mean square is significant, then at least one comparison will be significant and, of course, there may be more.

QUESTIONS AND PROBLEMS

In this chapter we shall give a limited number of problems. Additional problems, if they are desired, can be obtained by applying the methods of this chapter to the analysis of variance problems of other chapters and to the illustrative examples presented in the text.

8.1 We have a randomized group design with $n = 6$ subjects assigned to each of eight treatments. The error mean square of the analysis of variance is 53.02 with 40 d.f. The treatment means are given below:

A	*B*	*C*	*D*	*E*	*F*	*G*	*H*
19.7	36.7	50.6	51.4	55.1	61.3	65.3	72.0

Use Duncan's multiple range test with $\alpha = 0.01$ to investigate the differences between the means.

8.2 Assume we have a control group and $k = 5$ treatment groups, with ten observations for each group. The error mean square of the analysis of variance is 36.00. The means for the groups are given below:

Control	*A*	*B*	*C*	*D*	*E*
18.6	20.5	23.4	19.6	28.3	26.2

Use Dunnett's test to determine which of the treatment means is significantly greater than the mean of the control group.

8.3 For $k = 5$ treatment means, find a set of $k - 1$ orthogonal comparisons. Demonstrate that the comparisons are mutually orthogonal.

8.4 Describe an experiment involving six treatments in which a set of five planned and mutually orthogonal comparisons would be of experimental interest. Demonstrate that the comparisons are mutually orthogonal.

8.5 We have a randomized group design in which the treatments consist of three equally spaced intervals of testing. One group is tested for retention of learned material after 12 hours, another after 24 hours, and the third after 36 hours. The means for the groups are 11.0, 9.0, and 5.0, respectively. We have $n = 10$ subjects in each group and $s^2 = 20.0$ with 27 d.f. (*a*) Use the analysis of variance to determine whether the means differ significantly. (*b*) Test the linear component of the trend of the means for significance.

8.6 We have a set of k means and wish to test the differences between various subsets of the means. (*a*) What is the essential condition in order that a given difference may be described as a comparison? (*b*) What is the essential condition if two comparisons are to be described as orthogonal?

8.7 In an experiment involving $k = 4$ treatments, $n = 10$ subjects were assigned at random to each treatment. For the experiment we have $MS_W = 16.0$. Find the standard error of each of the following comparisons:

$$(a)\quad (\bar{X}_{1\cdot} + \bar{X}_{2\cdot}) - (\bar{X}_{3\cdot} + \bar{X}_{4\cdot})$$

$$(b)\quad \bar{X}_{1\cdot} - \bar{X}_{2\cdot}$$

$$(c)\quad 3\bar{X}_{1\cdot} - (\bar{X}_{2\cdot} + \bar{X}_{3\cdot} + \bar{X}_{4\cdot})$$

CHAPTER 9

THE 2^n FACTORIAL EXPERIMENT

9.1 Introduction

Many experiments are concerned with the influence of two or more independent variables, usually called *factors*, on a dependent variable. The number of ways in which a factor is varied is called the number of *levels* of the factor. A factor that is varied in two ways would be said to have two levels and a factor that is varied in three ways would be said to have three levels. With two or more factors each with two or more levels, a treatment consists of a combination of one level for each factor. When the treatments consist of all possible different combinations of one level from each factor, and we have an equal number of observations for each treatment, the experiment is described as a *complete factorial experiment with equal replications*.[1]

In this chapter we shall be concerned with the analysis of variance of a 2^n factorial experiment with a randomized group design. A 2^n factorial experiment is one in which we have n factors with two levels for each factor. Although our discussion will be confined to a 2^3 factorial experiment, it can readily be generalized to any 2^n factorial experiment.

9.2 An Example of a 2^3 Factorial Experiment

As an illustration let us suppose that the dependent variable is a measure of retention of verbal material. One factor of interest is the *number* of times the material is presented, and this is varied in two

[1] In factorial experiments, using the methods of analysis described in this chapter, we should have equal n's for each treatment combination. In Chapter 12 we describe a method that may be used in analyzing the results of factorial experiments in which, for one reason or another, we may happen to have unequal n's for the various treatment combinations.

ways by presenting the material once and by presenting the material twice. We shall designate this factor as A and the two levels by A_1, corresponding to one presentation, and A_2, corresponding to two presentations. A second factor of interest is the *mode* of presentation and this factor is also varied in two ways. In one case a passage is read to subjects, and we shall call this the *auditory mode* of presentation. In the other case, the subjects themselves read the passage, and we shall refer to this as the *visual mode* of presentation. We designate the mode of presentation as the B factor and the two levels as B_1, corresponding to the visual mode, and B_2, corresponding to the auditory mode. Still a third factor of interest is the *time* of testing and this factor is also to be varied in two ways. We designate this factor as C and let C_1 correspond to an immediate test and C_2 to a delayed test.

A given treatment will be obtained by selecting one level from each of the three factors. For example, one treatment will be $A_1B_1C_1$ and will represent a treatment consisting of one presentation, using the visual mode, and an immediate test. The total number of different treatments will be $2 \times 2 \times 2 = 8$, and they are shown in Table 9.1. The treatments are used with a randomized group design with $n = 10$ observations for each treatment. The outcomes of this hypothetical experiment are given in Table 9.2.

In our discussion of the analysis of variance for this experiment, we shall regard the levels of the factors as having been *selected* for investigation because they were of experimental interest. This is to emphasize that the levels of the factors are to be regarded as *fixed* and not as representing a *random* sampling from a larger population of levels.[2] We are not concerned, for example, with being able to generalize beyond the particular number of

Table 9.1

THE EIGHT TREATMENT COMBINATIONS OF THE 2^3 FACTORIAL EXPERIMENT

Treatment	Number	Mode	Time
$A_1B_1C_1$	one	visual	immediate
$A_1B_1C_2$	one	visual	delayed
$A_1B_2C_1$	one	auditory	immediate
$A_1B_2C_2$	one	auditory	delayed
$A_2B_1C_1$	two	visual	immediate
$A_2B_1C_2$	two	visual	delayed
$A_2B_2C_1$	two	auditory	immediate
$A_2B_2C_2$	two	auditory	delayed

[2] The fixed effect model of the analysis of variance is discussed in detail in Chapter 11.

Table 9.2
OUTCOME OF A $2 \times 2 \times 2$ FACTORIAL EXPERIMENT WITH A RANDOMIZED GROUP DESIGN

	A_1				A_2			
	B_1		B_2		B_1		B_2	
	C_1	C_2	C_1	C_2	C_1	C_2	C_1	C_2
	76	36	43	37	94	74	67	67
	66	45	75	22	85	74	64	60
	43	47	66	22	80	64	70	54
	62	23	46	25	81	86	65	51
	65	43	56	11	80	68	60	49
	43	43	62	27	80	72	55	38
	42	54	51	23	69	62	57	55
	60	45	63	24	80	64	66	56
	78	41	52	25	63	78	79	68
	66	40	50	31	58	61	80	58
Σ	601	417	564	247	770	703	663	556

presentations, the particular modes, or the particular times actually investigated. Under these circumstances and with a randomized group design, *the appropriate error mean square for all tests of significance will be the within treatment mean square.*

9.3 Sums of Squares: Total, Within, and Treatment

We begin our analysis in a manner already familiar. We first find the total sum of squares, then the treatment sum of squares, and then the within treatment sum of squares. For the data of Table 9.2, we have

$$\text{Total} = (76)^2 + (66)^2 + \cdots + (58)^2 - \frac{(4521)^2}{80} = 25{,}886.0$$

$$\text{Treatment} = \frac{(601)^2}{10} + \frac{(417)^2}{10} + \cdots + \frac{(556)^2}{10} - \frac{(4521)^2}{80} = 19{,}507.9$$

$$\text{Within} = 25{,}886.0 - 19{,}507.9 = 6378.1$$

As a check on the arithmetic, we calculate the sum of squares within each of the eight treatment groups. Then the sum of these sums of squares should be equal to the sum of squares within groups that we obtained by subtraction. For these eight sums of squares we have

$$\sum x_1^2 = (76)^2 + (66)^2 + \cdots + (66)^2 - \frac{(601)^2}{10} = 1582.9$$

$$\sum x_2^2 = (36)^2 + (45)^2 + \cdots + (40)^2 - \frac{(417)^2}{10} = 590.1$$

$$\sum x_3^2 = (43)^2 + (75)^2 + \cdots + (50)^2 - \frac{(564)^2}{10} = 890.4$$

$$\sum x_4^2 = (37)^2 + (22)^2 + \cdots + (31)^2 - \frac{(247)^2}{10} = 402.1$$

$$\sum x_5^2 = (94)^2 + (85)^2 + \cdots + (58)^2 - \frac{(770)^2}{10} = 1026.0$$

$$\sum x_6^2 = (74)^2 + (74)^2 + \cdots + (61)^2 - \frac{(703)^2}{10} = 576.1$$

$$\sum x_7^2 = (67)^2 + (64)^2 + \cdots + (80)^2 - \frac{(663)^2}{10} = 624.1$$

$$\sum x_8^2 = (67)^2 + (60)^2 + \cdots + (58)^2 - \frac{(556)^2}{10} = 686.4$$

Adding the above sums of squares, we have 1582.9 + 590.1 + 890.4 + 402.1 + 1026.0 + 576.1 + 624.1 + 686.4 = 6378.1 for the sum of squares within groups. This is the same value we obtained by subtraction.

9.4 Assumption of Homogeneity of Variance

Each of the sums of squares within each of the various treatment groups, when divided by the number of degrees of freedom, will provide a variance estimate s^2. Under the assumption that the population variance is the same for all treatment groups, the separate variance estimates will all be estimates of the same parameter. Dividing each of the sums of squares by 9, we obtain as the separate estimates: 175.9, 65.6, 98.9, 44.7, 114.0, 64.0, 69.3, and 76.3. It may seem that these estimates vary quite a bit to be estimates of a common population variance, and the cautious experimenter might wish to test this null hypothesis before proceeding with the analysis of variance.

If the null hypothesis were tested, it would be found that the outcome of the test is one in which the probability of the test statistic is about 0.50 when the null hypothesis of homogeneity of variance is true. The data, in other words, offer no significant evidence against the hypothesis that the sample variances are all estimates of a common population variance.

We have emphasized before that the F test in the analysis of variance is quite insensitive to heterogeneity of variance, provided that we have an equal number of observations for each treatment—as we do in the example under consideration. To make a test for heterogeneity of variance when one has an equal number of observations for each treatment before proceeding with the analysis of variance is, as Box (1953) has stated, somewhat like putting to sea in a rowing boat in order to find out if the water is safe for an ocean liner.

9.5 Significance of the Treatment Mean Square

The analysis of variance up to this point has resulted in a partitioning of the total sum of squares and degrees of freedom into two parts. One part is associated with differences among the eight treatment means and is based on $k - 1 = 7$ d.f. The other part is associated with the variation within each of the treatment groups and has $k(n - 1) = 72$ d.f. This analysis is shown in Table 9.3. Testing the treatment mean square for significance, we have $F = 2786.8/88.6 = 31.5$ with 7 and 72 d.f. From the table of F we find that for 7 and 72 d.f., $F = 31.5$ is significant with probability less than 0.01 and we conclude that the treatment means differ significantly.[3]

9.6 Partitioning the Treatment Sum of Squares: Main Effects

In the experiment we have described, the treatment sum of squares has 7 d.f. We now consider a possible division of the treatment sum of squares into seven component parts, each with 1 d.f. One of these components will be based on a comparison of the sums for one and two presentations of the material and will be called the A sum of squares. Another will be based on a comparison of the sums for the visual and auditory modes of presentation and will be called the B sum of squares. A third comparison will be based on the sums for the immediate and delayed tests and will be called the C sum of squares. Each of these com-

[3] Failure to obtain a significant treatment mean square is not necessarily a terminal test with a factorial experiment. Rather, the subsequent partitioning of the treatment sum of squares and tests of significance should be based on the structure of the factorial experiment and the comparisons that have been planned.

Table 9.3

ANALYSIS OF VARIANCE SHOWING THE TREATMENT SUM OF SQUARES AND THE SUM OF SQUARES WITHIN TREATMENTS FOR THE DATA OF TABLE 9.2

Source of Variation	Sum of Squares	d.f.	Mean Square	*F*
Treatments	19,507.9	7	2,786.8	31.5
Within treatments	6,378.1	72	88.6	
Total	25,886.0	79		

ponents will represent a comparison between the two levels of a given factor.

For the first comparison, we find the sum for $A_1 = 601 + 417 + 564 + 247 = 1829$ and the sum for $A_2 = 770 + 703 + 663 + 556 = 2692$. Each of these sums is based on $(4)(10) = 40$ observations. Then the sum of squares for A will be given by

$$A = \frac{(1829)^2}{40} + \frac{(2692)^2}{40} - \frac{(4521)^2}{80} = 9309.6$$

and because this sum of squares has 1 d.f., it is also a mean square.

For the second comparison, we have $B_1 = 601 + 417 + 770 + 703 = 2491$ and $B_2 = 564 + 247 + 663 + 556 = 2030$. Each of these sums is based on $(4)(10) = 40$ observations. Then the sum of squares for B will be given by

$$B = \frac{(2491)^2}{40} + \frac{(2030)^2}{40} - \frac{(4521)^2}{80} = 2656.5$$

and because this sum of squares has 1 d.f., it is also a mean square.

To find the sum of squares for C, we first find $C_1 = 601 + 564 + 770 + 663 = 2598$ and $C_2 = 417 + 247 + 703 + 556 = 1923$. Each of these sums is based on $(4)(10) = 40$ observations. Then the sum of squares for C will be given by

$$C = \frac{(2598)^2}{40} + \frac{(1923)^2}{40} - \frac{(4521)^2}{80} = 5695.3$$

and this sum of squares is also a mean square because it has 1 d.f.

We have accounted for 3 of the 7 d.f. associated with the treatment sum of squares. The A sum of squares corresponds to a comparison between A_1 and A_2 or between one and two presentations. The B sum of squares corresponds to a comparison between B_1 and B_2 or between the visual and auditory modes of presentation. The C sum of squares corresponds to a comparison between C_1 and C_2 or between the immediate and delayed tests. The mean squares for the levels of factors are often called the *main effects* of the factors.

9.7 *Partitioning the Treatment Sum of Squares: Interaction Effects*

In addition to the main effects of the factors, we have the possibility that there are *interactions* between the factors. We shall illustrate a method for calculating the interaction sum of squares between two factors when each factor has two levels, and postpone the discussion of the meaning of an interaction until later.

The sum of squares for the interaction of A and B, designated $A \times B$, may be found by setting up a two-way table for the factors, as shown in Table 9.4. Then the interaction sum of squares may be obtained by entering the sums corresponding to the cells of the table in the formula below:

$$\text{Interaction} = \frac{[(a + d) - (b + c)]^2}{4n} \qquad (9.1)$$

where n is the number of observations contributing to each of the sums in the cells of the table. In the present problem each of the cell sums is based on $(2)(10) = 20$ observations.

Table 9.5 gives the cell sums for the two-way tables for A and B, A and C, and B and C. For the two-way table for A and B, for example, the cell sums correspond to

$$a = \text{the sum for } A_1B_1 = 601 + 417 = 1018$$

$$b = \text{the sum for } A_1B_2 = 564 + 247 = 811$$

$$c = \text{the sum for } A_2B_1 = 770 + 703 = 1473$$

$$d = \text{the sum for } A_2B_2 = 663 + 556 = 1219$$

The cell sums for the other two-way tables have a similar interpretation.

Table 9.4
SCHEMATIC REPRESENTATION OF THE TWO-WAY TABLE FOR COMPUTING AN INTERACTION SUM OF SQUARES WITH 1 d.f.

	B_1	B_2
A_1	a	b
A_2	c	d

Table 9.5

THE TWO-WAY TABLES FOR $A \times B$, $A \times C$, AND $B \times C$ INTERACTIONS

(a) Two-Way Table for A and B

	B_1	B_2	Σ
A_1	1018	811	1829
A_2	1473	1219	2692
Σ	2491	2030	4521

(b) Two-Way Table for A and C

	C_1	C_2	Σ
A_1	1165	664	1829
A_2	1433	1259	2692
Σ	2598	1923	4521

(c) Two-Way Table for B and C

	C_1	C_2	Σ
B_1	1371	1120	2491
B_2	1227	803	2030
Σ	2598	1923	4521

Substituting with the appropriate values from Table 9.5 in (9.1) we have

$$A \times B = \frac{[(1018 + 1219) - (811 + 1473)]^2}{(4)(20)} = 27.6$$

$$A \times C = \frac{[(1165 + 1259) - (664 + 1433)]^2}{(4)(20)} = 1336.6$$

$$B \times C = \frac{[(1371 + 803) - (1120 + 1227)]^2}{(4)(20)} = 374.1$$

Each of the interaction sums of squares we have just calculated will have 1 d.f. and each is therefore a mean square. A general rule for determining the degrees of freedom for *any* interaction sum of squares is to multiply the degrees of freedom associated with the factors for which the interaction is being computed. In the present problem, we have 1 d.f. associated with A, 1 with B, and 1 with C, and the product of the degrees of freedom for A and B, for example, is 1 also.

The interaction sums of squares, $A \times B$, $A \times C$, and $B \times C$, each with 1 d.f., will account for 3 of the 4 d.f. that were left after we found the sums of squares for the main effects of A, B, and C. The single remaining degree of freedom is associated with the sum of squares for the interaction of the three factors, designated by $A \times B \times C$. The degrees of freedom associated with any $A \times B \times C$ interaction will be given by the product of the degrees of freedom associated with the factors involved in the interaction. Because we have 1 d.f. for each of the three factors, we will also have 1 d.f. for the $A \times B \times C$ interaction sum of squares.

The $A \times B \times C$ interaction sum of squares may be calculated directly, and we shall show methods for doing this later. In the present problem, it can be obtained most easily by subtraction. The treatment sum of squares is equal to the sum of the sums of squares for A, B, C, $A \times B$, $A \times C$, $B \times C$, and $A \times B \times C$. Because we have already calculated all of these sums of squares except $A \times B \times C$, the latter can be obtained by subtracting the other six sums of squares from the treatment sum of squares. The sum of the six sums of squares calculated so far is equal to 19,399.7, and by subtraction from the treatment sum of squares, we obtain

$$19{,}507.9 - 19{,}399.7 = 108.2$$

as the $A \times B \times C$ interaction sum of squares.

9.8 Summary of the Analysis of Variance for the 2^3 Factorial Experiment

The summary of the complete analysis of variance is presented in Table 9.6. The values of F that have been entered in the table were obtained by dividing each of the mean squares that is to be tested for significance by the error mean square, that is, the within treatment mean square. Thus, each F in the table is based on 1 and 72 d.f. No value of F was calculated for the $A \times B$ interaction because this mean square is obviously not significantly larger than the error mean square.

9.9 The Main Effects

From the table of F, we find that for 1 and 72 d.f., a value of F that is approximately equal to 4.0 will be significant with $\alpha = 0.05$. For the main effects, A, B, and C, we have significant F's. The A mean square corresponds to a comparison between the means for one and two presentations *averaged over the two levels of B and the two levels of C*. The mean for one presentation or the first level of A can be obtained from Table 9.5 and is equal to $1829/40 = 45.725$. The mean for two presentations or the second level of A can also be obtained from Table 9.5 and is equal to

Table 9.6

COMPLETE ANALYSIS OF VARIANCE FOR THE FACTORIAL EXPERIMENT OF TABLE 9.2

Source of Variation		Sum of Squares	d.f.	Mean Square	F
A:	Number	9,309.6	1	9,309.6	105.07
B:	Mode	2,656.5	1	2,656.5	29.98
C:	Time	5,695.3	1	5,695.3	64.28
$A \times B$:	Number $\times$ Mode	27.6	1	27.6	
$A \times C$:	Number $\times$ Time	1,336.6	1	1,336.6	15.09
$B \times C$:	Mode $\times$ Time	374.1	1	374.1	4.22
$A \times B \times C$:	Number $\times$ Mode $\times$ Time	108.2	1	108.2	1.22
Error:	Within treatments	6,378.1	72	88.6	
	Total	25,886.0	79		

2692/40 = 67.300. The fact that the A mean square is significant leads us to conclude that these two means differ significantly. Two presentations definitely result in a superior average retention compared with one presentation of the material.

Similarly, the main effect of B represents a comparison between the means for B_1, the visual mode, and B_2, the auditory mode, *averaged over the two levels of A and the two levels of C*. The mean for B_1 can be obtained from Table 9.5 and is equal to 2491/40 = 62.275 and corresponds to the mean for the visual mode of presentation. The mean for B_2 is equal to 2030/40 = 50.750 and corresponds to the mean for the auditory mode of presentation. Because the mean square for B is significant in the analysis of variance, we conclude that the means for B_1 and B_2 differ significantly. The visual mode of presentation results in greater average retention than the auditory mode of presentation.

The main effect of C represents a comparison between the means for C_1, the immediate test, and C_2, the delayed test, *averaged over the two levels of A and the two levels of B*. These two means can be obtained from Table 9.5. The mean for C_1 is equal to 2598/40 = 64.950 and corresponds to the mean for the immediate test. The mean for C_2 is equal to 1923/40 = 48.075 and corresponds to the mean for the delayed test. Because the C mean square of the analysis of variance is significant, we conclude that these two means differ significantly. We have greater average retention on the immediate test than on the delayed test.

9.10 The Two-Factor Interaction Effects

We come now to the interaction effects. Let us consider first the $A \times B$ interaction mean square which is *not* significant. The fact that this interaction mean square is not significant indicates that the difference between the means of A_1 and A_2 for the first level of B is not significantly different from the difference between the means of A_1 and A_2 for the second level of B. If the $A \times B$ sum of squares were exactly equal to zero, then the difference between the means of A_1 and A_2 for B_1 would be exactly equal to the difference between the means of A_1 and A_2 for B_2. With a nonsignificant $A \times B$ interaction, we can say that the A effect, the difference between A_1 and A_2, is *independent* of B; that is, we have approximately the same difference between A_1 and A_2, regardless of the levels of B.

We can see why, in order for the interaction sum of squares to be zero, what we have said above would have to be true. For example, setting the numerator of (9.1) equal to zero, we have $a - c = b - d$, and this equality would have to hold if the interaction sum of squares is to be zero. The left-hand side of the above expression represents the difference between A_1 and A_2 for B_1, and the right-hand side represents the difference between A_1 and A_2 for B_2. If the $A \times B$ interaction mean square is significant, it means that the A effect is not the same for the different levels of B; that is, $a - c$ and $b - d$ differ significantly.

Dividing each of the cell sums of Table 9.5(a) by 20, the number of observations contributing to the sums, we have as the mean difference between A_1 and A_2 for B_1

$$B_1\colon A_1 - A_2 = \frac{1018}{20} - \frac{1473}{20} = 50.90 - 73.65 = -22.75$$

and for the mean difference between A_1 and A_2 for B_2, we have

$$B_2\colon A_1 - A_2 = \frac{811}{20} - \frac{1219}{20} = 40.55 - 60.95 = -20.40$$

and it is the fact that these two differences are much the same that results in a nonsignificant $A \times B$ mean square.

Now, let us look at the $A \times C$ mean square, which is highly significant. Dividing each of the cell entries of Table 9.5(b) by 20, we have as the difference between the means of A_1 and A_2 for C_1

$$C_1\colon A_1 - A_2 = \frac{1165}{20} - \frac{1433}{20} = 58.25 - 71.65 = -13.40$$

and as the difference between the means for A_1 and A_2 for C_2

$$C_2\colon A_1 - A_2 = \frac{664}{20} - \frac{1259}{20} = 33.20 - 62.95 = -29.75$$

and we observe that these two differences are not at all comparable. Because the $A \times C$ mean square is significant, we know that the A effect is not independent of the C factor. In other words, the magnitude of the difference between A_1 and A_2 is not the same, within the limits of random variation, for C_1 and C_2. This is the meaning of the significant $A \times C$ mean square.

Dividing each of the cell entries of Table 9.5(c) by 20, we have as the difference between the means of C_1 and C_2 for B_1

$$B_1\colon C_1 - C_2 = \frac{1371}{20} - \frac{1120}{20} = 68.55 - 56.00 = 12.55$$

and for the difference between the means of C_1 and C_2 for B_2

$$B_2\colon C_1 - C_2 = \frac{1227}{20} - \frac{803}{20} = 61.35 - 40.15 = 21.20$$

and it is the fact that these two differences are not more alike that results in the significant $B \times C$ interaction.

9.11 Graphs for the Interaction Effects

Another way of examining the nature of an interaction is to present it graphically. We take one of the factors, say B, for the X axis and graph the means (or sums) for each level of A. For the

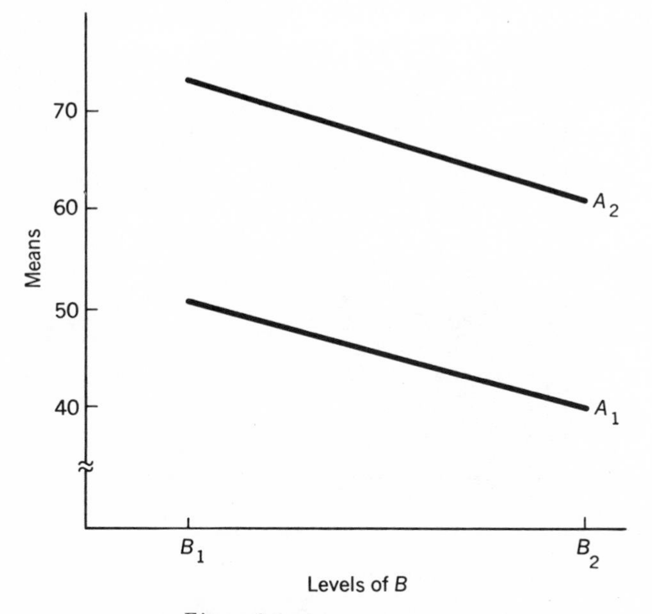

Figure 9.1
Means for levels of A at each level of B. A_1 and A_2 correspond to one and two presentations, respectively. B_1 and B_2 correspond to a visual and an auditory mode of presentation, respectively.

example under discussion, the graphs for A_1 and A_2 are given in Figure 9.1. Each of the lines in the figure corresponds to a different level of A. If the lines for A_1 and A_2 were exactly parallel, then the $A \times B$ interaction would be zero. The fact that the lines are very nearly parallel, within the limits of random variation, corresponds to the fact that the $A \times B$ interaction is not significant. Compare, however, the corresponding graph for the $A \times C$ interaction in Figure 9.2. Here we have taken C for the X axis and plotted the means for A_1 and A_2. Note that the lines for A_1 and A_2 are not parallel. The fact that the $A \times C$ interaction is significant is equivalent to stating that the lines A_1 and A_2 cannot be said to be parallel within the limits of random variation.

Figure 9.3 gives the graph for the $B \times C$ interaction, where we have chosen B for the X axis. We have a significant $B \times C$ interaction mean square, and this is shown graphically by the failure of the two lines, C_1 and C_2, in the figure to be parallel within the limits of random variation.

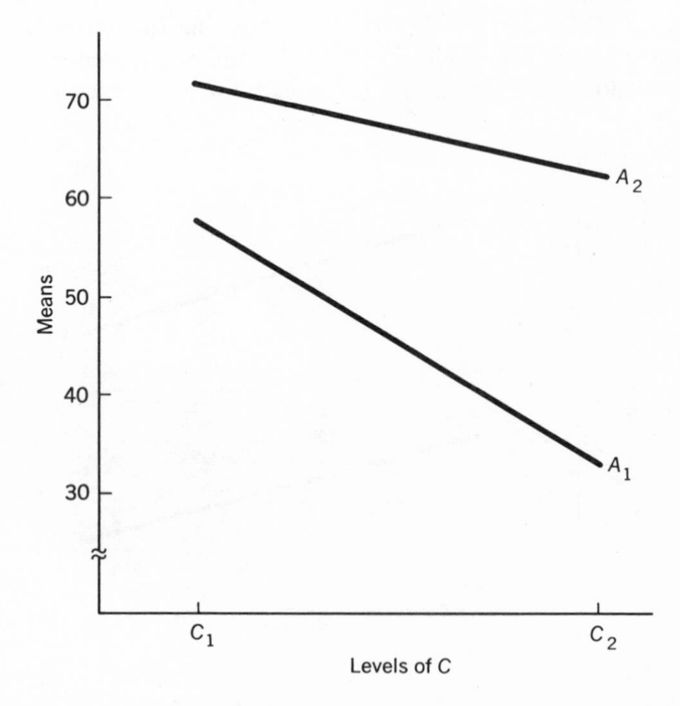

Figure 9.2
Means for levels of A at each level of C. A_1 and A_2 correspond to one and two presentations, respectively. C_1 and C_2 correspond to an immediate and a delayed test, respectively.

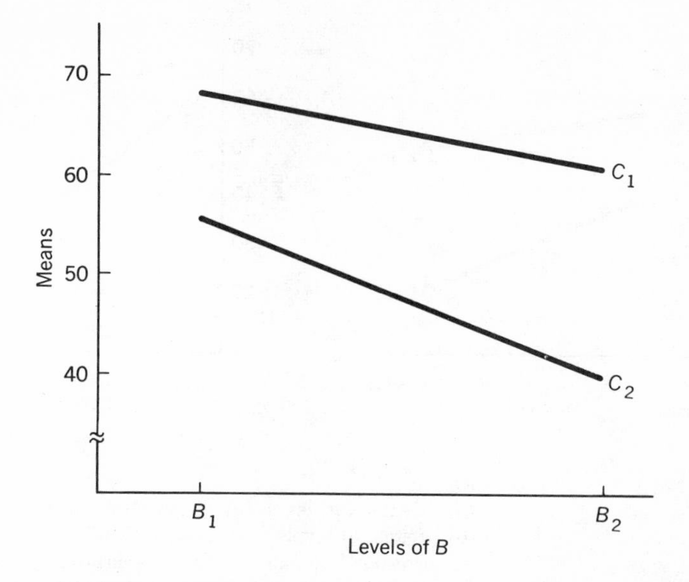

Figure 9.3
Means for levels of C at each level of B. C_1 and C_2 correspond to an immediate and a delayed test, respectively. B_1 and B_2 correspond to a visual and an auditory mode of presentation, respectively.

9.12 The Three-Factor Interaction

The $A \times B \times C$ interaction mean square is not significant. But to examine the nature of the $A \times B \times C$ interaction, we consider the $A \times C$ interaction separately for each level of B, as shown in Table 9.7. The graphs for A_1 and A_2 against C for B_1 are shown in Figure 9.4(a) and the graphs for A_1 and A_2 against C for B_2 are shown in Figure 9.4(b).

Significance or lack of significance of a two-factor interaction, $A \times C$, for example, tells us whether the A effect is the same for all levels of C.

Table 9.7
TWO-WAY TABLE OF MEANS FOR A AND C FOR EACH LEVEL OF B

	B_1			B_2	
	C_1	C_2		C_1	C_2
A_1	60.1	41.7	A_1	56.4	24.7
A_2	77.0	70.3	A_2	66.3	55.6

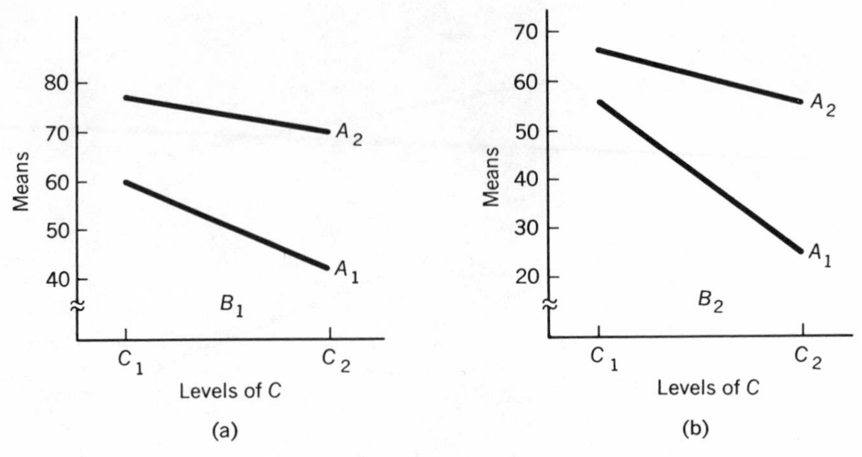

Figure 9.4
(a) Means for levels of A at each level of C for B_1. (b) Means for levels of A at each level of C for B_2. A_1 and A_2 correspond to one and two presentations, respectively. C_1 and C_2 correspond to an immediate and a delayed test, respectively. B_1 is a visual mode of presentation and B_2 is an auditory mode.

Similarly, if we examine the $A \times C$ interaction separately for each level of B, and if these interactions are of the same form for each level of B, then the $A \times B \times C$ interaction will not be significant. A significant $A \times B \times C$ interaction, in other words, means that the $A \times C$ interaction is not the same for the different levels of B.

We note that the forms of the graphs in Figures 9.4(a) and 9.4(b) are fairly similar and this finding is consistent with the nonsignificance of the $A \times B \times C$ mean square.

9.13 *Summary of Conclusions for the Example of a 2^3 Factorial Experiment*

Let us summarize the conclusions based on the analysis of variance for the experiment described. The significant A mean square tells us that the means for A_1 and A_2, averaged over the levels of B and C, differ significantly. Examination of the means shows that two presentations are superior to one. The significant B mean square tells us that the means for B_1 and B_2, averaged over the levels of A and C, differ significantly. Examination of these two means shows that the visual mode of presentation is superior to the auditory. The significant C mean square tells us that the means for C_1 and C_2, averaged over the levels of A and B, differ significantly. Examination of these two means shows that retention is greater on the immediate test than on the delayed test.

The $A \times B$ interaction is not significant. Therefore, the A effect, that is, the difference between A_1 and A_2, or between one and two presentations, is not dependent on the particular mode of presentation employed. The $A \times B$ interaction is identical with the $B \times A$ interaction and our statement about the difference between A_1 and A_2 being independent of B is also equivalent to stating that the difference between B_1 and B_2 is independent of A.

We do have a significant $A \times C$ interaction. This tells us that the difference between A_1 and A_2 is not independent of the levels of C or, equivalently, that the difference between C_1 and C_2 is not independent of the levels of A. In other words, a statement about the A effect must be qualified by the particular level of C involved, or, equivalently, a statement about the C effect must be qualified by the particular level of A involved. The nature of the interaction can be shown by graphic methods, as described previously.

The interpretation of the significant $B \times C$ interaction with respect to the main effects of B and C is similar to that described above for the $A \times C$ interaction with respect to the main effects of A and C.

The $A \times B \times C$ interaction was discussed by considering the $A \times C$ interaction separately for each level of B. We could just as well have considered the $A \times B$ interaction separately for each level of C or the $B \times C$ interaction separately for each level of A. Just as the $A \times B$ interaction is a symmetrical property of A and B, so also the $A \times B \times C$ interaction is a symmetrical property for the three factors A, B, and C. In other words, the nonsignificance of the $A \times B \times C$ interaction means that the $A \times B$ interactions for the separate levels of C are of the same form; that the $A \times C$ interactions for the separate levels of B are of the same form; and that the $B \times C$ interactions for the separate levels of A are of the same form.

If the $A \times B \times C$ interaction is significant, the nature of the interaction can be examined by the graphic methods described previously. Because the $A \times B \times C$ interaction is a symmetrical property of A, B, and C, we may consider graphing any one of the two-factor interactions separately for the levels of the third factor. This is to say that we may examine the nature of the $A \times B \times C$ interaction by graphing the $A \times B$ interaction separately for the levels of C, or by graphing the $A \times C$ interaction separately for the levels of B, or by graphing the $B \times C$ interaction separately for the levels of A.

9.14 Orthogonal Comparisons for the 2^3 Factorial Experiment

The comparisons we have made with respect to the $2 \times 2 \times 2$ factorial experiment correspond to what we have described earlier as *orthogonal* comparisons, each with 1 d.f.[4] That this is so can easily be seen in Table 9.8 where we show the nature of the comparisons in the same manner described previously when we discussed orthogonal comparisons. We note, for example, that the sum of the coefficients in each row is zero and that the sum of the products of the coefficients in each pair of rows is also zero. Multiplying the treatment sums by the corresponding coefficients, we obtain the values of D for each comparison. Each treatment sum is based on $n = 10$ observations and the sum of the squares of the coefficients in each row is 8. Then squaring D and dividing by $n\Sigma a._i^2 = 80$, we have the mean squares shown in the last column. We note that the mean squares for the comparisons shown in Table 9.8 are the same as those shown in the summary analysis of variance table, Table 9.6, except for rounding errors.

The source of the coefficients for the main effects is apparent. The coefficients for a two-factor interaction are obtained by multiplying the corresponding coefficients for the factors involved in the interaction. For example, the coefficients for the $A \times B$ interaction are obtained by multiplying the coefficients in rows A and B and entering the product with the appropriate sign in row $A \times B$. To obtain the coefficients for the $A \times B \times C$ interaction, we multiply the corresponding coefficients in the rows for the A, B, and C comparisons.

In partitioning the treatment sum of squares into the seven orthogonal comparisons, we assumed that the comparisons were the ones of experimental interest and that they were planned in advance. It would also be possible to examine specific comparisons between the treatment means or sums using Scheffé's test for multiple comparisons. In this way we could test certain comparisons suggested by the data.

Confidence limits may also be established for any and all of the comparisons in the manner described previously.

9.15 Notation and Sums of Squares for a Factorial Experiment

We consider now a general notation for the factorial experiment. We let a = the number of levels of A, b = the number of levels of B, c = the number of levels of C, and n = the number of observations for each treatment combination. The number of treatment combinations will be $k = abc$ and the total number of observations will be kn. Then we let a general observation be X_{abcn} with the understanding that when

[4] Although the numerators of the F ratios for orthogonal comparisons are independently distributed, the F ratios themselves are not. The reason for this is that for each ratio we have a common denominator or estimate of experimental error.

Table 9.8

ORTHOGONAL COMPARISONS FOR THE $2 \times 2 \times 2$ FACTORIAL

	Treatment Sums										
	$A_1B_1C_1$	$A_1B_1C_2$	$A_1B_2C_1$	$A_1B_2C_2$	$A_2B_1C_1$	$A_2B_1C_2$	$A_2B_2C_1$	$A_2B_2C_2$			
Comparison	601	417	564	247	770	703	663	556	$\sum a_{\cdot i}^2$	D_i	MS_{D_i}
A	1	1	1	1	−1	−1	−1	−1	8	−863	9309.6
B	1	1	−1	−1	1	1	−1	−1	8	461	2656.5
C	1	−1	1	−1	1	−1	1	−1	8	675	5695.3
$A \times B$	1	1	−1	−1	−1	−1	1	1	8	−47	27.6
$A \times C$	1	−1	1	−1	−1	1	−1	1	8	327	1336.6
$B \times C$	1	−1	−1	1	1	−1	−1	1	8	−173	374.1
$A \times B \times C$	1	−1	−1	1	−1	1	1	−1	8	93	108.1

a, b, c, and n are used as subscripts they represent variables. Thus, with A at two levels, B at two levels, C at two levels, and with $n = 10$, a as a subscript can take values of 1 or 2, b can take values of 1 or 2, c can take values of 1 or 2, and n can take values of 1 to 10. Thus, X_{1126} would correspond to the sixth observation of the first level of A, the first level of B, and the second level of C.

Then, using the dot notation, we can write the following identity

$$
\begin{aligned}
X_{abcn} - \bar{X}_{....} = {} & (X_{abcn} - \bar{X}_{abc\cdot}) \\
& + (\bar{X}_{a\cdots} - \bar{X}_{....}) \\
& + (\bar{X}_{\cdot b\cdot\cdot} - \bar{X}_{....}) \\
& + (\bar{X}_{\cdot\cdot c\cdot} - \bar{X}_{....}) \\
& + (\bar{X}_{ab\cdot\cdot} - \bar{X}_{a\cdots} - \bar{X}_{\cdot b\cdot\cdot} + \bar{X}_{....}) \\
& + (\bar{X}_{a\cdot c\cdot} - \bar{X}_{a\cdots} - \bar{X}_{\cdot\cdot c\cdot} + \bar{X}_{....}) \\
& + (\bar{X}_{\cdot bc\cdot} - \bar{X}_{\cdot b\cdot\cdot} - \bar{X}_{\cdot\cdot c\cdot} + \bar{X}_{....}) \\
& + (\bar{X}_{abc\cdot} + \bar{X}_{a\cdots} + \bar{X}_{\cdot b\cdot\cdot} + \bar{X}_{\cdot\cdot c\cdot} - \bar{X}_{ab\cdot\cdot} - \bar{X}_{a\cdot c\cdot} \\
& \quad - \bar{X}_{\cdot bc\cdot} - \bar{X}_{....})
\end{aligned}
$$

which states that the deviation of an observation from the overall mean can be expressed as the sum of the eight terms on the right.

If we square both sides of the above expression and sum over all observations, we find that the products of all terms on the right sum to zero. Thus,

Table 9.9

TWO-WAY TABLE OF MEANS FOR A AND B WITH ALL RESIDUALS EQUAL TO ZERO

	B_1	B_2	$\bar{X}$
A_1	5.0	10.0	7.5
A_2	15.0	20.0	17.5
$\bar{X}$	10.0	15.0	12.5

$\bar{X}_{ab\cdot\cdot} - \bar{X}_{a\cdots} - \bar{X}_{\cdot b\cdot\cdot} + \bar{X}_{....} =$ Residual

$5.0 - 7.5 - 10.0 + 12.5 = 0$

$10.0 - 7.5 - 15.0 + 12.5 = 0$

$15.0 - 17.5 - 10.0 + 12.5 = 0$

$20.0 - 17.5 - 15.0 + 12.5 = 0$

with $k = abc$, we have

$$\sum_{1}^{kn} (X_{abcn} - \bar{X}_{....})^2 = \sum_{1}^{kn} (X_{abcn} - \bar{X}_{abc.})^2$$

$$+ bcn \sum_{1}^{a} (\bar{X}_{a...} - \bar{X}_{....})^2$$

$$+ acn \sum_{1}^{b} (\bar{X}_{.b..} - \bar{X}_{....})^2$$

$$+ abn \sum_{1}^{c} (\bar{X}_{..c.} - \bar{X}_{....})^2$$

$$+ cn \sum_{1}^{ab} (\bar{X}_{ab..} - \bar{X}_{a...} - \bar{X}_{.b..} + \bar{X}_{....})^2$$

$$+ bn \sum_{1}^{ac} (\bar{X}_{a.c.} - \bar{X}_{a...} - \bar{X}_{..c.} + \bar{X}_{....})^2$$

$$+ an \sum_{1}^{bc} (\bar{X}_{.bc.} - \bar{X}_{.b..} - \bar{X}_{..c.} + \bar{X}_{....})^2$$

$$+ n \sum_{1}^{k} (\bar{X}_{abc.} + \bar{X}_{a...} + \bar{X}_{.b..} + \bar{X}_{..c.}$$

$$- \bar{X}_{ab..} - \bar{X}_{a.c.} - \bar{X}_{.bc.} - \bar{X}_{....})^2$$

The term on the left is the total sum of squares. The successive terms on the right give the sum of squares within treatments, the sum of squares for A, the sum of squares for B, the sum of squares for C, the $A \times B$ sum of squares, the $A \times C$ sum of squares, the $B \times C$ sum of squares, and the $A \times B \times C$ sum of squares.

9.16 *An Example Where $A \times B = 0$*

Consider the expression for any one of the two-factor interaction sums of squares. If we set up the two-way table involving these two factors, then the condition for a zero interaction is that each of the residuals for the table be equal to zero. That is, we will have a zero interaction for $A \times B$, for example, if

$$\bar{X}_{ab..} - \bar{X}_{a...} - \bar{X}_{.b..} + \bar{X}_{....} = 0$$

for all possible values. This condition is met by the data of Table 9.9, where we show the table of means at the top and the corresponding residuals at the bottom of the table.

9.17 An Example Where $A \times B \times C = 0$ and $A \times B \neq 0$

A sufficient condition for a three-factor interaction sum of squares to be equal to zero is that the tables of residuals for the two-factor interactions be exactly the same for each level of the third factor and consequently the same as the table of residuals for the observations averaged over the levels of the third factor. This condition is met by the data of Table 9.10. There we show the means for the $A \times B$ interactions

Table 9.10

TWO-WAY TABLES OF MEANS FOR A AND B FOR EACH LEVEL OF C AND AVERAGED OVER THE LEVELS OF C WITH IDENTICAL RESIDUALS FOR EACH TABLE

	C_1			Residuals	
	B_1	B_2	$\bar{X}$	B_1	B_2
A_1	10.0	20.0	15.0	−5.0	5.0
A_2	40.0	30.0	35.0	5.0	−5.0
$\bar{X}$	25.0	25.0	25.0		

	C_2			Residuals	
	B_1	B_2	$\bar{X}$	B_1	B_2
A_1	20.0	30.0	25.0	−5.0	5.0
A_2	50.0	40.0	45.0	5.0	−5.0
$\bar{X}$	35.0	35.0	35.0		

	$C_1 + C_2$			Residuals	
	B_1	B_2	$\bar{X}$	B_1	B_2
A_1	15.0	25.0	20.0	−5.0	5.0
A_2	45.0	35.0	40.0	5.0	−5.0
$\bar{X}$	30.0	30.0	30.0		

separately for each level of C and also averaged over the levels of C. We note that the residuals for each of the three tables are identical, and consequently, the $A \times B \times C$ interaction sum of squares will be equal to zero. However, note that the $A \times B$ interaction sum of squares is *not* equal to zero.

Figure 9.5 shows the graphs of A_1 and A_2 against the levels of B separately for C_1 and C_2 and we observe that the forms of the two graphs are the same. The fact that the two graphs have exactly the same form tells us that the nature of the $A \times B$ interaction is the same for C_1 and C_2. This, in turn, is equivalent to stating that the $A \times B \times C$ interaction is equal to zero. Note also, in the lower figure, that when we average over the levels of C_1 and C_2, the lines of A_1 and A_2 have the same form as for each of the separate C levels. The lower figure also shows that the $A \times B$ interaction is not equal to zero.

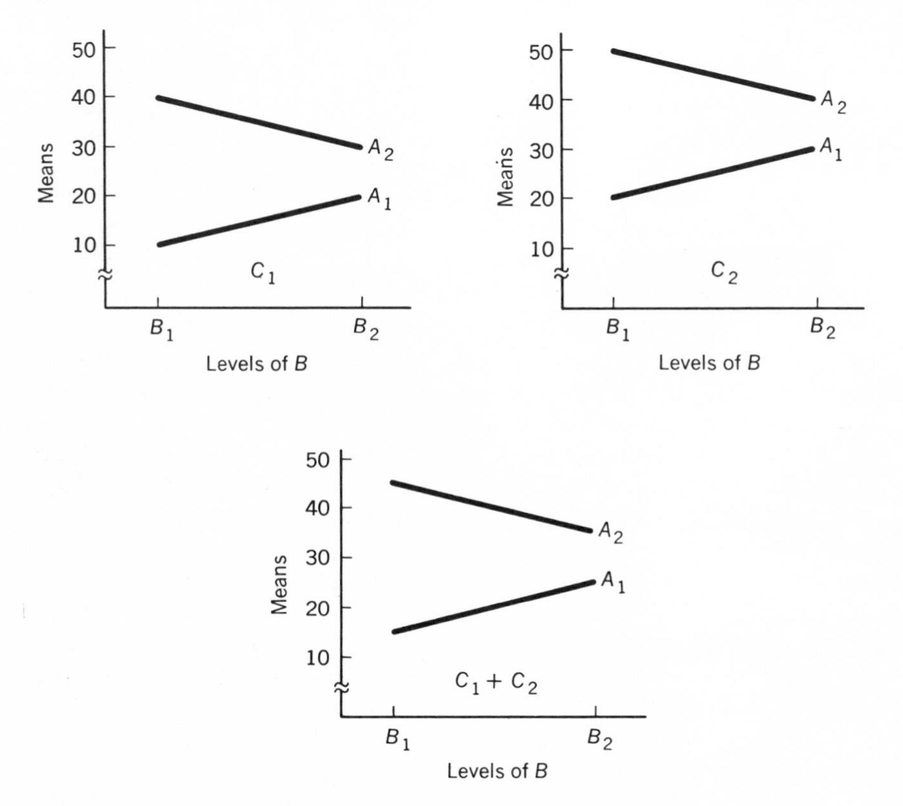

Figure 9.5
Means for levels of A at each level of B for C_1 and C_2 and averaged over the levels of C. The $A \times B \times C$ interaction sum of squares is equal to zero, but the $A \times B$ interaction sum of squares is not equal to zero.

Table 9.11

TWO-WAY TABLES OF MEANS FOR A AND B FOR EACH LEVEL OF C AND AVERAGED OVER THE LEVELS OF C WITH $A \times B = 0$ AND $A \times B \times C \neq 0$

C_1		
	B_1	B_2
A_1	5.0	15.0
A_2	25.0	10.0

C_2		
	B_1	B_2
A_1	5.0	5.0
A_2	5.0	30.0

$C_1 + C_2$		
	B_1	B_2
A_1	5.0	10.0
A_2	15.0	20.0

9.18 An Example Where $A \times B \times C \neq 0$ and $A \times B = 0$

Table 9.11 gives the means for a $2 \times 2 \times 2$ factorial experiment where the $A \times B$ interaction *is* equal to zero, but where the $A \times B \times C$ interaction is *not* equal to zero. Note that the forms of the graphs, as shown in Figure 9.6, for A_1 and A_2 against the levels of B separately for C_1 and C_2 are quite different. The fact that these two graphs differ in form tells us that the $A \times B$ interaction is not the same for C_1 and C_2. Yet, when we average over C_1 and C_2, as shown in the lower figure, we see that the lines A_1 and A_2 are parallel indicating that the $A \times B$ interaction is zero. Thus, two-factor interactions, even when not significant, must always be interpreted in accordance with whether the three-factor interaction is significant. If the three-factor interaction is significant, it means that the two-factor interactions are not the same for the different levels of the third factor. This can be true, as we have just seen, even when the two-factor interaction averaged over the third factor is equal to zero or is nonsignificant.

9.19 Other 2^n Factorial Experiments

We have considered only the $2 \times 2 \times 2$ factorial experiment. We could, of course, have a factorial experiment with only two factors or with more than three factors, with each factor having two levels.

With two factors, each varied in two ways, we would have four treatment groups, with 3 d.f. for the treatment sum of squares. The treatment sum of squares could then be analyzed into the sum of squares for A, the sum of squares for B, and the $A \times B$ interaction sum of squares, each with 1 d.f. The procedures for the analysis of variance of the 2×2 factorial experiment would be the same as those we have described for the $2 \times 2 \times 2$ factorial experiment.

If we have a factorial experiment with four factors, each varied in two ways, then we shall have sixteen treatment combinations with 15 d.f. for the treatment sum of squares. To find the sums of squares for the main effects and interactions, we can make use of a table of orthogonal com-

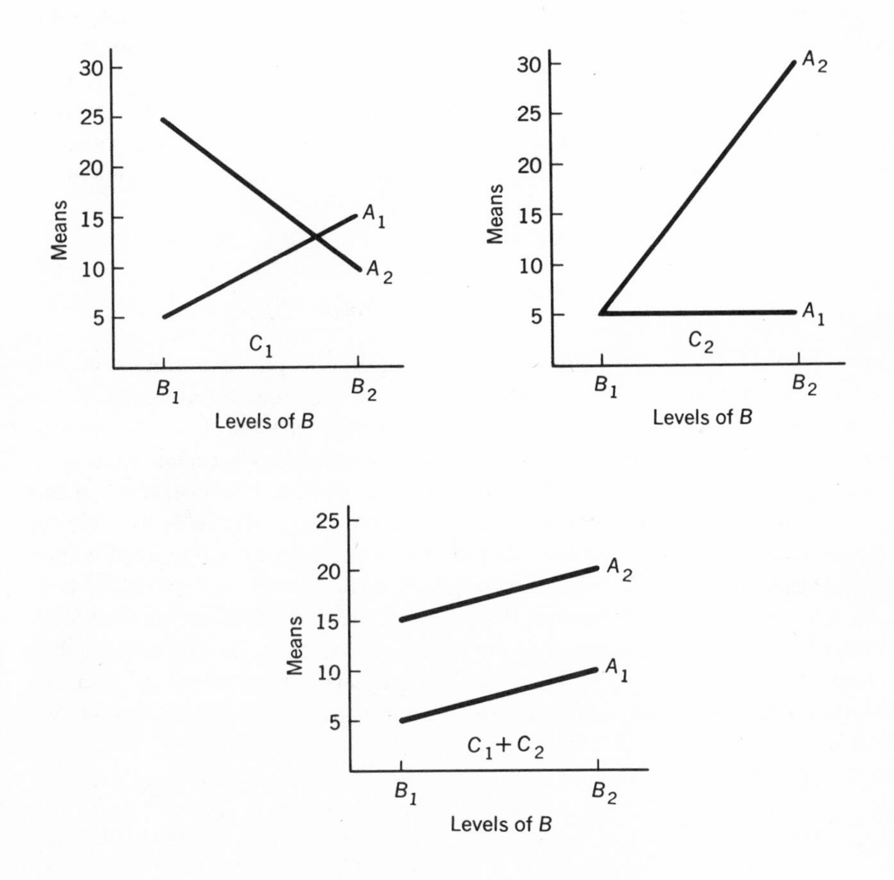

Figure 9.6
Means for levels of A at each level of B for C_1 and C_2 and averaged over the levels of C. The $A \times B \times C$ interaction sum of squares is not equal to zero, but the $A \times B$ interaction sum of squares is equal to zero.

parisons, as shown in Table 9.8 for the $2 \times 2 \times 2$ factorial. The coefficients for the main effects are easy to enter into the table. Then the coefficients for the $A \times B$ interaction can be obtained by multiplying each of the coefficients in the row for the A effect by the corresponding coefficients in the row for the B effect. These products will give the coefficients for the $A \times B$ comparison. The coefficients of the other two-factor interactions can be obtained in the same manner. The coefficients for the $A \times B \times C$ interaction can be obtained by multiplying the corresponding coefficients of the A, B, and C comparisons. In the same manner, we can obtain the coefficients for any three-factor interaction and the coefficients for any four-factor interaction.

If we have a 2^5 factorial experiment, we will have thirty-two treatment combinations, and with a 2^6 factorial experiment, we will have sixty-four treatment combinations. These two experiments with $n = 10$ observations for each treatment would require 320 and 640 subjects, respectively. In a later chapter, we shall discuss some techniques that can be used to reduce the total number of observations in an experiment involving a large number of treatment combinations, provided we can assume that the higher order interactions are negligible.

9.20 Advantages of Factorial Experiments

The factorial experiment has a number of advantages over a single-factor experiment. It may be noted that, in the illustrative example, the full number of observations, that is, eighty, entered into every comparison made, despite the fact that each treatment group consisted of only ten observations. This is true because, to take but one comparison—that between the number of presentations—the eighty observations could be split into a group of forty that were alike in that they were based on a single presentation and another group of forty that were alike in that they were based on two presentations of the material. The forty observations in each of the two sets were not alike in other respects, but they were alike in sets of ten, differing only with respect to A_1 and A_2. For example, the treatment groups contributing to A_1 and A_2 are shown below:

A_1	A_2
$A_1B_1C_1$	$A_2B_1C_1$
$A_1B_1C_2$	$A_2B_1C_2$
$A_1B_2C_1$	$A_2B_2C_1$
$A_1B_2C_2$	$A_2B_2C_2$

We note that for each of these corresponding sets the experimental conditions are constant except for the levels of the one factor, A. In comparing A_1 and A_2 we are testing for the difference between these two levels averaged over levels of B and C that are the same for A_1 and A_2.

It should also be observed that the estimate of experimental error, the within treatment mean square, is based on 72 d.f. If the experiment had been confined to a single factor with two levels, and if an estimate of experimental error is to be obtained with the same number of degrees of freedom, we would have to have thirty-seven observations for each level of the factor or a total of seventy-four observations. And this experiment would provide only information about the single factor investigated. In the factorial experiment, on the other hand, with eighty observations, we not only have information about the *main* effects of *three* factors, but also about the interactions between these factors.

If the interactions involving a given factor are not significant, then we obviously have a broader basis for generalizing about the main effect of the factor, because it has been tested in conjunction with variations of other factors rather than holding the other factors constant at arbitrary levels. If, on the other hand, we have a significant two-factor interaction, examination of the nature of the interaction may provide us with additional insight as to how each factor operates.

QUESTIONS AND PROBLEMS

9.1 The following data have been modified from an experiment by Glanville, Kreezer, and Dallenbach (1946). The problem was an investigation of the accuracy of apprehension of printed words under various experimental conditions. Three factors were selected for investigation: time of exposure, type size, and background. Let these factors be represented by A, B, and C, respectively, with each at two levels. We have A_1 corresponding to a 60-millisecond exposure and A_2 to a 120-millisecond exposure. We have B_1 corresponding to 6-point and B_2 to 12-point type. We also have C_1 corresponding to a blank and C_2 to a printed background.

Unfortunately, for our purposes, a list of only 100 words was used in the test conditions, and for the treatments with the longer exposure time the means were all close to the upper limit with the associated result of very small variances for these treatments. Not only are the variances heterogeneous, but, because the means for the treatments with the longer exposure time approach the upper limit, the distributions for these treatments were probably markedly skewed. A transformation of scale is probably in order. However, we shall do no violence to the conclusions arrived at by the experimenters if we assume slightly different experimental conditions than those actually used.

Let us assume that subjects were assigned at random to the eight experimental conditions and that fifty subjects were used in each treatment group. Let us further assume that the lists contained more than 100 words and that the conditions of normality of distribution and homogeneity of variance are satisfied for all treatments. With these assumed conditions, we have the following sums (unchanged from the original experiment).

Exposure Time (in milliseconds)	Type (in points)	Background	Sum of Scores
60	6	Blank	1319
120	6	Blank	4592
60	6	Printed	1196
120	6	Printed	4365
60	12	Blank	3682
120	12	Blank	4939
60	12	Printed	3357
120	12	Printed	4885

The within treatment sum of squares is given as equal to 84,397; the total sum of squares is given as equal to 405,084; and, by calculation, you will find that the treatment sum of squares is equal to 320,687. Complete the analysis of variance. If any interactions are significant, examine them by the graphic methods described in the chapter.

9.2 We have two factors, A and B, each at two levels. For each treatment combination, we have $n = 8$ subjects assigned at random. The data are given in the table below. Complete the analysis of variance.

A_1		A_2	
B_1	B_2	B_1	B_2
8	5	10	5
6	8	9	7
9	10	4	3
9	7	8	5
8	10	8	3
7	7	4	5
6	8	3	5
3	5	6	8

9.3 In a $2 \times 2 \times 2$ factorial experiment we have the following measures:

A_1				A_2			
B_1		B_2		B_1		B_2	
C_1	C_2	C_1	C_2	C_1	C_2	C_1	C_2
8	5	10	5	7	6	5	2
6	8	9	7	10	8	7	7
9	10	4	3	6	7	4	5
9	7	8	5	7	6	7	7
8	10	8	3	5	8	6	5
7	7	4	5	7	9	8	9
6	8	3	5	6	8	10	6
3	5	6	8	10	9	6	6

Complete the analysis of variance.

9.4 Describe an experiment in which a significant two-factor interaction might be expected. Describe the nature of the interaction and state why it would be expected.

9.5 Consult a recent issue of the *Journal of Experimental Psychology* or some other journal which publishes the results of experiments. Find a study in which a significant two-factor interaction is reported. What is the nature of the interaction? Can you offer some explanation as to why it occurred?

9.6 We have a $2 \times 2 \times 2$ factorial experiment with $n = 10$ subjects assigned at random to each treatment combination. Complete the analysis of variance.

A_1				A_2			
B_1		B_2		B_1		B_2	
C_1	C_2	C_1	C_2	C_1	C_2	C_1	C_2
7	6	10	3	10	1	1	9
7	8	1	2	2	4	10	6
4	1	3	2	8	3	10	5
1	7	9	5	10	4	6	6
5	7	9	2	3	6	7	3
9	7	4	4	1	9	3	9
6	8	1	7	4	3	8	5
8	7	3	1	3	4	8	3
9	9	10	8	4	4	3	7
8	3	3	9	1	10	5	2

9.7 We have two factors, A and B, each varied in two ways, and with $n = 5$ subjects assigned at random to each treatment combination. The pooled sum of squares within treatments is 80.00. The treatment sums are: $A_1B_1 = 15$, $A_1B_2 = 20$, $A_2B_1 = 25$, and $A_2B_2 = 40$. (*a*) Test the treatment mean square for significance. (*b*) Partition the treatment sum of squares into the two main effects and the interaction and test each mean square for significance.

9.8 The outcome of a 2×2 factorial experiment with $n = 3$ observations for each treatment combination is given below:

A_1B_1	A_1B_2	A_2B_1	A_2B_2
2	3	1	2
3	2	2	0
4	1	0	1

Complete the analysis of variance.

9.9 The outcome of a 2×2 factorial experiment with $n = 5$ observations for each treatment combination is given below:

A_1B_1	A_1B_2	A_2B_1	A_2B_2
2	8	10	10
7	11	8	15
3	11	5	12
2	10	5	13
6	15	7	10

Complete the analysis of variance.

9.10 The outcome of a $2 \times 2 \times 2$ factorial experiment with $n = 3$ observations for each treatment combination is given below:

$A_1B_1C_1$	$A_1B_1C_2$	$A_1B_2C_1$	$A_1B_2C_2$	$A_2B_1C_1$	$A_2B_1C_2$	$A_2B_2C_1$	$A_2B_2C_2$
2	2	11	15	5	10	13	12
7	6	11	10	5	15	10	14
3	8	10	8	7	12	11	15

Complete the analysis of variance.

9.11 A table of random numbers was used to assign values of X from 1 to 10 at random to each treatment combination in a $2 \times 2 \times 2$ factorial experiment. All observations involving A_1 were increased by adding one and all those involving A_2 were decreased by subtracting one. All observations involving B_1 were increased by adding two and all those involving B_2 were decreased by subtracting two. Similarly, all observations involving C_1 were increased by adding three and all those involving C_2 were decreased by subtracting three. The sum of these arbitrary constants for each treatment combination is shown at the bottom of the following table:

	A_1				A_2			
	B_1		B_2		B_1		B_2	
	C_1	C_2	C_1	C_2	C_1	C_2	C_1	C_2
	11	5	10	5	8	6	3	−2
	11	10	11	4	9	8	10	−5
	12	10	10	6	13	1	3	−4
	13	8	5	−1	5	7	7	1
	12	8	7	−2	8	5	5	3
	12	4	6	−2	9	1	2	−1
	8	3	7	−3	11	6	10	−2
	8	5	12	2	12	2	5	−1
	7	1	6	−1	5	−1	2	−1
	13	7	3	2	8	4	4	−5
$\sum X_{k\cdot}$	107	61	77	10	88	39	51	−17
A:	1	1	1	1	−1	−1	−1	−1
B:	2	2	−2	−2	2	2	−2	−2
C:	3	−3	3	−3	3	−3	3	−3
Sum:	6	0	2	−4	4	−2	0	−6

Note that the constants were added or subtracted to simulate the main effects for A, B, and C and not interactions. Consequently, there is no reason to believe that any of the two-factor interactions or the three-factor interaction will be significant. (*a*) Is it reasonable to believe that you will find $MS_C > MS_B > MS_A$? Explain why. (*b*) Complete the analysis of variance. (*c*) The additive constants will have no effect on MS_W. Because the values of X were drawn from a uniform or rectangular population in which $P = 1/10$ for each of the possible values, 1, 2, 3, . . ., 10, we have $\sigma^2 = (10^2 - 1)/12 = 8.25$. Is MS_W a reasonably accurate estimate of σ^2?

CHAPTER 10

FACTORIAL EXPERIMENTS: FACTORS WITH MORE THAN TWO LEVELS

10.1 Introduction

A factorial experiment is not limited to the investigation of factors at only two levels, as in the examples cited in the last chapter. Factorial experiments may involve factors at several levels. If a factor has three or more levels, then the sum of squares for this factor will have more than 1 d.f. Then it also follows that the interactions of this factor with other factors will also have more than 1 d.f. The rule for determining the degrees of freedom for an interaction sum of squares, as stated previously, is to find the product of the degrees of freedom associated with the factors involved in the interaction.

Because the method of calculating the sum of squares for a two-factor interaction with 1 d.f. is a special case of a more general method, we shall examine a factorial experiment that involves interactions with more than 1 d.f. We shall also show methods for the direct calculation of any interaction, regardless of the number of factors or the number of levels of the factors involved in the interaction. It should be possible for the reader to generalize from the examples described, then, to any particular factorial experiment in which he is interested.

10.2 An Example of a $4 \times 3 \times 2$ Factorial Experiment

Let us take as an example an experiment in which three factors are involved, namely, A, B, and C. Suppose A has four levels, B has three levels, and C has two levels. Then we shall have $(4)(3)(2) = 24$

treatment combinations. We shall assume that a randomized group design is used, with $n = 5$ observations for each treatment, so that we have a total of 120 observations.[1] Then the total sum of squares with 119 d.f. can be partitioned into the following components with degrees of freedom as shown at the right:

Main effects:	A	3
	B	2
	C	1
Two-factor interactions:	$A \times B$	6
	$A \times C$	3
	$B \times C$	2
Three-factor interaction:	$A \times B \times C$	6
Within treatment:	Error	96

10.3 *Calculation of the Sums of Squares for the $4 \times 3 \times 2$ Factorial Experiment*

The method of calculating the within treatment sum of squares would be exactly the same as in the examples previously described. We could calculate the total sum of squares and the treatment sum of squares and obtain the within treatment sum of squares by sub-

Table 10.1
OUTCOME OF A $4 \times 3 \times 2$ FACTORIAL EXPERIMENT

Each cell entry is the sum of five observations.

		A_1	A_2	A_3	A_4	Σ
	B_1	60	90	94	86	330
C_1	B_2	54	92	98	96	340
	B_3	70	76	80	60	286
	B_1	58	72	78	84	292
C_2	B_2	76	82	74	64	296
	B_3	66	56	72	78	272
	Σ	384	468	496	468	1816

[1] Again we emphasize that the methods of analysis described in this chapter require that we have an equal number of observations for each treatment combination. In Chapter 12 we discuss some possible ways in which to deal with the problem of unequal n's.

traction. Or we could calculate the sum of squares within each treatment group separately and the sum of these sums of squares would be equal to the within treatment sum of squares. Let us assume that the within treatment sum of squares has already been calculated and has been found to be equal to 1198.00 and that we have added together the values of the observations for each treatment to obtain the treatment sums shown in Table 10.1. Each sum in the table is based on $n = 5$ observations. Then the treatment sum of squares will be given by

$$\text{Treatment} = \frac{(60)^2}{5} + \frac{(54)^2}{5} + \cdots + \frac{(78)^2}{5} - \frac{(1816)^2}{120} = 783.47$$

It is this sum of squares, 783.47, with 23 d.f., that is to be analyzed into the sums of squares for the main effects and interactions.

From the data of Table 10.1 we may set up a table for factors A and C summed over the levels of B. The resulting sums are shown in Table 10.2. Because B has three levels, each sum in the table will be based on $(3)(5) = 15$ observations. The two sums, 956 and 860, are the sums for C_1 and C_2, respectively. Each of these two sums is based on $(4)(15) = 60$ observations. Then, as the sum of squares for C, we have

$$C = \frac{(956)^2}{60} + \frac{(860)^2}{60} - \frac{(1816)^2}{120} = 76.80$$

The sum of squares for A will be based on the sums for A_1, A_2, A_3, and A_4 given at the bottom of the table. Each of these sums is based on $(2)(15) = 30$ observations, and the sum of squares for A will be

$$A = \frac{(384)^2}{30} + \frac{(468)^2}{30} + \frac{(496)^2}{30} + \frac{(468)^2}{30} - \frac{(1816)^2}{120} = 235.20$$

The $A \times C$ interaction sum of squares may also be obtained from Table

Table 10.2

THE $A \times C$ TABLE WITH THE CELL ENTRIES SUMMED OVER THE LEVELS OF B

Each cell entry is the sum of fifteen observations.

	A_1	A_2	A_3	A_4	Σ
C_1	184	258	272	242	956
C_2	200	210	224	226	860
Σ	384	468	496	468	1816

10.2. We calculate first the sum of squares between the eight sums entered in the cells of the table, keeping in mind that each of these sums is based on fifteen observations. Then the cell sum of squares will be

$$\text{Cell} = \frac{(184)^2}{15} + \frac{(200)^2}{15} + \cdots + \frac{(226)^2}{15} - \frac{(1816)^2}{120} = 405.87$$

Then the $A \times C$ interaction sum of squares may be obtained by subtracting the A and C sums of squares, which we have already calculated, from the cell sum of squares. As a general formula, if we have a two-way table with rows corresponding to the levels of one factor and columns corresponding to the levels of another factor, the interaction sum of squares for the row and column factors will be given by

$$R \times C = \text{Cell} - \text{row} - \text{column} \tag{10.1}$$

Then, remembering that the $R \times C$ interaction is the same as the $C \times R$ interaction, we have as the $A \times C$ interaction sum of squares

$$A \times C = 405.87 - 76.80 - 235.20 = 93.87$$

We now go back to the data of Table 10.1 and set up another two-way table for factors A and B summed over the levels of C. In this way we obtain the entries in Table 10.3. Because C has two levels, each sum in the table will be based on $(2)(5) = 10$ observations. The sums for B_1, B_2, and B_3 are the row sums, 622, 636, and 558, respectively. Each of these sums is based on $(4)(10) = 40$ observations. Then for the B sum of squares we have

$$B = \frac{(622)^2}{40} + \frac{(636)^2}{40} + \frac{(558)^2}{40} - \frac{(1816)^2}{120} = 86.47$$

To obtain the $A \times B$ interaction sum of squares, we first calculate the

Table 10.3
THE $A \times B$ TABLE WITH THE CELL ENTRIES SUMMED OVER THE LEVELS OF C

Each cell entry is the sum of ten observations.

	A_1	A_2	A_3	A_4	$\sum$
B_1	118	162	172	170	622
B_2	130	174	172	160	636
B_3	136	132	152	138	558
$\sum$	384	468	496	468	1816

Table 10.4
THE $B \times C$ TABLE WITH THE CELL ENTRIES SUMMED OVER THE LEVELS OF A

Each cell entry is the sum of twenty observations.

	B_1	B_2	B_3	Σ
C_1	330	340	286	956
C_2	292	296	272	860
Σ	622	636	558	1816

cell sum of squares for the entries in Table 10.3. Thus,

$$\text{Cell} = \frac{(118)^2}{10} + \frac{(130)^2}{10} + \cdots + \frac{(138)^2}{10} - \frac{(1816)^2}{120} = 425.87$$

We have already calculated the A (column) sum of squares and the B (row) sum of squares so that, by substitution in (10.1), we have

$$A \times B = 425.87 - 86.47 - 235.20 = 104.20$$

To find the $B \times C$ interaction sum of squares, we set up still another two-way table for factors B and C summed over the levels of A. Thus, from Table 10.1, we obtain Table 10.4. Because A has four levels, each sum in the table will be based on (4) (5) = 20 observations. For this table we have already calculated the C (row) sum of squares and the B (column) sum of squares. Therefore, all that we need to do is to find the sum of squares for cells and we can then find the $B \times C$ interaction sum of squares using (10.1). For the cell sum of squares we have

$$\text{Cell} = \frac{(330)^2}{20} + \frac{(292)^2}{20} + \cdots + \frac{(272)^2}{20} - \frac{(1816)^2}{120} = 175.87$$

Then, by subtraction we have

$$B \times C = 175.87 - 76.80 - 86.47 = 12.60$$

We shall show how to calculate the $A \times B \times C$ interaction sum of squares later. For the moment, we shall obtain it by subtraction. We know that the sum of the sums of squares for A, B, C, $A \times B$, $A \times C$, $B \times C$, and $A \times B \times C$ must equal the treatment sum of squares. Because we have calculated the first six of the seven sums of squares, we can obtain the last one by subtraction. The sum of the sums of squares we have calculated is equal to $235.20 + 86.47 + 76.80 + 104.20 + 93.87 + 12.60 = 609.14$. Subtracting this sum from the treatment sum of squares, we have

$$A \times B \times C = 783.47 - 609.14 = 174.33$$

10.4 Summary of the Analysis of Variance for the 4 × 3 × 2 Factorial Experiment

We have assumed that the within treatment sum of squares has already been calculated and found to be equal to 1198.00. The sums of squares we have just calculated and the within treatment sum of squares are shown in Table 10.5, which summarizes the analysis.

We again assume that the levels of the factors are *fixed* and do not represent random selections from any larger populations. Our conclusions, therefore, are to be restricted to the conditions we have actually investigated. In this case, the within treatment sum of squares, when divided by its degrees of freedom, provides us with an estimate of experimental error for testing the significance of the other mean squares.

The values of F shown in the summary analysis of variance table must be interpreted in terms of the number of degrees of freedom involved, and these vary depending on the mean square being tested for significance. To simplify matters, we have starred the values of F that are significant with $\alpha = 0.05$. The interpretation of the significant value of F follows the same pattern as that described in the previous chapter.

As a matter of interest, because it is the smallest interaction mean square, we have graphed the $B \times C$ interaction to show that the lines for the means of C_1 and C_2 plotted against the levels of B have much the same form. This graph is shown at the bottom of Figure 10.1. The $A \times B \times C$ interaction is significant, indicating that the two-factor interactions are not the same for the levels of the third factor. For purposes of comparison, we have also graphed in Figure 10.1 the $B \times C$ interactions for each level of A.

Table 10.5
ANALYSIS OF VARIANCE OF THE 4 × 3 × 2 FACTORIAL EXPERIMENT OF TABLE 10.1

Source of Variation	Sum of Squares	d.f.	Mean Square	F
A	235.20	3	78.40	6.28*
B	86.47	2	43.24	3.46*
C	76.80	1	76.80	6.15*
$A \times B$	104.20	6	17.37	1.39
$A \times C$	93.87	3	31.29	2.51
$B \times C$	12.60	2	6.30	
$A \times B \times C$	174.33	6	29.06	2.33*
Within treatments	1198.00	96	12.48	
Total	1981.47	119		

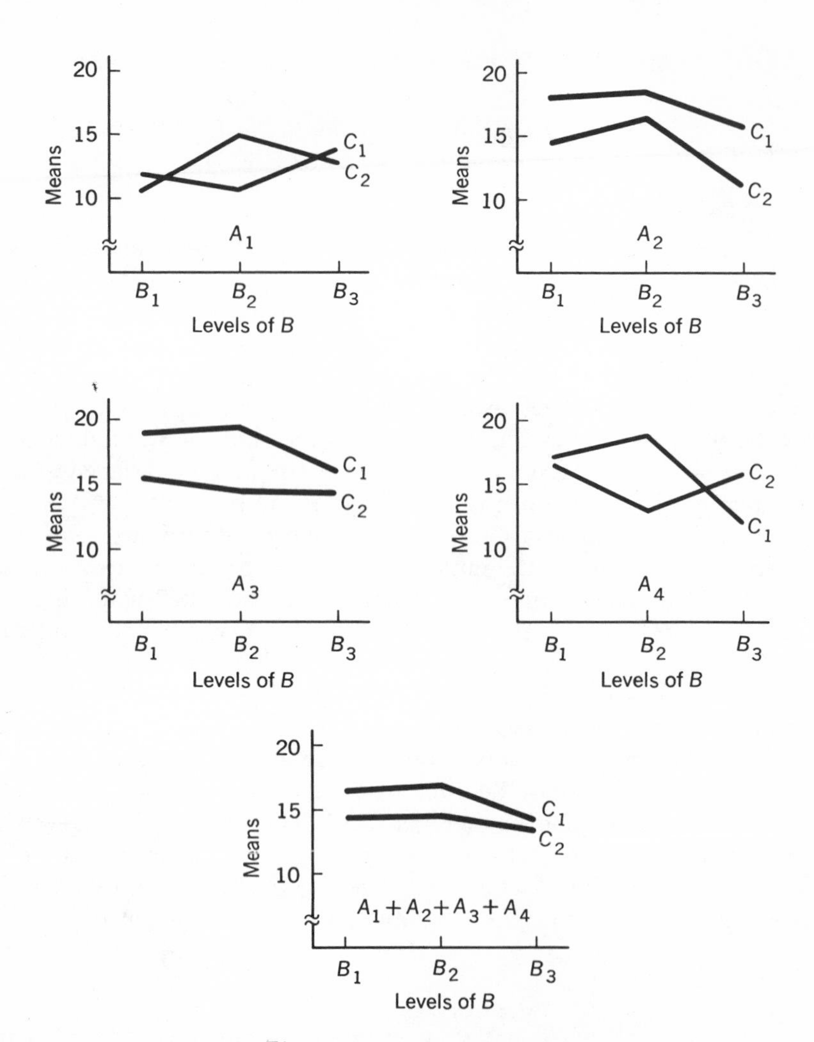

Figure 10.1
Means for levels of C at each level of B for A_1, A_2, A_3, and A_4, and also averaged over the levels of A. The $B \times C$ interaction averaged over the levels of A is not significant. The fact that the $A \times B \times C$ interaction is significant means that the $B \times C$ interactions are not of the same form for the different levels of A. The nature of these interactions is shown for each level of A.

10.5 Calculation of a Three-Factor Interaction Sum of Squares

In some factorial experiments it will be necessary to calculate directly the sum of squares for a three-factor interaction. We now illustrate a method of calculating the sum of squares for a three-factor interaction. The procedure is perfectly general and can be applied to obtain any interaction sum of squares, regardless of the number of factors and the number of levels of the factors involved in the interaction.

The interaction sum of squares to be calculated is that of $A \times B \times C$. Examine the data of Table 10.1. The cell entries there are the sums of $n = 5$ observations for each treatment. The sums of Table 10.1 may be rearranged in the manner of Table 10.6. We have first separated the treatments according to the four levels of A and then according to the levels of B and C.

Consider only Table (1). For the data in this table we can calculate a sum of squares between the six cells of the table. We can also calculate a

Table 10.6

TABLES FOR THE CALCULATION OF THE $A \times B \times C$ INTERACTION SUM OF SQUARES

Table (1)

	A_1			Σ
	B_1	B_2	B_3	
C_1	60	54	70	184
C_2	58	76	66	200
Σ	118	130	136	384

Table (2)

	A_2			Σ
	B_1	B_2	B_3	
C_1	90	92	76	258
C_2	72	82	56	210
Σ	162	174	132	468

Table (3)

	A_3			Σ
	B_1	B_2	B_3	
C_1	94	98	80	272
C_2	78	74	72	224
Σ	172	172	152	496

Table (4)

	A_4			Σ
	B_1	B_2	B_3	
C_1	86	96	60	242
C_2	84	64	78	226
Σ	170	160	138	468

Table (5)

	$A = A_1 + A_2 + A_3 + A_4$			Σ
	B_1	B_2	B_3	
C_1	330	340	286	956
C_2	292	296	272	860
Σ	622	636	558	1816

sum of squares for rows (C) and for columns (B). The row sum of squares would be the sum of squares for C averaged over the levels of B with the level of A (A_1) held constant. If we subtract the row and column sums of squares from the cell sum of squares, we have an interaction sum of squares. This procedure involves nothing new. We have used this method of calculation before in obtaining a two-factor interaction sum of squares. The interaction obtained from Table (1), however, is the $B \times C$ interaction with the level of A held constant and we designate this interaction sum of squares by $A_1(B \times C)$.

The process described could be repeated for Tables (2), (3), and (4). We would then have the interactions: $A_1(B \times C)$, $A_2(B \times C)$, $A_3(B \times C)$, and $A_4(B \times C)$. The necessary calculations for these interactions are as follows:

Table (1):

$$\text{Cell} = \frac{(60)^2}{5} + \frac{(58)^2}{5} + \cdots + \frac{(66)^2}{5} - \frac{(384)^2}{30} = 67.20$$

$$\text{Row} = \frac{(184)^2}{15} + \frac{(200)^2}{15} - \frac{(384)^2}{30} = 8.53$$

$$\text{Column} = \frac{(118)^2}{10} + \frac{(130)^2}{10} + \frac{(136)^2}{10} - \frac{(384)^2}{30} = 16.80$$

$$A_1(B \times C) = 67.20 - 8.53 - 16.80 = 41.87$$

Table (2):

$$\text{Cell} = \frac{(90)^2}{5} + \frac{(72)^2}{5} + \cdots + \frac{(56)^2}{5} - \frac{(468)^2}{30} = 176.00$$

$$\text{Row} = \frac{(258)^2}{15} + \frac{(210)^2}{15} - \frac{(468)^2}{30} = 76.80$$

$$\text{Column} = \frac{(162)^2}{10} + \frac{(174)^2}{10} + \frac{(132)^2}{10} - \frac{(468)^2}{30} = 93.60$$

$$A_2(B \times C) = 176.00 - 76.80 - 93.60 = 5.60$$

Table (3):

$$\text{Cell} = \frac{(94)^2}{5} + \frac{(78)^2}{5} + \cdots + \frac{(72)^2}{5} - \frac{(496)^2}{30} = 116.27$$

$$\text{Row} = \frac{(272)^2}{15} + \frac{(224)^2}{15} - \frac{(496)^2}{30} = 76.80$$

$$\text{Column} = \frac{(172)^2}{10} + \frac{(172)^2}{10} + \frac{(152)^2}{10} - \frac{(496)^2}{30} = 26.67$$

$$A_3(B \times C) = 116.27 - 76.80 - 26.67 = 12.80$$

Table (4):

$$\text{Cell} = \frac{(86)^2}{5} + \frac{(84)^2}{5} + \cdots + \frac{(78)^2}{5} - \frac{(468)^2}{30} = 188.80$$

$$\text{Row} = \frac{(242)^2}{15} + \frac{(226)^2}{15} - \frac{(468)^2}{30} = 8.53$$

$$\text{Column} = \frac{(170)^2}{10} + \frac{(160)^2}{10} + \frac{(138)^2}{10} - \frac{(468)^2}{30} = 53.60$$

$$A_4(B \times C) = 188.80 - 8.53 - 53.60 = 126.67$$

Summing the interactions of $B \times C$ for each level of A, we have

$$\sum A(B \times C) = 41.87 + 5.60 + 12.80 + 126.67 = 186.94$$

We have already calculated the $B \times C$ interaction averaged over the levels of A, that is, the $B \times C$ interaction for Table (5). Table (5), for example, is identical with Table 10.4 and for the $B \times C$ interaction we obtained a sum of squares equal to 12.60. Then the sum of squares for the three-factor interaction, $A \times B \times C$, may be obtained by subtracting the $B \times C$ interaction sum of squares from the sum of the interactions of $B \times C$ for the separate levels of A. Thus,

$$A \times B \times C = \sum A(B \times C) - B \times C \qquad (10.2)$$

Substituting in (10.2), we obtain

$$A \times B \times C = 186.94 - 12.60 = 174.34$$

which checks, within rounding errors, with the value previously found for the $A \times B \times C$ sum of squares.

10.6 General Methods for Calculating Interaction Sums of Squares for Three or More Factors

The method described above for calculating the sum of squares for a three-factor interaction can be varied to fit the needs of a particular factorial experiment. For example, the three-factor interaction sum of squares might have been obtained by calculating the $A \times C$ interactions for each level of B, summing, and then subtracting the $A \times C$ interaction obtained from the two-way table in which the $A \times C$ entries are summed over the levels of B. Thus, in general, a three-factor interaction sum of squares will be given by

$$\begin{aligned} A \times B \times C &= \sum A(B \times C) - B \times C \\ &= \sum B(A \times C) - A \times C \\ &= \sum C(A \times B) - A \times B \qquad (10.3) \end{aligned}$$

and a four-factor interaction sum of squares will be given by

$$\begin{aligned} A \times B \times C \times D &= \sum A(B \times C \times D) - B \times C \times D \\ &= \sum B(A \times C \times D) - A \times C \times D \\ &= \sum C(A \times B \times D) - A \times B \times D \\ &= \sum D(A \times B \times C) - A \times B \times C \end{aligned} \tag{10.4}$$

and a five-factor interaction sum of squares will be given by

$$\begin{aligned} A \times B \times C \times D \times E &= \sum A(B \times C \times D \times E) - B \times C \times D \times E \\ &= \sum B(A \times C \times D \times E) - A \times C \times D \times E \\ &= \sum C(A \times B \times D \times E) - A \times B \times D \times E \\ &= \sum D(A \times B \times C \times E) - A \times B \times C \times E \\ &= \sum E(A \times B \times C \times D) - A \times B \times C \times D \end{aligned} \tag{10.5}$$

A similar series of equations may be written for any interaction involving six factors, and so on. A general proof of these equations is given by Edwards and Horst (1950).

Because a three-or-more-factor interaction may be found in a variety of ways, it is worthwhile to examine the data to determine the set of tables that will require the least effort as far as calculations are concerned. We have not, for example, taken the most economical set of tables in the present problems. The calculations would be reduced considerably if we had taken $A \times B \times C = \Sigma C(A \times B) - A \times B$ instead of taking $A \times B \times C = \Sigma A(B \times C) - B \times C$. The first equation would require only two tables, one for $C_1(A \times B)$ and one for $C_2(A \times B)$, whereas the second equation, as we have seen, requires four tables.

10.7 *Orthogonal Comparisons for Interaction Effects*

If we have factors at more than two levels, it is always possible to obtain a set of orthogonal comparisons separately for the levels of each factor such that each comparison has 1 d.f. For example, in the experiment described, A has four levels and the sum of squares for A has 3 d.f. Therefore, we may analyze the sum of squares for A into a set of three orthogonal comparisons, each with 1 d.f. We designate these three comparisons by D_1, D_2, and D_3. If the three comparisons are made, then the $A \times B$ interaction sum of squares with 6 d.f. can be analyzed into the following comparisons: $D_1 \times B$, $D_2 \times B$, and $D_3 \times B$, each with 2 d.f. The $A \times C$ sum of squares with 3 d.f. can also be analyzed into the following comparisons: $D_1 \times C$, $D_2 \times C$, and $D_3 \times C$, each with 1 d.f.

Table 10.7

THREE ORTHOGONAL COMPARISONS ON THE A SUMS

	A_1	A_2	A_3	A_4	D_i^2
	384	468	496	468	$n \sum a_{\cdot i}^2$
D_1	1	1	−1	−1	104.53
D_2	1	−1	0	0	117.60
D_3	0	0	1	−1	13.07

In Table 10.7 we show three orthogonal comparisons on the A sums and we note that the sum of the sums of squares for the three comparisons is $104.53 + 117.60 + 13.07 = 235.20$, and this is equal to the A sum of squares. Using the treatment sums given in Table 10.3, we obtain the entries shown in Table 10.8. The three subtables in Table 10.8 can be used to find the $D_1 \times B$, $D_2 \times B$, and $D_3 \times B$ interaction sums of squares.

Each cell entry in the first subtable is based on $n = 20$ observations and for this subtable the cell sum of squares will be

$$\text{Cell} = \frac{(280)^2}{20} + \frac{(304)^2}{20} + \cdots + \frac{(290)^2}{20} - \frac{(1816)^2}{120} = 214.27$$

The column sum of squares for this subtable is the sum of squares for D_1 and is equal to 104.53. The row sum of squares is the sum of squares for B and is equal to 86.47. Then by subtraction

$$D_1 \times B = 214.27 - 86.47 - 104.53 = 23.27$$

with 2 d.f.

Each cell entry in the second subtable is based on $n = 10$ observations and for this subtable the cell sum of squares will be

$$\text{Cell} = \frac{(118)^2}{10} + \frac{(130)^2}{10} + \cdots + \frac{(132)^2}{10} - \frac{(852)^2}{60} = 228.00$$

The column sum of squares is the sum of squares for comparison D_2 and is equal to 117.60. For the row sum of squares, we have

$$\text{Row} = \frac{(280)^2}{20} + \frac{(304)^2}{20} + \frac{(268)^2}{20} - \frac{(852)^2}{60} = 33.60$$

and by subtraction we obtain

$$D_2 \times B = 228.00 - 33.60 - 117.60 = 76.80$$

with 2 d.f.

For the third subtable, each cell entry is based on $n = 10$ observations and the cell sum of squares will be given by

$$\text{Cell} = \frac{(172)^2}{10} + \frac{(172)^2}{10} + \cdots + \frac{(138)^2}{10} - \frac{(964)^2}{60} = 93.33$$

Table 10.8

PARTITIONING OF THE $A \times B$ INTERACTION SUM OF SQUARES INTO THREE COMPONENTS

$D_1(n = 20)$

	$A_1 + A_2$	A_3+A_4	Σ
B_1	280	342	622
B_2	304	332	636
B_3	268	290	558
Σ	852	964	1816

$D_2(n = 10)$

	A_1	A_2	Σ
B_1	118	162	280
B_2	130	174	304
B_3	136	132	268
Σ	384	468	852

$D_3(n = 10)$

	A_3	A_4	Σ
B_1	172	170	342
B_2	172	160	332
B_3	152	138	290
Σ	496	468	964

and the row sum of squares by

$$\text{Row} = \frac{(342)^2}{20} + \frac{(332)^2}{20} + \frac{(290)^2}{20} - \frac{(964)^2}{60} = 76.13$$

The column sum of squares is the sum of squares for comparison D_3 and is equal to 13.07. Then by subtraction we obtain

$$D_3 \times B = 93.33 - 76.13 - 13.07 = 4.13$$

with 2 d.f.

We note that the sum of the three sums of squares we have just calculated is

$$D_1 \times B + D_2 \times B + D_3 \times B = 23.27 + 76.80 + 4.13 = 104.20$$

and this is equal to the $A \times B$ sum of squares. In the same manner, we could find the $D_1 \times C$, $D_2 \times C$, and $D_3 \times C$ interaction sums of squares and the sum of these three sums of squares would be equal to the $A \times C$ interaction sum of squares. Similarly, the $A \times B \times C$ sum of squares could be analyzed into the comparisons: $D_1 \times B \times C$, $D_2 \times B \times C$, and $D_3 \times B \times C$ and each of these sums of squares would have 2 d.f. The sum of these sums of squares would be equal to the $A \times B \times C$ interaction sum of squares.

The B sum of squares has 2 d.f. and it would be possible to analyze this sum of squares into two orthogonal comparisons, each with 1 d.f. If we designate the two comparisons on the B sums as D_4 and D_5 and if these two comparisons are made along with the three comparisons on the A sums, then the various interaction sums of squares could be analyzed into the following comparisons:

$A \times B$:	$D_1 \times D_4$	$D_1 \times D_5$	
	$D_2 \times D_4$	$D_2 \times D_5$	
	$D_3 \times D_4$	$D_3 \times D_5$	
$A \times C$:	$D_1 \times C$	$D_2 \times C$	$D_3 \times C$
$B \times C$:	$D_4 \times C$	$D_5 \times C$	
$A \times B \times C$:	$D_1 \times D_4 \times C$	$D_1 \times D_5 \times C$	
	$D_2 \times D_4 \times C$	$D_2 \times D_5 \times C$	
	$D_3 \times D_4 \times C$	$D_3 \times D_5 \times C$	

each with 1 d.f.

It is obvious that a different set of orthogonal comparisons on the A sums could have been made that would not be the same as those shown in Table 10.7. Similarly, we have a choice as to the orthogonal comparisons to be made on the B sums. With either a different set of comparisons for A or for B, we would obtain a different set of orthogonal comparisons for the interactions also.

We see that, in this experiment, if three orthogonal comparisons are made on the A sums and two orthogonal comparisons are made on the B sums, then all of the interactions, $A \times B$, $A \times C$, $B \times C$, and $A \times B \times C$, can also be analyzed into orthogonal comparisons, each with 1 d.f. The total number of comparisons for both main effects and interactions would be equal to twenty-three, the number of degrees of freedom associated with the treatment sum of squares. It is, of course, not necessary that we make a complete set of twenty-three orthogonal comparisons. We may be interested only in a selected few of the comparisons. It is also obvious that many other possible comparisons, not necessarily orthogonal comparisons, could be made on the set of twenty-four treatment sums or means. If we wish to explore the data thoroughly, then the recommended test procedure is Scheffé's test, described previously.

QUESTIONS AND PROBLEMS

10.1 An experiment involves factor A, which is varied in three ways, factor B, which is varied in two ways, factor C, which is varied in two ways, and factor D, which is varied in three ways. The experiment is replicated with $n = 5$ subjects for each treatment combination. (a) Set up the summary analysis of variance table showing the sources of variation and the number of degrees of freedom associated with each. (b) How would you calculate the $A \times B \times D$ interaction sum of squares?

10.2 A factorial experiment involves two factors, A and B, with A varied in four ways and B varied in three ways. The treatment combinations are replicated with $n = 5$ observations for each. Results are given below:

A_1			A_2			A_3			A_4		
B_1	B_2	B_3	B_1	B_2	B_3	B_1	B_2	B_3	B_1	B_2	B_3
38	54	65	24	21	35	36	35	35	45	45	34
45	34	86	43	67	45	81	36	65	55	98	65
22	54	62	56	98	76	22	54	67	34	65	65
23	23	26	75	46	89	23	65	76	34	34	43
45	32	42	43	55	98	45	78	55	45	54	36

Use the analysis of variance to analyze the results of the experiment.

10.3 We have a factorial experiment in which A is varied in two ways and B is varied in three ways. Results are given below:

A_1			A_2		
B_1	B_2	B_3	B_1	B_2	B_3
19	43	53	30	49	64
10	42	51	24	43	61
21	41	57	25	49	68
15	44	57	30	53	56
20	42	68	28	44	60
24	49	60	34	46	55
16	46	48	31	46	54
22	39	47	32	56	68
18	48	60	27	54	59
18	39	60	32	53	57

(*a*) Analyze the data using the analysis of variance. (*b*) Examine the $A \times B$ interaction, regardless of whether it is significant, by graphic methods.

10.4 Suppose we have a factorial experiment with a levels of A and b levels of B. Then, as we have pointed out, it is always possible to analyze the sum of squares for A into a set of $a - 1$ mutually orthogonal comparisons, and it is also possible to analyze the sum of squares for B into a set of $b - 1$ mutually orthogonal comparisons. Consider a simple example, with two levels of A and three levels of B. The sum for each treatment combination is given below:

	A_1			A_2		
	B_1	B_2	B_3	B_1	B_2	B_3
Comparison	10	15	20	15	10	30
1	1	1	1	−1	−1	−1
2	2	−1	−1	2	−1	−1
3	0	1	−1	0	1	−1
4	2	−1	−1	−2	1	1
5	0	1	−1	0	−1	1

(*a*) Assuming that $n = 10$ subjects have been randomly assigned to each treatment combination, find the treatment sum of squares with 5 d.f. (*b*) Analyze the treatment sum of squares into the A, B, and $A \times B$ sums of squares. (*c*) Now examine the comparisons shown in the above table. Are they mutually orthogonal? (*d*) Each of the comparisons shown in the table will have 1 d.f. Find the mean square for each comparison.

Note that the sum of the sums of squares for comparisons (2) and (3) is equal to the B sum of squares. Note also that the coefficients for comparison (4) are obtained by multiplying the coefficients of (1) and (2). Similarly, the coefficients for comparison (5) are obtained by multiplying the coefficients of (1) and (3). Furthermore, the sum of the sums of squares for comparisons (4) and (5) is equal to the $A \times B$ interaction sum of squares. In other words, we have partitioned the $A \times B$ interaction sum of squares with 2 d.f. into the two orthogonal comparisons (4) and (5), each with 1 d.f.

10.5 As another example, assume that the thirst, hunger, and sex drives of rats are each at three levels or intensities, 1, 2, and 3. (*a*) Assuming $n = 10$ rats have been randomly assigned to each treatment combination, find the treatment sum of squares with 8 d.f. (*b*) Partition the treatment sum of squares into the sum of squares for drive, intensity, and drive $\times$ intensity. (*c*) Now examine the comparisons shown in the table. Are they mutually orthogonal? (*d*) Find the mean squares for each of the comparisons shown in the table. (*e*) Into what comparisons has the sum of squares for drive been partitioned? Into what comparisons has the sum of squares for intensity been partitioned? Into what comparisons has the sum of squares for drive × intensity been partitioned?

	Thirst			Hunger			Sex		
	1	2	3	1	2	3	1	2	3
Comparison	10	18	28	12	20	30	10	12	15
1	1	1	1	1	1	1	−2	−2	−2
2	1	1	1	−1	−1	−1	0	0	0
3	−1	−1	2	−1	−1	2	−1	−1	2
4	−1	1	0	−1	1	0	−1	1	0
5	−1	−1	2	−1	−1	2	2	2	−4
6	−1	1	0	−1	1	0	2	−2	0
7	−1	−1	2	1	1	−2	0	0	0
8	−1	1	0	1	−1	0	0	0	0

10.6 We have a complete factorial experiment with $n = 5$ subjects assigned at random to each treatment combination. We have $a = 3$, $b = 2$, $c = 4$, and $d = 2$ levels for A, B, C, and D, respectively. Give the number of observations on which each of the following means will be based:

(*a*) $\bar{X}_{a\cdots}$ (*e*) $\bar{X}_{ab\cdot\cdot}$ (*i*) $\bar{X}_{abc\cdot}$

(*b*) $\bar{X}_{\cdot b\cdot\cdot}$ (*f*) $\bar{X}_{a\cdot c\cdot}$ (*j*) $\bar{X}_{a\cdot cd}$

(*c*) $\bar{X}_{\cdot\cdot c\cdot}$ (*g*) $\bar{X}_{\cdot bc\cdot}$ (*k*) $\bar{X}_{\cdot bcd}$

(*d*) $\bar{X}_{\cdots d}$ (*h*) $\bar{X}_{a\cdot\cdot d}$ (*l*) $\bar{X}_{abcd}$

10.7 If a significant three-factor interaction is obtained in an experiment, how could one go about examining the nature of the interaction?

10.8 Describe an experiment in which one might expect to find a significant three-factor interaction. Explain why you would expect this result.

CHAPTER 11

MODELS FOR FACTORIAL EXPERIMENTS WITH A RANDOMIZED GROUP DESIGN

11.1 Introduction

In our discussion of factorial experiments in Chapters 9 and 10, we assumed that the levels of each factor included in the experiment were selected by the experimenter because they were the ones in which he was interested. Generalizations based on tests of significance in these factorial experiments were therefore confined to the particular levels of the factors and combinations of levels actually investigated. In other words, the levels of a factor were *not* considered to be a *random* selection from a larger population of levels. For example, one of the factors in an experiment may be shock with three levels. The experimenter obviously has a choice in selecting the levels or intensities of shock to be investigated. It is not likely, however, that he will decide to choose the three levels by random selection from a larger population of intensities. When the treatments, or levels of factors, are *not* randomly selected, the analysis of variance model is referred to as Model I, or as a *fixed-effect* model.

Assume now that we have several factors and that the levels of each factor have been *randomly* selected from some larger populations. The analysis of variance model, in this instance, is referred to as Model II, or as a *random-effect* model. If the levels of some factors have been randomly selected and those of others have not, the analysis of variance model is referred to as a *mixed* model. Thus, the mixed model involves both fixed effects and random effects.

If all of the treatments or levels about which inferences are to be made are included in the experiment, then the treatments or levels may be regarded as fixed and Model I is appropriate for the analysis of variance. On the other hand, if generalizations and inferences about treatments or levels not included in the experiment are to be made, then the treatments or levels investigated must be *randomly* selected from the population of interest. When this is the case, the treatments or levels are regarded as random and Model II is appropriate for the analysis of variance.

If it can be assumed that Model II applies to a given experiment, then generalizations based on the tests of significance are also assumed to hold for the levels or treatments in the population of interest. This assumption is warranted, of course, only if the treatments or levels actually investigated do represent a random selection from the population of interest. There may be instances in which Model II can be justified in a behavioral science experiment but, in general, this model seems unrealistic. The fixed-effect and the mixed model appear to be much closer to the realities of experimental procedures in the behavioral sciences.

11.2 Assumptions of the Fixed-Effect Model: Model I

In this section we consider the assumptions for a fixed-effect model for a factorial experiment with a randomized group design. For simplicity we illustrate the model for a two-factor experiment. Later we shall consider the model for factorial experiments with three or more factors.

We assume that

$$X_{abn} = \mu + t_a + t_b + t_{ab} + e_{abn} \qquad (11.1)$$

where $\mu = (\mu_1 + \mu_2 + \cdots + \mu_k)/k$, $t_a = \mu_a - \mu$ represents a treatment effect due to the A factor, $t_b = \mu_b - \mu$ represents a treatment effect due to the B factor, $t_{ab} = \mu_{ab} - \mu_a - \mu_b + \mu$ represents a treatment effect due to the interaction of A and B, and e_{abn} represents a random error associated with a given experimental unit or observation. The e_{abn}'s are assumed to be independently and normally distributed with $E(e_{abn}) = 0$ and $E(e_{abn}^2) = \sigma^2$.

For the fixed-effect model we have $E(t_a) = t_a$ and $E(t_a^2) = t_a^2$, $E(t_b) = t_b$ and $E(t_b^2) = t_b^2$, and $E(t_{ab}) = t_{ab}$ and $E(t_{ab}^2) = t_{ab}^2$. Without loss of generality, we also have the following restrictions on t_a, t_b, and t_{ab}:

$$\sum_1^a t_a = \sum_1^b t_b = \sum_1^a t_{ab} = \sum_1^b t_{ab} = 0 \qquad (11.2)$$

11.3 Expected Value of the Correction Term: Model I

To find the expected value of the correction term we sum (11.1) over all abn observations, square the resulting sum, divide by abn, and take the expectation. Thus, for the expected value of the correction term, we have

$$E\left[\frac{(\Sigma X_{...})^2}{abn}\right] = \frac{1}{abn} E\left(abn\mu + bn\sum_1^a t_a + an\sum_1^b t_b + n\sum_1^{ab} t_{ab} + \sum_1^{abn} e_{abn}\right)^2$$

$$= abn\mu^2 + \sigma^2 \qquad (11.3)$$

11.4 Expected Value of the Total Sum of Squares: Model I

We also have

$$E\left(\sum_1^{abn} X_{abn}^2\right) = E\left[\sum_1^{abn} (\mu + t_a + t_b + t_{ab} + e_{abn})^2\right]$$

$$= abn\mu^2 + bn\sum_1^a t_a^2 + an\sum_1^b t_b^2 + n\sum_1^{ab} t_{ab}^2 + abn\sigma^2 \qquad (11.4)$$

because all cross product terms on the right disappear when we sum over a and b. Subtracting (11.3) from (11.4) we obtain the expected value of the total sum of squares or

$$E(SS_{tot}) = bn\sum_1^a t_a^2 + an\sum_1^b t_b^2 + n\sum_1^{ab} t_{ab}^2 + (abn - 1)\sigma^2 \qquad (11.5)$$

11.5 Expected Value of the Treatment Sum of Squares: Model I

To obtain the expected value of the treatment sum of squares, we first find

$$E\left[\sum_1^{ab} \frac{(\Sigma X_{ab.})^2}{n}\right] = \frac{1}{n} E\left[\sum_1^{ab} (n\mu + nt_a + nt_b + nt_{ab} + \sum_1^{n} e_{abn})^2\right]$$

$$= abn\mu^2 + bn\sum_1^a t_a^2 + an\sum_1^b t_b^2 + n\sum_1^{ab} t_{ab}^2 + ab\sigma^2 \qquad (11.6)$$

Subtracting the correction term, as given by (11.3), from (11.6) we have the expected value of the treatment sum of squares or

$$E(SS_T) = bn \sum_1^a t_a^2 + an \sum_1^b t_b^2 + n \sum_1^{ab} t_{ab}^2 + (ab - 1)\sigma^2 \qquad (11.7)$$

11.6 *Expected Value of the Within Treatment Mean Square: Model I*

The expected value of the within treatment sum of squares can then be obtained by subtraction. Thus,

$$E(SS_W) = E(SS_{tot}) - E(SS_T) \qquad (11.8)$$

Subtracting (11.7) from (11.5), we see that

$$E(SS_W) = (abn - 1)\sigma^2 - (ab - 1)\sigma^2$$

$$= ab(n - 1)\sigma^2 \qquad (11.9)$$

Dividing the expected value of the within treatment sum of squares by its degrees of freedom, we have

$$E(MS_W) = \sigma^2 \qquad (11.10)$$

and we see that MS_W is an unbiased estimate of the common population variance σ^2.

11.7 *Expected Values of the A and B Mean Squares: Model I*

To find the expected value of the A sum of squares, we first find

$$E\left[\sum_1^a \frac{(\Sigma X_{a\cdot\cdot})^2}{bn}\right] = \frac{1}{bn} E\left[\sum_1^a \left(bn\mu + bnt_a + n\sum_1^b t_b + n\sum_1^b t_{ab} + \sum_1^{bn} e_{abn}\right)^2\right]$$

$$= abn\mu^2 + bn \sum_1^a t_a^2 + a\sigma^2 \qquad (11.11)$$

Subtracting the correction term, as given by (11.3), from (11.11) we have the expected value of the A sum of squares or

$$E(SS_A) = bn \sum_1^a t_a^2 + (a - 1)\sigma^2 \qquad (11.12)$$

and the expected value of the A mean square will be

$$E(MS_A) = \sigma^2 + bn \frac{\sum_1^a t_a^2}{a - 1} \quad (11.13)$$

The expected value of the B sum of squares can be obtained following the same procedures we used in finding the expected value of the A sum of squares. Carrying out these operations, we find that the expected value of the B sum of squares is

$$E(SS_B) = an \sum_1^b t_b^2 + (b - 1)\sigma^2 \quad (11.14)$$

and the expected value of the B mean square will be

$$E(MS_B) = \sigma^2 + an \frac{\sum_1^b t_b^2}{b - 1} \quad (11.15)$$

11.8 Expected Value of the A × B Mean Square: Model I

Having obtained the expected values of the A and B sums of squares, we can subtract these two values from the expected value of the treatment sum of squares to obtain the expected value of the $A \times B$ sum of squares. Thus,

$$E(SS_{AB}) = E(SS_T) - E(SS_A) - E(SS_B) \quad (11.16)$$

Subtracting (11.12) and (11.14) from (11.7), we obtain

$$E(SS_{AB}) = n \sum_1^{ab} t_{ab}^2 + (a - 1)(b - 1)\sigma^2 \quad (11.17)$$

Dividing (11.17) by the degrees of freedom associated with the $A \times B$ sum of squares, we have

$$E(MS_{AB}) = \sigma^2 + n \frac{\sum_1^{ab} t_{ab}^2}{(a - 1)(b - 1)}$$

11.9 Tests of Significance with the Fixed-Effect Model

The expectations of the mean squares for a two-factor experiment are summarized in Table 11.1 where, as a matter of

Table 11.1
EXPECTATION OF MEAN SQUARES FOR A TWO-FACTOR EXPERIMENT: MODEL I

Source	Expectation of Mean Square
A	$\sigma^2 + bn\theta_a^2$
B	$\sigma^2 + an\theta_b^2$
$A \times B$	$\sigma^2 + n\theta_{ab}^2$
Within	σ^2

convenience, we let θ^2 represent a fixed effect so that

$$\theta_a^2 = \frac{\sum_1^a t_a^2}{a - 1}, \qquad \theta_b^2 = \frac{\sum_1^b t_b^2}{b - 1}, \qquad \text{and} \qquad \theta_{ab}^2 = \frac{\sum_1^{ab} t_{ab}^2}{(a - 1)(b - 1)}$$

Under the assumption that

$$\theta_a^2 = \theta_b^2 = \theta_{ab}^2 = 0$$

$F = MS_A/MS_W$ will have an F distribution with $a - 1$ and $k(n - 1)$ d.f., $F = MS_B/MS_W$ will have an F distribution with $b - 1$ and $k(n - 1)$ d.f., and $F = MS_{AB}/MS_W$ will have an F distribution with $(a - 1)(b - 1)$ and $k(n - 1)$ d.f., where k is the number of treatment groups or $k = ab$. In other words, these F ratios, respectively, provide tests of the hypotheses that

$$\sum_1^a t_a^2 = 0, \qquad \sum_1^b t_b^2 = 0, \qquad \text{and} \qquad \sum_1^{ab} t_{ab}^2 = 0$$

11.10 *Assumptions of the Random-Effect Model: Model II*

For a random-effect model, we also have

$$X_{abn} = \mu + t_a + t_b + t_{ab} + e_{abn}$$

For a random-effect model, however, t_a, t_b, and t_{ab}, as well as e_{abn} are assumed to be independent and normally distributed *random variables* such that

$$E(t_a) = 0 \quad \text{and} \quad E(t_a^2) = \sigma_a^2$$
$$E(t_b) = 0 \quad \text{and} \quad E(t_b^2) = \sigma_b^2$$
$$E(t_{ab}) = 0 \quad \text{and} \quad E(t_{ab}^2) = \sigma_{ab}^2$$
$$E(e_{abn}) = 0 \quad \text{and} \quad E(e_{abn}^2) = \sigma^2$$

11.11 *Expected Value of the Correction Term: Model II*

For the expectation of the correction term, we have

$$E\left[\frac{(\Sigma X...)^2}{abn}\right] = \frac{1}{abn} E\left(abn\mu + bn\sum_1^a t_a + an\sum_1^b t_b + n\sum_1^{ab} t_{ab} + \sum_1^{abn} e_{abn}\right)^2$$

$$= abn\mu^2 + bn\sigma_a{}^2 + an\sigma_b{}^2 + n\sigma_{ab}{}^2 + \sigma^2 \qquad (11.18)$$

11.12 *Expected Value of the Total Sum of Squares: Model II*

We also have

$$E(X_{abn}{}^2) = \mu^2 + \sigma_a{}^2 + \sigma_b{}^2 + \sigma_{ab}{}^2 + \sigma^2 \qquad (11.19)$$

and

$$E\left(\sum_1^{abn} X_{abn}{}^2\right) = abn\mu^2 + abn\sigma_a{}^2 + abn\sigma_{ab}{}^2 + abn\sigma^2 \qquad (11.20)$$

Subtracting the correction term, as given by (11.18), from (11.20) we have the expected value of the total sum of squares or

$$E(SS_{tot}) = bn(a-1)\sigma_a{}^2 + an(b-1)\sigma_b{}^2 + n(ab-1)\sigma_{ab}{}^2 + (abn-1)\sigma^2 \qquad (11.21)$$

11.13 *Expected Value of the Treatment Sum of Squares: Model II*

To find the expected value of the treatment sum of squares, we first find

$$E\left[\sum_1^{ab}\frac{(\Sigma X_{ab}.)^2}{n}\right] = \frac{1}{n}E\left[\sum_1^{ab}(n\mu + nt_a + nt_b + nt_{ab} + \sum_1^n e_{abn})^2\right]$$

$$= abn\mu^2 + abn\sigma_a{}^2 + abn\sigma_b{}^2 + abn\sigma_{ab}{}^2 + ab\sigma^2 \qquad (11.22)$$

and subtracting the correction term from (11.22), we obtain the expected value of the treatment sum of squares or

$$E(SS_T) = bn(a-1)\sigma_a{}^2 + an(b-1)\sigma_b{}^2 + n(ab-1)\sigma_{ab}{}^2 + (ab-1)\sigma^2 \qquad (11.23)$$

11.14 Expected Value of the Within Treatment Mean Square: Model II

The expected value of the within treatment sum of squares can be obtained by subtracting the expected value of the treatment sum of squares from the expected value of the total sum of squares. Thus,

$$E(SS_W) = E(SS_{tot}) - E(SS_T) \tag{11.24}$$

Subtracting (11.23) from (11.21) we obtain

$$\begin{aligned} E(SS_W) &= (abn - 1)\sigma^2 - (ab - 1)\sigma^2 \\ &= ab(n - 1)\sigma^2 \end{aligned} \tag{11.25}$$

and dividing (11.25) by the degrees of freedom associated with the within treatment sum of squares, we have

$$E(MS_W) = \sigma^2 \tag{11.26}$$

11.15 Expected Values of the A and B Mean Squares: Model II

To obtain the expected value of the A sum of squares, we note that

$$\begin{aligned} E\left[\sum_1^a \frac{(\Sigma X_{a..})^2}{bn}\right] &= \frac{1}{bn} E\left[\sum_1^a \left(bn\mu + bnt_a + n\sum_1^b t_b + n\sum_1^b t_{ab} + \sum_1^{bn} e_{abn}\right)^2\right] \\ &= abn\mu^2 + abn\sigma_a^2 + an\sigma_b^2 + an\sigma_b^2 + a\sigma^2 \end{aligned} \tag{11.27}$$

Subtracting the correction term from (11.27) we have

$$E(SS_A) = bn(a - 1)\sigma_a^2 + n(a - 1)\sigma_{ab}^2 + (a - 1)\sigma_a^2 \tag{11.28}$$

Dividing (11.28) by $a - 1$, the degrees of freedom associated with the A sum of squares, we have

$$E(MS_A) = bn\sigma_a^2 + n\sigma_{ab}^2 + \sigma^2 \tag{11.29}$$

In a similar manner, we find that the expected value of the B sum of squares is

$$E(SS_B) = an(b - 1)\sigma_b^2 + n(b - 1)\sigma_{ab}^2 + (b - 1)\sigma^2 \tag{11.30}$$

and

$$E(MS_B) = an\sigma_b^2 + n\sigma_{ab}^2 + \sigma^2 \tag{11.31}$$

11.16 Expected Value of the A × B Mean Square: Model II

Subtracting the expected values of the A and B sums of squares from the expected value of the treatment sum of squares,

Table 11.2
EXPECTATION OF MEAN SQUARES FOR A TWO-FACTOR EXPERIMENT: MODEL II

Source	Expectation of Mean Square
A	$\sigma^2 + n\sigma_{ab}^2 + bn\sigma_a^2$
B	$\sigma^2 + n\sigma_{ab}^2 + an\sigma_b^2$
$A \times B$	$\sigma^2 + n\sigma_{ab}^2$
Within	σ^2

we have the expected value of the $A \times B$ sum of squares or

$$E(SS_{AB}) = E(SS_T) - E(SS_A) - E(SS_B) \tag{11.32}$$

Thus, subtracting (11.28) and (11.30) from (11.23), we obtain

$$E(SS_{AB}) = n(a-1)(b-1)\sigma_{ab}^2 + (a-1)(b-1)\sigma^2 \tag{11.33}$$

and for the expected value of the $A \times B$ mean square we have

$$E(MS_{AB}) = n\sigma_{ab}^2 + \sigma^2 \tag{11.34}$$

11.17 Tests of Significance: Model II

The expectations of the mean squares for a two-factor experiment with a random-effect model are summarized in Table 11.2. For the random-effect model, $F = MS_{AB}/MS_W$ will be distributed as F with $(a-1)(b-1)$ and $ab(n-1)$ d.f., if $\sigma_{ab}^2 = 0$. The obtained value of F thus provides a test of the hypothesis that $\sigma_{ab}^2 = 0$. If $\sigma_a^2 = 0$, then $F = MS_A/MS_{AB}$ will be distributed as F with $(a-1)$ and $(a-1)(b-1)$ d.f., and this F ratio provides a test of the hypothesis that $\sigma_a^2 = 0$. Similarly, if $\sigma_b^2 = 0$, then $F = MS_B/MS_{AB}$ will be distributed as F with $(b-1)$ and $(a-1)(b-1)$ d.f., and this F ratio provides a test of the hypothesis that $\sigma_b^2 = 0$.

11.18 Rules for Obtaining Expected Values of Mean Squares: Model II

For a factorial experiment involving three factors, A, B, and C, with corresponding levels of a, b, and c, respectively, we assume that

$$X_{abcn} = \mu + t_a + t_b + t_c + t_{ab} + t_{ac} + t_{bc} + t_{abc} + e_{abn} \tag{11.35}$$

We could find the expected values of the mean squares for the three-factor experiment for either a fixed-effect or a random-effect model following the procedures we used for a two-factor experiment. Fortunately, however, there are rules that enable us to write the expectations of the mean squares without going through the tedious algebra involved in deriving them.[1] In this section we give the rules for obtaining the expected values of the mean squares for a *random-effect* model. We shall then show how to obtain the expected values for a *mixed* model and for a *fixed-effect* model. We shall illustrate the rules for a three-factor experiment, but they can easily be generalized to factorial experiments with any number of factors.

In Table 11.3 we give the expectations of the mean squares for a factorial experiment with three factors, A, B, and C, with n observations for each treatment combination. The expectations of the mean squares are based on Model II, the random-effect model. The entries in each row of Table 11.3 are called *components of variance*. For each factor A, B, and C, we have assigned a letter a, b, and c, respectively, that is used to designate a source of variation when used as a subscript and also to designate the number of levels of the source when used as a coefficient. The expectation of the mean square for a given source always includes the error variance σ^2, and also a component directly attributable to that source. Thus, in Table 11.3, the expectation of the mean square for A includes σ^2 and σ_a^2, the former denoting the error variance and the latter the variance directly attributable to A. We observe in the table, however, that associated with σ_a^2 are the coefficients bcn. The coefficients of a component for a given source consist of all those letters that are not used as identifying subscripts for the source. Thus, the coefficients of σ_a^2 are bcn, because b, c, and n are not used as identifying subscripts.[2] The component of variance directly attributable to each source and the coefficients of the component are the last ones entered in each row of Table 11.3.

In addition to the error variance and the component directly attributable to a source, the other components of variance in the expectation of a given mean square consist of all other components whose identifying subscripts contain all of the letters necessary to describe the source under consideration.

[1] Methods for obtaining the expectations of mean squares have been described by Anderson and Bancroft (1952), Crump (1946), Edwards (1964), Schultz (1955), and Villars (1951).

[2] The rule we have given regarding the coefficients of a given component of variance assumes that the total number of observations is equal to the product of the subscripts necessary to identify a given observation. In a three-factor experiment, for example, with n observations for each treatment combination, we have X_{abcn} and the total number of observations will be equal to $abcn$. In some experimental designs, to be discussed later, the total number of observations is *not* equal to the product of the subscripts necessary to identify a given observation. In these designs a more general method for determining the coefficients of a given component of variance is necessary.

We illustrate the method for the three-factor experiment with n observations for each treatment combination. We note that $t_a = \mu_a - \mu$ and that $\bar{X}_{a...}$, the sample estimate of μ_a, is based on bcn observations and these are also the coefficients of σ_a^2. Similarly, for $t_b = \mu_b - \mu$ we observe that $\bar{X}_{b...}$, the sample estimate of μ_b, is based on acn observations and these are also the coefficients of σ_b^2. For the two-factor interaction, $t_{ab} = \mu_{ab} - \mu_a - \mu_b + \mu$, we have $\bar{X}_{ab..}$ as the sample estimate of μ_{ab} and $\bar{X}_{ab..}$ is based on cn observations. The coefficients of σ_{ab}^2 are, therefore, cn.

Table 11.3

EXPECTATION OF MEAN SQUARES FOR A THREE-FACTOR EXPERIMENT: MODEL II

Source	Expectation of Mean Square
A	$\sigma^2 + n\sigma_{abc}^2 + bn\sigma_{ac}^2 + cn\sigma_{ab}^2 + bcn\sigma_a^2$
B	$\sigma^2 + n\sigma_{abc}^2 + an\sigma_{bc}^2 + cn\sigma_{ab}^2 + acn\sigma_b^2$
C	$\sigma^2 + n\sigma_{abc}^2 + an\sigma_{bc}^2 + bn\sigma_{ac}^2 + abn\sigma_c^2$
$A \times B$	$\sigma^2 + n\sigma_{abc}^2 + cn\sigma_{ab}^2$
$A \times C$	$\sigma^2 + n\sigma_{abc}^2 + bn\sigma_{ac}^2$
$B \times C$	$\sigma^2 + n\sigma_{abc}^2 + an\sigma_{bc}^2$
$A \times B \times C$	$\sigma^2 + n\sigma_{abc}^2$
Within	σ^2

Thus, for A, we also have $n\sigma_{abc}^2$, $cn\sigma_{ab}^2$, and $bn\sigma_{ac}^2$, because the subscripts for each of these components contain a, this being the only letter necessary to describe completely the source of variance under consideration. The expectations of the other mean squares are obtained in the same manner as we obtained the expectation of the A mean square.

The *appropriate error term* for a test of significance of a given effect is the mean square whose expectation contains all of the components that are in the expectation of the mean square for the given effect *except* the component directly attributable to the given effect. In Table 11.3 it is obvious that, for the random-effect model, the within treatment mean square provides an appropriate error term for the $A \times B \times C$ interaction because, in this instance, we would have

$$\frac{E(MS_{ABC})}{E(MS_W)} = \frac{\sigma^2 + n\sigma_{abc}^2}{\sigma^2}$$

To test any of the two-factor interactions for significance, the appropriate error term is the $A \times B \times C$ interaction mean square. For example, as a test of significance of the $B \times C$ interaction effect, we would have

$$\frac{E(MS_{BC})}{E(MS_{ABC})} = \frac{\sigma^2 + n\sigma_{abc}^2 + na\sigma_{bc}^2}{\sigma^2 + n\sigma_{abc}^2}$$

For the main effects, there is no appropriate error mean square in the table, but approximate tests for situations of this kind have been devised by Cochran (1951) and Satterthwaite (1946).

11.19 Rules for Obtaining Expected Values of Mean Squares: A Mixed Model

In Table 11.4 we show the expectations of the mean squares for a mixed model, where the levels of C are regarded as random and the levels of A and B are regarded as fixed. As before, we use

θ^2, with appropriate subscripts, to represent a component due directly to a fixed effect. For example, because the levels of A are fixed, we have

$$\theta_a^2 = \sum_1^a t_a^2/(a - 1)$$

We can obtain the expectations of the mean squares in Table 11.4 from those in Table 11.3. When we have fixed effects, certain components are deleted from the expectations of the mean squares of Model II. To determine which components are to be deleted, we first delete the one or more subscript letters in the components of Table 11.3 that are necessary to describe the source in which the component is listed. Then, if any one of the remaining subscripts specifies a fixed effect, we delete the component from the expectation of the mean square for the source. For example, consider the expectation for A in Table 11.3. For the component σ_{abc}^2 we delete the subscript a because it is necessary to describe the source under consideration. The two remaining subscripts are then b and c. Because b specifies a

Table 11.4

EXPECTATION OF MEAN SQUARES FOR A FACTORIAL EXPERIMENT WITH A AND B FIXED AND C RANDOM

Source	Expectation of Mean Square						
A	$\sigma^2 +$		$bn\sigma_{ac}^2 +$				$bnc\theta_a^2$
B	$\sigma^2 +$	$an\sigma_{bc}^2 +$				$acn\theta_b^2$	
C	$\sigma^2 +$				$anb\sigma_c^2$		
$A \times B$	$\sigma^2 + n\sigma_{abc}^2 +$			$cn\theta_{ab}^2$			
$A \times C$	$\sigma^2 +$		$bn\sigma_{ac}^2$				
$B \times C$	$\sigma^2 +$	$an\sigma_{bc}^2$					
$A \times B \times C$	$\sigma^2 + n\sigma_{abc}^2$						
Within	σ^2						

Source	Error Mean Square for Test of Significance
A	$A \times C$
B	$B \times C$
C	Within
$A \times B$	$A \times B \times C$
$A \times C$	Within
$B \times C$	Within
$A \times B \times C$	Within

fixed effect, this component is deleted from the expectation of the mean square for A in Table 11.4. Similarly, with respect to σ_{ab}^2 we delete the subscript a which is necessary to describe the source under consideration. The remaining subscript is b and because this specifies a fixed effect, the component is deleted from the expectation of A in Table 11.4. For σ_{ac}^2 we again delete a which is necessary to describe the source. The remaining subscript is c and because this subscript does *not* specify a fixed effect, this component with coefficients bn remains as part of the expectation for A in Table 11.4.

We note that when we have a fixed effect which cross-classifies with a random effect, the interaction results in a component that is random in only one direction. In other words, the component appears as part of the expectation of the mean square for the fixed effect, but not as part of the expectation of the mean square for the random effect. Consider, for example, the component σ_{bc}^2 of Table 11.4, where B is fixed and C is random. The component σ_{bc}^2 appears as part of the expectation of the mean square for B, the fixed effect, but not as part of the expectation of the mean square for C, the random effect. The reason for this is that the B effect is measured over a random sampling of the levels of C and the B effect may be expected to show random variation for different random samples of the levels of C. Thus, σ_{bc}^2 appears as a component of the expectation of the mean square for B. On the other hand, the C effect is measured over fixed levels of B. Thus, there is no uncertainty associated with the C effect as a result of being measured over random levels of B, because the levels of B are not random, but fixed.

Following the procedures described above, we obtain the expectations of the other mean squares in Table 11.4 from those in Table 11.3. The appropriate error terms for the mean squares of the mixed model of Table 11.4 are given at the bottom of the table.

Table 11.5

EXPECTATION OF MEAN SQUARES FOR A THREE-FACTOR EXPERIMENT WITH A, B, AND C FIXED

Source	Expectation of Mean Square
A	$\sigma^2 + bcn\theta_a^2$
B	$\sigma^2 + acn\theta_b^2$
C	$\sigma^2 + abn\theta_c^2$
$A \times B$	$\sigma^2 + cn\theta_{ab}^2$
$A \times C$	$\sigma^2 + bn\theta_{ac}^2$
$B \times C$	$\sigma^2 + an\theta_{bc}^2$
$A \times B \times C$	$\sigma^2 + n\theta_{abc}^2$
Within	σ^2

11.20 Expected Values of Mean Squares: The Fixed-Effect Model

If the levels of C are fixed, as well as those of A and B, then we have the expectations of the mean squares shown in Table 11.5. In this case the appropriate error mean square for testing each of the main effects and interactions for significance is the within treatment mean square. With A, B, and C all representing fixed effects, the appropriate analysis of variance model is Model I. Regardless of the number of factors involved, for a fixed-effect model the appropriate error mean square for testing any main effects and interactions for significance is the within treatment mean square.

QUESTIONS AND PROBLEMS

11.1 Discuss the difference between a fixed-effect model and a random-effect model.

11.2 If the levels of factor A are fixed, what assumptions are made regarding t_a?

11.3 If the levels of factor A are random, what assumptions are made regarding t_a?

11.4 For a two-factor experiment with a fixed-effect model and with n observations for each treatment combination, show that $E(SS_{tot}) - E(SS_T) = ab(n-1)\sigma^2$.

11.5 For a two-factor experiment with a fixed-effect model and with n observations for each treatment combination, show that $E(SS_B) =$

$$(b-1)\sigma^2 + an \sum_1^b t_b^2.$$

11.6 For a two-factor experiment with a random-effect model and with n observations for each treatment combination, show that the expected value of the within treatment sum of squares is equal to $ab(n-1)\sigma^2$.

11.7 Assume we have a four-factor experiment with n observations for each treatment combination. (a) If all factors are random, what are the expectations of the mean squares? (b) If A is fixed and the other factors are random, what are the expectations of the mean squares? (c) If A and B are fixed and the other factors are random, what are the expectations of the mean squares? (d) If A, B, and C are fixed and D is random, what are the expectations of the mean squares?

11.8 For a four-factor experiment with n observations for each treatment combination, assume that A and C are fixed and that B and D are random. What are the expectations of the mean squares for this mixed model?

11.9 How does one determine the appropriate error term for the test of significance of a given effect?

CHAPTER 12

FACTORIAL EXPERIMENTS: FURTHER CONSIDERATIONS

12.1 Obtaining the Same Number of Observations for Each Treatment

The methods for calculating sums of squares in factorial experiments that we have described in previous chapters require that we have an equal number of observations for each treatment combination. When the supply of subjects is considerably larger than the number we plan to use in an experiment, it is not difficult to complete the experiment in such a way that we will have an equal number of observations for each treatment.

Suppose, for example, that we have k treatments and plan on n observations for each treatment so that the experiment will require a total of kn subjects. Let N be the available supply of subjects with N being considerably larger than kn. If we select a subject at random from the N available, then this subject can be randomly assigned to one of the k treatments. After this subject has been tested, we select another subject at random from the $N - 1$ available subjects and assign this subject at random to one of the $k - 1$ remaining treatments. We continue in this way until we have one complete replication of the experiment, that is, until we have tested one subject for each of the k treatments. We then repeat the process for a second replication of the experiment and continue in this way until we have obtained the desired number of replications.

12.2 Discarding Observations at Random to Obtain Equal n's

When we have a limited supply of subjects and when it is necessary to divide the subjects at random into k treatment groups before undertaking the testing of subjects, it can and may happen that we will end up with an unequal number of subjects in the treatment groups. A rat assigned to a given treatment may die before it can be tested and a human subject assigned to a given treatment may fail to keep a laboratory appointment. We shall assume that the loss of subjects is random and in no way related to the experimental conditions or treatments themselves. The fact is, however, that if we do not have the same number of observations for each treatment, the analysis of variance methods described previously for factorial experiments are not applicable.

The situation is not hopeless. Assume that in planning the experiment we anticipated a possible loss of subjects and assigned to each treatment a relatively larger n than we might otherwise have done. Then, if the number of obtained observations in the treatment group with the smallest number is still judged to be adequate, we might use a table of random numbers to discard, at random, observations from the other treatment groups in order to make the number of observations for all treatments equal to those for the treatment with the smallest number. Then, after discarding observations, we can proceed to analyze the results of the experiment in terms of the methods described, because we will have an equal number of observations for each treatment. This is one possible solution to the problem of unequal n's.

Suppose, however, that if we have to discard at random observations from the $k - 1$ treatment groups in order to equalize the n's for these groups with the n for the group with the smallest number of observations, we believe we would have considerably less precision in the means for the $k - 1$ treatment groups. For example, the n for the treatment group with the smallest number of observations may be quite small compared with the n's for the other $k - 1$ treatments. Thus, in order to obtain equal n's for all the treatments, we might have to discard a relatively larger number of observations from each of the other $k - 1$ treatments. Not only would this result in a loss of accuracy of the treatment means for which

Table 12.1
SUMS FOR A 2×2 FACTORIAL EXPERIMENT

Each cell entry is the sum of $n = 10$ observations.

	B_1	B_2	$\sum$
A_1	120	180	300
A_2	130	100	230
$\sum$	250	280	530

Table 12.2
ANALYSIS OF VARIANCE FOR THE 2 × 2 FACTORIAL EXPERIMENT OF TABLE 12.1

Source of Variation	Sum of Squares	d.f.	Mean Square	F
A	122.5	1	122.5	4.9
B	22.5	1	22.5	
$A \times B$	202.5	1	202.5	8.1
Within treatments	900.0	36	25.0	
Total	1247.5	39		

we have to discard observations, but it would also result in a loss of degrees of freedom for our estimate of experimental error.

In the next section we consider a method of analysis for factorial experiments with unequal n's that uses the information provided by all of the observations that are made in the experiment.

12.3 The Analysis of Variance Using Treatment Means as Single Observations

In Table 12.1 we give the treatment sums for a 2 × 2 factorial experiment with $n = 10$ subjects assigned to each treatment combination. For the treatment sum of squares we have

$$\text{Treatment} = \frac{(120)^2}{10} + \frac{(130)^2}{10} + \frac{(180)^2}{10} + \frac{(100)^2}{10} - \frac{(530)^2}{40} = 347.5$$

and this sum of squares can be analyzed into the sum of squares for A, B, and $A \times B$. Thus,

$$A = \frac{(300)^2}{20} + \frac{(230)^2}{20} - \frac{(530)^2}{40} = 122.5$$

$$B = \frac{(250)^2}{20} + \frac{(280)^2}{20} - \frac{(530)^2}{40} = 22.5$$

$$A \times B = 347.5 - 122.5 - 22.5 \qquad = 202.5$$

Let us assume that the pooled within treatment sum of squares, based on the variation within each of the four groups, is equal to 900.0. This pooled sum of squares will have $k(n - 1) = 4(10 - 1) = 36$ d.f. Table 12.2 summarizes the analysis of variance of the 2 × 2 factorial experiment with $n = 10$ observations for each treatment group.

Table 12.3
MEANS FOR EACH TREATMENT COMBINATION IN THE 2×2 FACTORIAL EXPERIMENT

	B_1	B_2	Σ
A_1	12	18	30
A_2	13	10	23
Σ	25	28	53

In Table 12.3 we give the treatment *means* for the 2×2 factorial experiment of Table 12.1. We shall assume that the within treatment sum of squares is the same as before, that is, 900.0, and with $n = 10$ observations assigned to each treatment we have $MS_W = 900.0/36 = 25.0$, as before. It is important to note, however, that in order to calculate the within treatment sum of squares in a factorial experiment, it is *not* necessary that each treatment have the same number of observations. Thus, we can obtain MS_W for any factorial experiment as long as we have two or more observations for each treatment.

Because the cell entries in Table 12.3 are *means*, each entry has an expected variance equal to

$$\sigma_{\bar{X}_k}{}^2 = \frac{\sigma_k{}^2}{n_k} \tag{12.1}$$

where $\sigma_k{}^2$ is the population variance and n_k is the number of observations on which a cell mean is based. We do *not* require that the n_k's be equal, as they happen to be in this instance. The average variance of the cell entries would then be

$$\frac{1}{k}\sum_1^k \frac{\sigma_k{}^2}{n_k} = \frac{\sum_1^k \sigma_{\bar{X}_k}{}^2}{k} \tag{12.2}$$

If we assume, as we ordinarily do, that all σ_k's are equal to the same population value σ^2, then (12.2) becomes

$$\sigma^2 \frac{1}{k}\sum_1^k \frac{1}{n_k} = \frac{\sum_1^k \sigma_{\bar{X}_k}{}^2}{k} \tag{12.3}$$

We let

$$c = \frac{1}{k}\sum_1^k \frac{1}{n_k} \tag{12.4}$$

and, as usual, our estimate of σ^2 is the within treatment mean square,

MS_W. Then, as an estimate of the variance defined by (12.3), we have

$$cMS_W = \frac{1}{k} \sum_{1}^{k} \frac{1}{n_k} MS_W \qquad (12.5)$$

In our example we have $n = 10$ observations for each of the cell means and, substituting in (12.4), we obtain

$$c = \frac{1}{4}\left(\frac{1}{10} + \frac{1}{10} + \frac{1}{10} + \frac{1}{10}\right) = 0.1$$

Then, using the value of c obtained above and substituting in (12.3) with $MS_W = 25.0$ as an estimate of σ^2, we obtain

$$cMS_W = (0.1)(25.0) = 2.5$$

We use $cMS_W = 2.5$ as our estimate of experimental error for the data of Table 12.3 where the cell entries are *means*, not sums.

We apply the analysis of variance to the data of Table 12.3 in the usual way except that we regard each cell entry (mean) as a *single* observation. Then we have the following sums of squares:

$$\text{Treatments} = (12)^2 + (18)^2 + (13)^2 + (10)^2 - \frac{(53)^2}{4} = 34.75$$

$$A = \frac{(30)^2}{2} + \frac{(23)^2}{2} - \frac{(53)^2}{4} = 12.25$$

$$B = \frac{(25)^2}{2} + \frac{(28)^2}{2} - \frac{(53)^2}{4} = 2.25$$

$$A \times B = 34.75 - 12.25 - 2.25 = 20.25$$

Table 12.4 shows the results of our calculations. As our estimate of experimental error we use $cMS_W = 2.5$. If you will compare the F ratios given in

Table 12.4

ANALYSIS OF VARIANCE OF THE MEANS GIVEN IN TABLE 12.3 WITH EACH CELL MEAN REGARDED AS A SINGLE OBSERVATION

Source of Variation	Sum of Squares	d.f.	Mean Square	F
A	12.25	1	12.25	4.9
B	2.25	1	2.25	
$A \times B$	20.25	1	20.25	8.1
Error		36	2.50	

this table with those given previously in Table 12.2, you will note that they are exactly the same, as they must be in the case of equal n's. However, the methods of analysis used in obtaining the results shown in Table 12.4 do *not* require that we have an equal number of observations for each treatment and, consequently, this method of analysis might be used for those cases where we have a factorial experiment with unequal n's for the various treatments.

To apply the methods of this section, we find the within treatment sum of squares for each of the treatment groups. We pool these sums of squares and divide by the pooled degrees of freedom, which will be equal to $\sum_{1}^{k} n_k - k$, to obtain the within treatment mean square. The within treatment mean square is then multiplied by c as defined by (12.4). We then set up a table in which the cell entries are the means for each treatment combination. The sums of squares for main effects and interactions are then calculated using the table of means, with each mean being regarded as if it were a *single* observation. Tests of significance of the mean squares for main effects and interactions are then made using (12.5) as the estimate of experimental error.[1]

Although we have used as an example a two-factor experiment, it is obvious that the procedures described in this section can be applied to any factorial experiment, regardless of the number of factors involved.

12.4 Single Replication of a Factorial Experiment

In the analysis of variance, a given set of observations consisting of one observation for each treatment is called a *replication*. Thus, if we have $k = 5$ treatments with $n = 10$ observations for each treatment, the experiment would be described as having ten replications. In factorial experiments, the number of treatment combinations increases as both the number of factors and the number of levels of the factors increase. For example, a 2^5 factorial experiment involves 32 treatments and a 3^5 factorial experiment involves 243 treatments. Although a 2^5 factorial experiment, involving 32 treatments, may not seem like a large number of treatments, a total of 320 subjects or experimental units will be needed if we have $n = 10$ replications. Similarly, if we have ten replications for a 3^5 factorial experiment, a total of 2430 subjects or experimental units will be needed.

[1] We emphasize that the analysis and tests of significance described are only approximate and that other more complicated methods of analysis have also been suggested. See, for example, Snedecor (1956) and Winer (1962).

In some factorial experiments, which involve a large number of treatments, only one replication is used; that is, for each treatment we have only a single observation. For example, in an experiment by Crutchfield (1938), five factors, each varied in three ways, were investigated. Animals were placed in a compartment in which a string led through a series of pulleys to a food pan. By pulling on the string the animals could pull the food pan next to the compartment and obtain food. A friction device was used to increase or decrease the force required for pulling the food pan. Factor A was the length of string attached to the food pan and this was varied by the use of a 60-cm, 120-cm, and a 240-cm length of string. Factor B was the force required to pull the food pan in on the training trials and this was varied by using a low, medium, and high setting of the friction devices. Factor C was the number of training trials and was varied by giving thirty, sixty, and ninety trials. Factor D consisted of the number of hours between the critical test trial and the last feeding period. This was varied with the intervals of 12, 24, and 48 hours. The final factor, E, was the force required to pull the food pan during the critical test trial and this was varied in the same way as during the training trials.

With each of the five factors varied in three ways, we have a total of $3^5 = 243$ treatment combinations. One animal was assigned to each of these treatment combinations.

In the analysis of variance, each of the five main effects will have 2 d.f. Each two-factor interaction will have 4 d.f. and the total number of such interactions is $5!/3!2! = 10$. Similarly, the total number of three-factor interactions is $5!/2!3! = 10$, and each of these interactions will have 8 d.f. The number of four-factor interactions is $5!/4!1! = 5$ and each of these interactions will have 16 d.f. We have only one five-factor interaction with 32 d.f.

In this experiment, it seems reasonable to regard the levels of the factors as fixed effects. Then, if the five-factor and four-factor interactions are all negligible, we know that the expected values of each of these mean squares will be approximately equal to σ^2. If this assumption is true, then we could pool the sums of squares and degrees of freedom for these interactions and use the resulting mean square with 112 d.f. as an estimate of error for testing the significance of the main effects and the two-factor and three-factor interactions.

The danger involved in using the pooled interactions as an estimate of error is that the various interaction effects may not be negligible. When this is the case, the estimate of error based on the pooled interactions may be considerably larger than an estimate that could have been obtained if the complete experiment had been replicated a number of times. For example, with additional replications of the experiment, we would have available a within treatment sum of squares and, in this case, the within treatment mean square would be used as the error mean square. If the higher order interactions are not negligible, then, as we have shown previously, the expected value of the within treatment mean square will be less than the expected value of the mean square based on the pooled higher order interactions. Thus, in the experiment described, if the five-factor and four-factor interactions are not negligible, the significance of the main effects and the two-factor and three-factor interactions will be underestimated. In other words, a main effect or two-factor or three-factor interaction that is not significant, when the pooled interaction mean square is used as an estimate of error, might prove to be significant if tested in terms of a within treatment mean square.

12.5 Fractional Replication of the 2^n Factorial Experiment

Another solution to the problem of a large number of treatments is based on the notion of fractional replication. For a 2^n factorial experiment, that is, with all factors at two levels, only a certain fraction, ½ or ¼ or ⅛, of the possible treatment combinations is used instead of the complete set. Fractional replication is also based on the assumption that particular interaction effects are unimportant or negligible. We shall consider only the case of a ½ fractional replication for the 2^n factorial experiment.

Consider, for example, the 2^3 factorial experiment of Table 9.8. We see there that the $A \times B \times C$ interaction is based on a comparison of the difference between four treatment sums that have plus signs and four treatment sums that have minus signs. The treatment sums with plus and minus signs are shown below:

Plus	*Minus*
$A_1B_1C_1$	$A_1B_1C_2$
$A_1B_2C_2$	$A_1B_2C_1$
$A_2B_1C_2$	$A_2B_1C_1$
$A_2B_2C_1$	$A_2B_2C_2$

Note that in this 2^3 factorial experiment the highest order interaction involves three factors and that 3 is an odd number and so is the sum of the subscripts for each treatment combination with a plus sign in the three-factor interaction comparison. In a 2^4 factorial experiment, the highest order interaction will involve four factors. Then the treatment combinations with plus signs in the four-factor interaction comparison will be those for which the subscripts sum to an even number. Similarly, in a 2^5

factorial experiment, the treatment combinations with plus signs in the five-factor interaction comparison will be those for which the subscripts sum to an odd number.

Now assume that we use a ½ fractional replication and replicate only those treatments that have plus signs in the $A \times B \times C$ interaction of a 2^3 factorial experiment. We thus sacrifice information on the $A \times B \times C$ interaction provided by complete replication. If the treatments with plus signs are replicated n times, then we will have a treatment sum of squares with 3 d.f. and a within treatment sum of squares with $4(n - 1)$ d.f.

The treatments that are replicated are shown in Table 12.5. We note that the first three rows of the table are comparisons for the A effect, the B effect, and the C effect, respectively, and that these three comparisons are mutually orthogonal. But we also observe that the A comparison is identical with the $B \times C$ comparison, the B comparison is identical with the $A \times C$ comparison, and that the C comparison is identical with the $A \times B$ comparison. In other words, the main effect of any given factor is completely *confounded* with the two-factor interaction of the other two factors. Only if the two-factor interactions can be assumed to be negligible will the ½ fractional replication of the 2^3 factorial experiment provide unbiased estimates of the A, B, and C effects. Of course, for the 2^3 factorial experiment we would ordinarily not use fractional replication, because the number of treatment combinations is only eight.

For a 2^4 factorial experiment, we have sixteen treatment combinations. Again we consider the treatments that will have plus signs and those that will have minus signs in the highest order interaction, $A \times B \times C \times D$. The eight treatments with plus signs are shown in Table 12.6.

We note in Table 12.6 that each main effect or comparison is confounded with a three-factor interaction and that each of the two-factor interaction comparisons is confounded with another two-factor interaction comparison;

Table 12.5

COMPARISONS ON THE SUMS OF A ½ FRACTIONAL REPLICATION OF A 2^3 FACTORIAL EXPERIMENT

Comparison	$A_1B_1C_1$	$A_1B_2C_2$	$A_2B_1C_2$	$A_2B_2C_1$
A	1	1	−1	−1
B	1	−1	1	−1
C	1	−1	−1	1
$A \times B$	1	−1	−1	1
$A \times C$	1	−1	1	−1
$B \times C$	1	1	−1	−1
$A \times B \times C$	1	1	1	1

Table 12.6

COMPARISONS ON THE SUMS OF A ½ FRACTIONAL REPLICATION OF A 2^4 FACTORIAL EXPERIMENT

	Treatment Combination							
	A_1 B_1 C_1 D_1	A_1 B_1 C_2 D_2	A_1 B_2 C_1 D_2	A_1 B_2 C_2 D_1	A_2 B_1 C_1 D_2	A_2 B_1 C_2 D_1	A_2 B_2 C_1 D_1	A_2 B_2 C_2 D_2
Comparison	1111	1122	1212	1221	2112	2121	2211	2222
A	1	1	1	1	−1	−1	−1	−1
B	1	1	−1	−1	1	1	−1	−1
C	1	−1	1	−1	1	−1	1	−1
D	1	−1	−1	1	−1	1	1	−1
$A \times B = C \times D$	1	1	−1	−1	−1	−1	1	1
$A \times C = B \times D$	1	−1	1	−1	−1	1	−1	1
$A \times D = B \times C$	1	−1	−1	1	1	−1	−1	1
$B \times C = A \times D$	1	−1	−1	1	1	−1	−1	1
$B \times D = A \times C$	1	−1	1	−1	−1	1	−1	1
$C \times D = A \times B$	1	1	−1	−1	−1	−1	1	1
$A \times B \times C = D$	1	−1	−1	1	−1	1	1	−1
$A \times B \times D = C$	1	−1	1	−1	1	−1	1	−1
$A \times C \times D = B$	1	1	−1	−1	1	1	−1	−1
$B \times C \times D = A$	1	1	1	1	−1	−1	−1	−1
$A \times B \times C \times D$	1	1	1	1	1	1	1	1

that is,

$$A = B \times C \times D$$

$$B = A \times C \times D$$

$$C = A \times B \times D$$

$$D = A \times B \times C$$

and

$$A \times B = C \times D$$

$$A \times C = B \times D$$

$$A \times D = B \times C$$

If we can assume that the three-factor interactions are negligible, then the ½ fractional replication of the 2^4 factorial experiment provides information about the main effects of all four factors.

If we use a ½ fractional replication of a 2^5 factorial experiment, then each main effect will be confounded with a four-factor interaction. For example, the main effect of A will be confounded with $B \times C \times D \times E$. Each two-factor interaction will be confounded with a three-factor interaction. For example, $A \times B$ will be confounded with $C \times D \times E$. If we can assume that all four- and three-factor interactions are negligible, then a ½ fractional replication of the 2^5 factorial experiment will provide information about all main effects of the factors and also about the two-factor interactions.

Additional details concerning the use of fractional replication for the 2^n series of factorial experiments can be found in Cochran and Cox (1957). The discussion of fractional replication by Kempthorne (1952) and Cox (1958) will also prove of value to the experimenter who wishes to consider seriously the use of fractional replication.

12.6 *Factorial Experiments and Control Groups*

In some factorial experiments, it may be desirable to include a control group, that is, a group of subjects who receive none of the treatment combinations. If a comparison is made of the difference between the mean for the control group and the average value of the means for all of the treatment groups, then this comparison will be orthogonal to the usual comparisons that would be made on the set of treatment means. For example, assume we have a two-factor experiment with each factor at two levels. We include in the experiment a control group that does not receive any of the treatment combinations. Then, if we make a comparison of the difference between the mean for the control group and the average of the means for the treatment groups, this comparison will be orthogonal to the A, B, and $A \times B$ comparisons.

In some cases the levels of a factor may simply consist of either the factor being present or the factor being absent. Assume, for example, that A and B are two drugs. Let A_1 be the presence of drug A in a standard dosage and A_0 be the absence of A. Similarly, let B_1 be the presence of drug B in a standard dosage and B_0 be the absence of B. Four pills are prepared representing the four treatment combinations: A_1B_1, A_1B_0, A_0B_1, and A_0B_0. Note that A_0B_0 contains neither of the two drugs and is commonly referred to as a placebo, A_1B_1 contains both drugs, and A_1B_0 contains only A and A_0B_1 contains only B. With n subjects assigned at random to each pill we might undertake the usual analysis of variance for a two-factor experiment, partitioning the treatment sum of squares into the A, B, and $A \times B$ interaction. The results of the usual tests of significance of these mean squares may, however, be misleading. For example, for the A effect or comparison we have

$$D_1 = (A_1B_1 + A_1B_0) - (A_0B_1 + A_0B_0)$$

and D_1 may be significant because the combination of A and B is highly effective and A and B alone are relatively ineffective. If the effect of the combination of A and B is such as to make both drugs ineffective, then D_1 may be nonsignificant even though A alone may be highly effective.

In designs of the kind described above, where the levels of a factor simply consist of either the factor being present or absent, there are more meaningful comparisons that may be made rather than the standard orthogonal comparisons regarding A, B, and $A \times B$. In Table 12.7 we show some of the comparisons that would ordinarily be of experimental interest. The group receiving the placebo, A_0B_0, obviously serves as a control group for the other three groups which have received either A, B, or A *and* B, and the first three comparisons in the table test the effectiveness of A, B, and AB in relation to the control group. If these were the only comparisons of interest, then for these tests we might use Dunnett's procedure, described previously, for testing each treatment mean against a control mean. However, there are other comparisons that are of interest and, consequently, we might make all tests using Scheffé's procedure, which we have also described previously in the chapter on multiple comparisons.

In Table 12.7 we observe that D_4 provides a test of the difference between the effectiveness of A and B, D_5 a test of the effectiveness of the combination AB against A alone, and D_6 a test of the effectiveness of the combination AB against B alone. These three comparisons would ordinarily also be of experimental interest.

As another example of a two-factor experiment in which the levels consist of the factor either being present or absent, let A_1 represent a pretest and A_0 the absence of a pretest. Similarly, let B_1 represent a treatment and B_0 the absence of the treatment. The dependent variable in an experiment of this kind consists of the scores of the subjects on a posttest on the *same* variable on which they were given a pretest. Then with four groups of

Table 12.7

SOME COMPARISONS ON THE MEANS OF FOUR GROUPS WHERE A_0 AND B_0 REPRESENT THE ABSENCE OF DRUGS A AND B, AND A_1 AND B_1 REPRESENT THE PRESENCE OF DRUGS A AND B

	Groups			
Comparison	A_1B_1	A_1B_0	A_0B_1	A_0B_0
D_1	0	1	0	−1
D_2	0	0	1	−1
D_3	1	0	0	−1
D_4	0	1	−1	0
D_5	1	−1	0	0
D_6	1	0	−1	0

subjects we would have the following treatment combinations:

Group	*Pretest*	*Treatment*
A_1B_1	Yes	Yes
A_1B_0	Yes	No
A_0B_1	No	Yes
A_0B_0	No	No

The usefulness of the above design in developmental studies has been described in detail by Solomon and Lessac (1968). For example, in some developmental studies the treatment consists of isolating subjects in early infancy or of depriving them of a normal environment. In these studies the control groups (A_1B_0 and A_0B_0) are raised in a normal environment and the experimental groups (A_1B_1 and A_0B_1) are raised in an isolated or restricted environment. All four groups of subjects are given a posttest. For one of the control groups (A_1B_0) and for one of the experimental groups (A_1B_1) we have scores on both the pretest and the posttest. Thus, we have six sets of scores or measures, as shown below:

Control			Experimental		
A_1B_0		A_0B_0	A_1B_1		A_0B_1
Pretest	Posttest	Posttest	Pretest	Posttest	Posttest

This design makes possible various comparisons on the six means which may be of value in clarifying different interpretations of the nature of developmental processes. For a discussion of these comparisons and the necessity of a design of the kind described, the reader is referred to the article by Solomon and Lessac (1968).

12.7 Factorial Experiments and Binomial Variables

It is obvious that the results of the Hsu and Feldt (1969) study, cited previously, regarding the analysis of variance of binomial variables, are not limited to a single factor experiment. If we have a factorial experiment in which the dependent variable is a binomial variable, then we may also use the analysis of variance to test the main effects and interactions for significance, provided we have a reasonably large number of subjects assigned to each of the treatment combinations.

QUESTIONS AND PROBLEMS

12.1 In Chapter 9 we presented the results of a 2^3 factorial experiment with $n = 10$ observations for each treatment combination. Using a table of random numbers, two treatments were selected and then three observations were discarded at random from each of the two treatments. Two other treatments were selected at random and two observations were discarded at random from these two treatments. Similarly, one observation was discarded at random from two other treatments. As a result, we have a 2^3 factorial experiment with an unequal number of observations for the treatments. The results are shown below:

A_1				A_2			
B_1		B_2		B_1		B_2	
C_1	C_2	C_1	C_2	C_1	C_2	C_1	C_2
76	36	43	37	94	74	67	60
43	45	75	22	80	64	64	54
65	47	66	22	81	68	70	51
42	23	56	25	80	72	65	49
60	43	62	11	80	62	60	38
78	43	51	27	69	78	55	55
66	54	63	23	80	61	57	56
	45	52	25	58		66	68
	41	50	31			79	
	40					80	

(*a*) Find the within treatment mean square and the value of c as defined by (12.4). (*b*) Find the mean for each treatment combination and regard these means as representing a single observation. Complete the analysis of variance in the manner described in the chapter for a factorial experiment with unequal n's. (*c*) Compare the results of this analysis with the analysis of variance of the complete factorial experiment with equal replications as given in Chapter 9.

12.2 The smallest number of observations for any treatment in the above problem is $n = 7$. Use the table of random numbers to discard at random observations from those treatments with $n > 7$ so that all treatments have $n = 7$ observations. Then analyze the remaining observations in terms of a 2^3 factorial experiment with $n = 7$ observations for each treatment. Compare the F ratios obtained with those found in Problem 12.1 and with those given in Chapter 9.

12.3 We have a 2^5 factorial experiment and decide to use a ½ fractional replication of those treatment combinations with plus signs in the five-factor interaction. Describe the nature of the confounding in this experiment.

12.4 Child (1946) designed an experiment to test the hypothesis that preference for a more distant goal object, when found, is the result of experience in previous situations. "The experiment was planned so that if this assumption was correct, certain influences of previous learning would be exhibited" (p. 3). The factors introduced were as follows: the sex of the children used as subjects in the experiment; the sex of the experimenter present during the test situation; the nature of the barrier introduced between the subject and the distant goal object; and the type of instructions given to the child. "The basic technique of these experiments was to place children in the position of having to choose between two desirable goals, one of which was more accessible than the other, and to observe their reactions" (p. 5).

Subjects were school children in grades 1 through 7. They were divided into groups of thirty-four to forty-five subjects each. The data given are in terms of the percentage choosing the more distant goal. Child states that the percentages are "close enough to 50, to suggest an adequate approximation to the assumption of normal distribution of sampling errors" (pp. 18–19). The analysis of variance was applied, however, making use of the inverse sine transformation. The values of F obtained with the transformation were slightly different, but no conclusions concerning significance were changed by the analysis of the data on the transformed scale. The results are given below:

	Male Subjects		Female Subjects	
	Cued Instructions	Noncued Instructions	Cued Instructions	Noncued Instructions
Male experimenter				
Table barrier	43	36	13	21
Ladder barrier	40	50	24	32
Female experimenter				
Table barrier	33	41	39	30
Ladder barrier	55	46	37	43

(*a*) Compute the various sums of squares. You may find the procedure of setting up a table with orthogonal coefficients a convenient method of calculation. (*b*) Note that in this experiment there is no mean square within treatments and that for tests of significance the higher order interactions must be used for an error mean square. For tests of significance, Child used the pooled sum of squares for all interactions with 11 d.f. What are some of the problems and assumptions involved in this procedure? (*c*) Note also that sex of the subjects corresponds to an organismic variable. If this mean square is significant, what interpretation may be made? (*d*) Sex of the experimenter corresponds to a treatment factor in which the levels (male and female) may be randomly assigned to subjects. As in the other factorial experiments we have discussed, however, levels of this factor do not represent a random sampling from a larger population of male and female experimenters. Thus, if significant, the conclusions should be restricted to the particular female and male experimenter involved in the experiment and not generalized beyond the two actually used.

12.5 Assume we use a ½ fractional replication of a 2^6 factorial experiment. (*a*) With what other effects will the main effects of the factors be confounded? (*b*) With what other effects will the two-factor interactions be confounded?

12.6 Kittens at age one month are divided at random into a treatment group and a control group which does not receive the treatment. The treatment is of such a nature that it extends over three months. (*a*) The control group is divided at random into two groups, one of which receives a pretest and one of which does not. Similarly, the treatment or experimental group is divided at random into two groups, one of which receives a pretest and one of which does not. At the end of the experiment all four groups are given a posttest. As alternatives to this design assume: (*b*) that neither the control nor the experimental group is given a pretest but that both groups are given the posttest; (*c*) that both groups are given a pretest and a posttest. What advantages, if any, does the design described by (*a*) have over that described by (*b*) and that described by (*c*)?

CHAPTER 13

RANDOMIZED BLOCK DESIGNS

13.1 Introduction

In research in the behavioral sciences the experimental unit to which a treatment is applied is most often a subject—a person, a rat, a dog, a cat, a pigeon. It is well known, of course, that an unselected group of subjects will vary with respect to almost any variable that we might care to measure. Subjects differ in their reaction times, their ability to solve problems, to learn, to recall, to perceive, and so forth. In many experiments, the dependent variable may be one in which there are widespread individual differences.

In this chapter we consider a design called a *randomized block design*. The randomized block design is based on the principle of grouping experimental units into blocks. The blocks are formed under the assumption that the units within each block will be more homogeneous in their response on some dependent variable of interest, in the absence of treatment effects, than units selected completely at random. By taking into account the differences existing between blocks in the analysis of variance, it is anticipated that a smaller error mean square will be obtained, for the same number of observations, than if a randomized group design had been used. The design, in other words, is one in which differences among blocks are eliminated from the estimate of experimental error.

The term "randomized blocks" comes from agricultural experiments in which the experimental unit to which a treatment is applied is a plot of land. A block consists of a strip of land consisting of a number of adjacent or neighboring plots. It is often true that plots that are near to one another in a field are more alike with respect to fertility and general soil condition than an equal number of randomly selected plots. The randomized block design, as used in agriculture, attempts to control for some of the existing differences between randomly selected plots by grouping the plots into blocks.

In psychological research, the experimental unit corresponding to a plot is a subject. A group of subjects relatively homogeneous with respect to some variable corresponds to a block. In essence, each block of subjects is matched with respect to a given variable and for this reason the randomized block design is also called a *matched group design*. It is anticipated that the subjects within each block will be relatively more homogeneous on the dependent variable, in the absence of treatment effects, than subjects selected completely at random.

13.2 A Randomized Block Design with $t = 5$ Treatments and $b = 5$ Blocks

Suppose that the dependent variable in an experiment is the number of arithmetic problems correctly solved in a fixed period of time and that subjects are to be tested under five treatments. Prior information available to the experimenter indicates that subjects vary considerably in the number of problems they can solve in a given period of time when tested under *uniform* conditions. Therefore, he decides to use a randomized block design in an attempt to eliminate from the estimate of experimental error some of the variance attributable to individual differences in the ability to solve arithmetic problems.

Let us assume that twenty-five subjects are to be used in the experiment, with five subjects assigned to each of the five treatments. On the basis of an initial test, administered under uniform conditions, a score is obtained for each subject that represents the number of problems solved in a given

Table 13.1

RANDOM PERMUTATIONS OF THE SUBJECTS IN EACH BLOCK IN A RANDOMIZED BLOCK DESIGN WITH FIVE BLOCKS AND FIVE TREATMENTS

	Treatments				
	1	2	3	4	5
Block 1	4	3	5	1	2
Block 2	1	4	2	3	5
Block 3	2	1	5	3	4
Block 4	5	4	3	1	2
Block 5	2	1	4	5	3

period of time. These initial scores are used to arrange the subjects into blocks. It will be convenient to let t represent the number of treatments in a randomized block design and b represent the number of blocks. In general, each block will consist of t subjects. With b blocks available, we will have a total of bt observations. If the subjects are arranged in rank order of their scores on the initial test, then the first t subjects will make up the first block or group, the next t subjects will make up the second block or group, and so on, until b such blocks or groups have been formed. In the present example, we would have five blocks of five subjects each.

Within each block the subjects are assigned at random to the t treatments. The randomization can be carried out by taking the ace, 2, 3, 4, and 5 from a deck of playing cards. If the cards are shuffled thoroughly and then turned face up in a row, the result will be a random permutation of the numbers 1 to 5. We need one such random permutation for each block.[1] Following the procedure described, the five random permutations shown in Table 13.1 were obtained. Now, if we have previously numbered the subjects from 1 to 5 in each block, then Table 13.1 tells us which subject in each block is to be assigned to which treatment. For example, in Block 1, Subject 4 is to be assigned to Treatment 1, Subject 3 to Treatment 2, Subject 5 to Treatment 3, Subject 1 to Treatment 4, and Subject 2 to Treatment 5.

13.3 Calculation of the Sums of Squares in a Randomized Block Design

Table 13.2 gives the outcome of an experiment with $t = 5$ treatments and $b = 5$ blocks. The analysis of variance[2] for the randomized block design begins with finding the total sum of squares. For the data of Table 13.2, we have

$$\text{Total} = (18)^2 + (17)^2 + \cdots + (16)^2 - \frac{(450)^2}{25} = 78.0$$

with $bt - 1$ d.f., where b is the number of blocks and t is the number of treatments.

[1] We can also use a table of random numbers to obtain random permutations. Suppose our point of entry in Table I in the Appendix is block 02, row 02, and column 10. Reading down, the first random permutation we obtain is 5, 4, 3, 1, and 2. Continuing to read down the table we can obtain the other necessary random permutations.

[2] In a randomized group design, if we should for one reason or another lose one or more observations for the treatments, the analysis of variance can still be used with unequal n's. With a randomized block design, however, the analysis of variance requires that we replace the missing value by some estimate. Estimating a single missing value is not difficult, but when several observations are missing, the situation is more complicated. Methods for obtaining estimates of missing values can be found in Cochran and Cox (1957), Federer (1955), Kempthorne (1952), and Snedecor (1956).

The treatment sum of squares can be found in the usual way. Thus,

$$\text{Treatment} = \frac{(83)^2}{5} + \frac{(88)^2}{5} + \cdots + \frac{(95)^2}{5} - \frac{(450)^2}{25} = 20.8$$

and the treatment sum of squares will have $t - 1$ d.f.

We now find the block sum of squares and this will be given by

$$\text{Block} = \frac{(100)^2}{5} + \frac{(95)^2}{5} + \cdots + \frac{(80)^2}{5} - \frac{(450)^2}{25} = 50.0$$

and the block sum of squares will have $b - 1$ d.f.

If we subtract the treatment and block sums of squares from the total sum of squares, we obtain a residual sum of squares that is called the *block* $\times$ *treatment* ($B \times T$) sum of squares. Thus,

$$\text{Block} \times \text{treatment} = \text{Total} - \text{treatment} - \text{block} \qquad (13.1)$$

For the present example, we have

$$\text{Block} \times \text{treatment} = 78.0 - 20.8 - 50.0 = 7.2$$

The degrees of freedom for the block $\times$ treatment sum of squares may also be obtained by subtracting the degrees of freedom for treatments and blocks from the total degrees of freedom. If we make this subtraction, we have

$$(bt - 1) - (t - 1) - (b - 1) = bt - t - b + 1 = (b - 1)(t - 1)$$

as the degrees of freedom for the block $\times$ treatment sum of squares.

In a randomized group design, the error sum of squares is the within treatment sum of squares. It is used as the error sum of squares because it

Table 13.2

OBSERVATIONS IN A RANDOMIZED BLOCK DESIGN WITH FIVE TREATMENTS AND FIVE BLOCKS

	Treatments					
Blocks	1	2	3	4	5	$\sum$
1	18	20	20	21	21	100
2	17	19	19	20	20	95
3	16	17	18	19	20	90
4	16	16	17	18	18	85
5	16	16	15	17	16	80
$\sum$	83	88	89	95	95	450

Table 13.3

ANALYSIS OF VARIANCE OF THE OBSERVATIONS IN TABLE 13.2

Source of Variation	Sum of Squares	d.f.	Mean Square	F
Treatments	20.8	4	5.20	11.56
Blocks	50.0	4	12.50	
$B \times T$	7.2	16	0.45	
Total	78.0	24		

is free from differences in the treatment means. In other words, the within treatment sum of squares represents the random variation to be expected when different subjects have received the same experimental treatment.

In the randomized block design, the block × treatment sum of squares is the error sum of squares. It represents the variation in the observations not only after differences in the treatment means have been eliminated but also after differences in the block means have been eliminated. Because, in the experiment described, all five treatments occur in each block, differences in the block means should not reflect differences in the treatments. However, if there are systematic differences in the average response to the treatments of subjects of varying levels of arithmetic ability, these differences will result in differences in the block means. Presumably, the block × treatment sum of squares, in the experiment described, represents the random variation to be expected when subjects are relatively homogeneous in arithmetic ability and also when all subjects have received the same treatment. If there are systematic differences in the block means, then the error sum of squares for a randomized block design will be smaller than the error sum of squares for a randomized group design.

13.4 Test of Significance with a Randomized Block Design

In Table 13.3 we show the sums of squares we have calculated and the degrees of freedom associated with these sums of squares. Dividing each sum of squares by its degrees of freedom, we obtain the mean squares given in the table. For the randomized block design, we have as a test of significance of the null hypothesis regarding the treatment means

$$F = \frac{MS_T}{MS_{BT}} \tag{13.2}$$

with $t - 1$ d.f. for the numerator and $(b - 1)(t - 1)$ d.f. for the denominator. In our example, we have $F = 11.56$ with 4 and 16 d.f. With $\alpha = 0.05$, $F = 11.56$ exceeds the tabled value of F and the null hypothesis would be rejected. We conclude that the treatment means do differ significantly.

Additional tests concerning the treatment means may be made in terms of the procedures discussed previously under the heading multiple comparisons. For the randomized block design, the block × treatment mean square is the error mean square for these tests.

13.5 Notation and Sums of Squares

In the randomized block design, the total sum of squares is partitioned into three component parts: the treatment sum of squares, the block sum of squares, and the block × treatment sum of squares. The nature of the treatment sum of squares is already familiar. The block sum of squares is based on the variation of the block means about the overall mean. We have obtained the block × treatment sum of squares by subtraction, but it can also be calculated directly.

For example, suppose we identify a given observation by X_{bt}, where b refers to a given block and t to a given treatment. Let b and t, when used as subscripts, be variables. Then, in the experiment described, b and t can each take values from 1 to 5. Thus, X_{32} would be the observation for Treatment 2 in Block 3. Let $X_{bt} - \bar{X}_{..}$ represent a deviation from the overall mean, $\bar{X}_{.t} - \bar{X}_{..}$ the deviation of a treatment mean from the overall mean, and $\bar{X}_{b.} - \bar{X}_{..}$ the deviation of a block mean from the overall mean. Then, by subtraction, we have

$$(X_{bt} - \bar{X}_{..}) - (\bar{X}_{.t} - \bar{X}_{..}) - (\bar{X}_{b.} - \bar{X}_{..}) = X_{bt} - \bar{X}_{.t} - \bar{X}_{b.} + \bar{X}_{..} \tag{13.3}$$

and the right-hand side of the above expression represents a residual.

Table 13.4 shows the residuals for each observation in Table 13.2. We note that the sum of the residuals in each column and in each row is zero. We have a total of bt residuals, but only $(b - 1)(t - 1)$ of them are free to vary, because the sum of the residuals in each column and each row of the table is zero. If we square and sum the residuals in the table, we will obtain

$$(-0.6)^2 + (-0.6)^2 + \cdots + (-1.0)^2 = 7.2$$

and this sum of squares is the block × treatment sum of squares of the analysis of variance, as shown previously in Table 13.3, with $(5 - 1)(5 - 1) = 16$ d.f.

Table 13.4
VALUE OF THE RESIDUALS $(X_{bt} - \bar{X}_{\cdot t} - \bar{X}_{b\cdot} + \bar{X}_{\cdot\cdot})$ FOR THE OBSERVATIONS IN TABLE 13.2

	Treatments					
Blocks	1	2	3	4	5	Σ
1	−0.6	0.4	0.2	0.0	0.0	0.0
2	−0.6	0.4	0.2	0.0	0.0	0.0
3	−0.6	−0.6	0.2	0.0	1.0	0.0
4	0.4	−0.6	0.2	0.0	0.0	0.0
5	1.4	0.4	−0.8	0.0	−1.0	0.0
Σ	0.0	0.0	0.0	0.0	0.0	0.0

Using the notation we have given, we can write the following identity:

$$X_{bt} - \bar{X}_{\cdot\cdot} = (\bar{X}_{\cdot t} - \bar{X}_{\cdot\cdot}) + (\bar{X}_{b\cdot} - \bar{X}_{\cdot\cdot}) + (X_{bt} - \bar{X}_{\cdot t} - \bar{X}_{b\cdot} + \bar{X}_{\cdot\cdot}) \tag{13.4}$$

which states that the deviation of a given value of X_{bt} from the overall mean can be expressed as a sum of the three component parts on the right. If we square both sides of this expression and sum over all bt observations, we find that all of the products between terms on the right sum to zero. Therefore,

$$\sum_1^{bt} (X_{bt} - \bar{X}_{\cdot\cdot})^2 = b \sum_1^{t} (\bar{X}_{\cdot t} - \bar{X}_{\cdot\cdot})^2 + t \sum_1^{b} (\bar{X}_{b\cdot} - \bar{X}_{\cdot\cdot})^2 + \sum_1^{bt} (X_{bt} - \bar{X}_{\cdot t} - \bar{X}_{b\cdot} + \bar{X}_{\cdot\cdot})^2 \tag{13.5}$$

The term on the left is the total sum of squares. The first term on the right is the treatment sum of squares and the second term is the block sum of squares. The last term is the block $\times$ treatment sum of squares.

13.6 Expectation of Mean Squares in a Randomized Block Design

Consider now a randomized block design such that within each block we have not a single subject for each treatment but n subjects. For example, suppose we have five blocks of eight subjects each, such that within each block the subjects are relatively homogeneous on a variable that we believe to be relevant to the measures to be obtained after the application of the treatments. Assume that we have only $t = 2$

treatments so that within each of the $b = 5$ blocks we can randomly assign $n = 4$ subjects to each of the treatments. For a given treatment in a given block, it will be possible to obtain a sum of squares based on the variation of the n observations for that treatment. This sum of squares will have $n - 1 = 3$ d.f. and the pooled sum of squares of all of these sums of squares will have $bt(n - 1) = (5)(2)(4 - 1) = 30$ d.f. We shall refer to this sum of squares as the *within cell* sum of squares. The analysis of variance for this experiment would result in the following sums of squares with degrees of freedom shown at the right:

Source	d.f.
Treatment	1
Block	4
Block × treatment	4
Within cell	30
Total	39

It is clear that if the blocks are regarded as a factor B with $b = 5$ levels and if treatments are regarded as a factor T with $t = 2$ levels, then this experimental design is similar to a two-factor experiment with $n = 4$ observations for each block-treatment combination. The expectations of the mean squares for the experiment would thus correspond to those for a two-factor experiment with n observations for each treatment combination. The within cell mean square, in this design, takes the place of the within treatment mean square.

Table 13.5 shows the expectations of the mean squares for Model II (random effect), Model I (fixed effect), and the mixed model where the blocks are assumed to be random and the treatments fixed. It is clear, for Model II, that the appropriate error mean square for testing the significance of the treatment mean square is the block × treatment interaction. This is also true for the mixed model where blocks are random and treatments fixed. For the fixed-effect model, where both blocks and treatments are regarded as fixed, the appropriate error mean square for testing the significance of the treatment mean square is the within cell mean square.

The expectations of the mean squares given in Table 13.5 are based on the assumption that

$$X_{btn} = \mu + t_t + t_b + t_{bt} + e_{btn} \tag{13.6}$$

where $t_t = \mu_t - \mu$, $t_b = \mu_b - \mu$, and $t_{bt} = \mu_{bt} - \mu_t - \mu_b + \mu$ represent a treatment, a block, and a block × treatment effect, respectively, and e_{btn} is a random error associated with each of the n observations in each of the block-treatment combinations. For the model defined by (13.6), we

Table 13.5

EXPECTATION OF MEAN SQUARES FOR A RANDOMIZED BLOCK DESIGN WITH n OBSERVATIONS IN EACH CELL

Model II: Blocks and Treatments Random

Source	Expectation of Mean Square
Treatments	$\sigma^2 + n\sigma_{bt}^2 + nb\sigma_t^2$
Blocks	$\sigma^2 + n\sigma_{bt}^2 + nt\sigma_b^2$
$B \times T$	$\sigma^2 + n\sigma_{bt}^2$
Within cell	σ^2

Model I: Blocks and Treatments Fixed

Source	Expectation of Mean Square
Treatments	$\sigma^2 + nb\theta_t^2$
Blocks	$\sigma^2 + nt\theta_b^2$
$B \times T$	$\sigma^2 + n\theta_{bt}^2$
Within cell	σ^2

Mixed Model: Blocks Random and Treatments Fixed

Source	Expectation of Mean Square
Treatments	$\sigma^2 + n\sigma_{bt}^2 + nb\theta_t^2$
Blocks	$\sigma^2 + nt\sigma_b^2$
$B \times T$	$\sigma^2 + n\sigma_{bt}^2$
Within cell	σ^2

have $E(MS_{WC}) = E(e_{btn}^2) = \sigma^2$, where MS_{WC} is the within cell mean square.

In the randomized block design, as we have described it earlier, we do not have an estimate of experimental error, σ^2, based on a within cell sum of squares because we have only $n = 1$ observation for each of the treatments within each of the blocks. With $n = 1$, we have the expectations of the mean squares shown in Table 13.6. It is of some importance to emphasize that simply because an experimental design does not provide an estimate of a variance component of a mean square, it does not follow that the variance component disappears from the expectation of the mean square. Thus, even though the randomized block design with $n = 1$ provides no estimate of $\sigma^2 = E(e_{btn}^2)$, this variance component is still part of the expectations of the mean squares shown in Table 13.6.

In general, in a randomized block design with $n = 1$, we shall not be able to regard the treatments as randomly selected and therefore we shall have

Table 13.6

EXPECTATION OF MEAN SQUARES IN A RANDOMIZED BLOCK DESIGN WITH $n = 1$ OBSERVATION IN EACH CELL

Model II: Blocks and Treatments Random

Source	Expectation of Mean Square
Treatments	$\sigma^2 + \sigma_{bt}^2 + \quad b\sigma_t^2$
Blocks	$\sigma^2 + \sigma_{bt}^2 + t\sigma_b^2$
$B \times T$	$\sigma^2 + \sigma_{bt}^2$

Model I: Blocks and Treatments Fixed

Source	Expectation of Mean Square
Treatments	$\sigma^2 + \quad b\theta_t^2$
Blocks	$\sigma^2 + \quad t\theta_b^2$
$B \times T$	$\sigma^2 + \theta_{bt}^2$

Mixed Model: Blocks Random and Treatments Fixed

Source	Expectation of Mean Square
Treatments	$\sigma^2 + \sigma_{bt}^2 + \quad b\theta_t^2$
Blocks	$\sigma^2 + \quad t\sigma_b^2$
$B \times T$	$\sigma^2 + \sigma_{bt}^2$

to assume that either the mixed model or the fixed-effect model is applicable. For the mixed model, with blocks random and treatments fixed, MS_{BT} provides an appropriate error term for testing the significance of MS_T. With a fixed-effect model, MS_{BT} is an appropriate error term for testing the significance of MS_T only if we can assume that the block $\times$ treatment component is negligible, that is, only if θ_{bt}^2 is negligible. If $\theta_{bt}^2 = 0$, then $E(MS_{BT}) = E(MS_{WC}) = \sigma^2$ and MS_{BT} can be used to test the significance of the treatment mean square.

13.7 Factorial Experiments with a Randomized Block Design

We have discussed factorial experiments with respect to a randomized group design. A factorial experiment may also be used with a randomized block design. Suppose, for example, we have two factors, A and C, each with two levels and that we have some reason for believing that subjects of comparable levels of intelligence will tend to be more homogeneous in their performance on the dependent variable, in the absence of treatment effects, than subjects selected completely at random. We have intelligence test scores for twenty subjects and on the basis of these scores the subjects are arranged into five blocks of four subjects each. Within each block subjects are assigned at random with one subject for each AC treatment combination.

Table 13.7 gives the results of the experiment. We find the various sums of squares in the usual manner. Thus,

$$\text{Total} = (5)^2 + (4)^2 + \cdots + (3)^2 - \frac{(70)^2}{20} = 65.0$$

$$\text{Block} = \frac{(20)^2}{4} + \frac{(17)^2}{4} + \cdots + \frac{(6)^2}{4} - \frac{(70)^2}{20} = 33.5$$

$$\text{Treatment} = \frac{(15)^2}{5} + \frac{(20)^2}{5} + \cdots + \frac{(25)^2}{5} - \frac{(70)^2}{20} = 25.0$$

Subtracting the block and treatment sums of squares from the total, we have the block $\times$ treatment sum of squares, or

$$\text{Block} \times \text{treatment} = 65.0 - 33.5 - 25.0 = 6.5$$

Table 13.7
A 2×2 FACTORIAL EXPERIMENT IN A RANDOMIZED BLOCK DESIGN

	Treatments				
Blocks	A_1C_1	A_1C_2	A_2C_1	A_2C_2	Σ
1	5	4	4	7	20
2	4	5	3	5	17
3	3	6	2	6	17
4	2	3	1	4	10
5	1	2	0	3	6
Σ	15	20	10	25	70

	C_1	C_2	Σ
A_1	15	20	35
A_2	10	25	35
Σ	25	45	70

Table 13.8
SUMMARY OF THE ANALYSIS OF VARIANCE FOR THE DATA OF TABLE 13.7

Source of Variation	Sum of Squares	d.f.	Mean Square	F
A	0.0	1		
C	20.0	1	20.00	37.0
$A \times C$	5.0	1	5.00	9.3
B: Blocks	33.5	4	8.38	
$B \times T$	6.5	12	0.54	
Total	65.0	19		

The treatment sum of squares, with 3 d.f., may be further analyzed into the three orthogonal comparisons: A, C, and $A \times C$, each with 1 d.f. However, because the sum for A_1 is equal to the sum for A_2, it is obvious that the A sum of squares must be equal to zero. For the C sum of squares, we have

$$C = \frac{(25)^2}{10} + \frac{(45)^2}{10} - \frac{(70)^2}{20} = 20.0$$

We know that the $A \times C$ sum of squares must be equal to 5.0, but we calculate it anyway, as a check, using (9.1). Then

$$A \times C = \frac{[(15 + 25) - (20 + 10)]^2}{(4)(5)} = 5.0$$

The results of the analysis are summarized in Table 13.8. The block $\times$ treatment mean square is often used to test the significance of the A, C, and $A \times C$ mean squares, and the F ratios for these tests are shown in the table. For a fixed-effect model, however, MS_{BT} is an appropriate error mean square for these tests only if it can be assumed that θ_{bt}^2 is equal to zero. For a mixed model, with blocks random and A and C fixed, MS_{BT} is an appropriate error term for testing the A, C, and $A \times C$ mean squares for significance only if $E(MS_{BA}) = E(MS_{BC}) = E(MS_{BAC})$, as we shall show in a later section.

13.8 Partitioning the Block × Treatment Sum of Squares

We have previously shown that if we have a two-way table, with columns corresponding to the levels of one factor and rows to the levels of another factor, and if we partition the column sum of squares into a set of orthogonal components or comparisons, then this also provides for a partitioning of the row $\times$ column interaction sum of squares. In the present example, we have partitioned the column (treatment) sum of squares into a set of orthogonal comparisons: A, C, and $A \times C$. If we regard the blocks (rows) as a factor, then it should also be possible to partition the block $\times$ treatment sum of squares into the three components:

$$\text{Block} \times A$$

$$\text{Block} \times C$$

$$\text{Block} \times A \times C$$

Because we have 4 d.f. for blocks and A and C each have 1 d.f., each of the above sums of squares will have 4 d.f.

In Table 13.9, obtained from Table 13.7, we show at the left the sums for A_1 and A_2 for each block. The sums for C_1 and C_2 for each block are shown at the right. Then, for the table at the left, the sum of squares between cells will be given by

$$\text{Cell} = \frac{(9)^2}{2} + \frac{(9)^2}{2} + \cdots + \frac{(3)^2}{2} - \frac{(70)^2}{20} = 35.0$$

We have already found that the block sum of squares is equal to 33.5 and the A sum of squares is equal to zero. Then

$$\text{Block} \times A = \text{Cell} - \text{block} - A$$

$$= 35.0 - 33.5 - 0.0 = 1.5$$

and this sum of squares will have (4)(1) = 4 d.f.

For the table at the right, the cell sum of squares will be given by

$$\text{Cell} = \frac{(9)^2}{2} + \frac{(7)^2}{2} + \cdots + \frac{(5)^2}{2} - \frac{(70)^2}{20} = 57.0$$

Table 13.9

SUMS FOR A_1 AND A_2 AND FOR C_1 AND C_2 FOR EACH BLOCK

The original data are given in Table 13.7.

Blocks	A_1	A_2	Σ	Blocks	C_1	C_2	Σ
1	9	11	20	1	9	11	20
2	9	8	17	2	7	10	17
3	9	8	17	3	5	12	17
4	5	5	10	4	3	7	10
5	3	3	6	5	1	5	6
Σ	35	35	70	Σ	25	45	70

Table 13.10

SUMMARY OF THE ANALYSIS OF VARIANCE FOR THE DATA OF TABLE 13.7 WITH A PARTITIONING OF THE BLOCK $\times$ TREATMENT SUM OF SQUARES

Source of Variation	Sum of Squares	d.f.	Mean Square	F
B: Blocks	33.5	4	8.375	
A	0.0	1		
$B \times A$	1.5	4	0.375	
C	20.0	1	20.000	22.86
$B \times C$	3.5	4	0.875	
$A \times C$	5.0	1	5.000	13.33
$B \times A \times C$	1.5	4	0.375	
Total	65.0	19		

The block sum of squares is equal to 33.5, the C sum of squares is equal to 20.0, and therefore,

$$\text{Block} \times C = 57.0 - 33.5 - 20.0 = 3.5$$

with 4 d.f.

The block $\times$ A $\times$ C sum of squares could be calculated directly in the manner described in a previous chapter, but it can also be obtained by subtraction, because

$$\text{Block} \times A \times C = \text{Block} \times \text{treatment} - \text{block} \times A - \text{block} \times C$$

and, in the present example, we have

$$\text{Block} \times A \times C = 6.5 - 1.5 - 3.5 = 1.5$$

with 4 d.f.

The results of the analysis of variance are shown in Table 13.10. In this table, the tests of significance of MS_A, MS_C, and MS_{AC} were made using MS_{BA}, MS_{BC}, and MS_{BAC}, respectively, as the error mean squares.

13.9 Expectation of the Mean Squares for a Two-Factor Experiment with a Randomized Block Design: Blocks Random and A and C Fixed

In Table 13.11 we give the expected values of the mean squares for a mixed model with blocks random and A and C fixed. The expectations of the mean squares are easily obtained by applying the rules described previously.

For the mixed model, with blocks random and A and C fixed, MS_{BA} is the appropriate error mean square for testing the A effect, MS_{BC} is the appropriate error mean square for testing the C effect, and MS_{BAC} is the appropriate error mean square for testing the $A \times C$ interaction effect. On the other hand, if

$$\sigma_{ba}^2 = \sigma_{bc}^2 = \sigma_{bac}^2 = 0$$

or if

$$c\sigma_{ba}^2 = a\sigma_{bc}^2 = \sigma_{bac}^2$$

then we will also have

$$E(MS_{BA}) = E(MS_{BC}) = E(MS_{BAC})$$

and, under this condition, we could pool the sums of squares for $B \times A$, $B \times C$, and $B \times A \times C$, and divide this pooled sum of squares by the pooled degrees of freedom. The result would be the block $\times$ treatment mean square and MS_{BT}, in this instance, would be the appropriate error mean square for testing the A, C, and $A \times C$ mean squares for significance.

In our example, we have

$$MS_{BA} = 0.375, \qquad MS_{BC} = 0.875, \qquad \text{and } MS_{BAC} = 0.375$$

each with 4 d.f., and it seems reasonable to regard these values as homogeneous. Consequently, MS_{BT}, with 12 d.f., would be used as the error mean square in the tests of significance of the A, C, and $A \times C$ mean squares.[3]

Table 13.11

EXPECTATION OF MEAN SQUARES FOR A TWO-FACTOR EXPERIMENT IN A RANDOMIZED BLOCK DESIGN WITH BLOCKS RANDOM AND A AND C FIXED

Source	Expectation of Mean Square
B: Blocks	$\sigma^2 + ac\sigma_b^2$
A	$\sigma^2 + c\sigma_{ba}^2 + bc\theta_a^2$
$B \times A$	$\sigma^2 + c\sigma_{ba}^2$
C	$\sigma^2 + a\sigma_{bc}^2 + ba\theta_c^2$
$B \times C$	$\sigma^2 + a\sigma_{bc}^2$
$A \times C$	$\sigma^2 + \sigma_{bac}^2 + b\theta_{ac}^2$
$B \times A \times C$	$\sigma^2 + \sigma_{bac}^2$

[3] For a fixed-effect model, with blocks as well as A and C fixed, if θ_{bt}^2 is equal to zero, then θ_{ba}^2, θ_{bc}^2, and θ_{bac}^2 will also be equal to zero. In this case, MS_{BT} is also the appropriate error mean square for the tests of significance.

13.10 Error Mean Squares for Comparisons

It should be noted that comparisons may be made on the treatment (column) means for any randomized block design and not just for randomized block designs in which the treatments correspond to a 2^n factorial experiment. If we make a comparison D_i on the treatment means, then we can always calculate the block $\times$ D_i mean square and this mean square can be used to test the comparison D_i for significance, assuming that we have a mixed model with blocks random and treatments fixed. The degrees of freedom for the block $\times$ D_i sums of squares will be equal to $b - 1$, where b is the number of blocks.

Under the assumption that the various block $\times$ D_i mean squares have the same expected values for all comparisons, we are better off using the block $\times$ treatment mean square in the denominator of the tests of significance of each of the comparisons, because the block $\times$ treatment mean

Table 13.12
OBSERVATIONS IN A RANDOMIZED BLOCK DESIGN WITH TWO TREATMENTS AND TWENTY BLOCKS

	Treatments			
Blocks	1	2	$\sum$	D
1	14	14	28	0
2	14	12	26	2
3	12	11	23	1
4	12	11	23	1
5	11	9	20	2
6	11	10	21	1
7	10	9	19	1
8	11	10	21	1
9	9	9	18	0
10	10	9	19	1
11	8	9	17	−1
12	10	9	19	1
13	8	8	16	0
14	7	8	15	−1
15	8	8	16	0
16	7	6	13	1
17	4	1	5	3
18	5	2	7	3
19	5	4	9	1
20	4	1	5	3
$\sum$	180	160	340	20

Table 13.13

ANALYSIS OF VARIANCE OF THE OBSERVATIONS IN TABLE 13.12

Source of Variation	Sum of Squares	d.f.	Mean Square	F
Treatments	10.0	1	10.000	14.6
Blocks	401.0	19	21.110	
$B \times T$	13.0	19	0.684	
Total	424.0	39		

square will be based on $(b - 1)(t - 1)$ degrees of freedom, whereas the error mean square for a specific comparison will have only $b - 1$ degrees of freedom.

13.11 A Randomized Block Design with $t = 2$ Treatments

In Table 13.12, we give the results of a randomized block design in which we have $t = 2$ treatments and $b = 20$ blocks. For the various sums of squares, we have

$$\text{Total} = (14)^2 + (14)^2 + \cdots + (1)^2 - \frac{(340)^2}{40} = 424.0$$

$$\text{Treatment} = \frac{(180)^2}{20} + \frac{(160)^2}{20} - \frac{(340)^2}{40} = 10.0$$

$$\text{Block} = \frac{(28)^2}{2} + \frac{(26)^2}{2} + \cdots + \frac{(5)^2}{2} - \frac{(340)^2}{40} = 401.0$$

$$\text{Block} \times \text{treatment} = 424.0 - 10.0 - 401.0 = 13.0$$

The analysis of variance for the experiment is shown in Table 13.13. Testing the treatment mean square for significance, we have $F = 14.6$ with 1 and 19 d.f. With $\alpha = 0.05$, this is a highly significant value.

13.12 The t Test for a Randomized Block Design with Two Treatments

In a randomized block design with two treatments, the test of significance of the difference between the two treatment means can also be made in terms of a t test. Let $D = X_{b1} - X_{b2}$ or the difference between the two observations in a given block. These differences are given in the last column of Table 13.12. Then the sum of squared deviations for the differences will be given by

$$\sum (D - \bar{D})^2 = \sum D^2 - \frac{(\sum D)^2}{b} \tag{13.7}$$

where b is the number of differences or blocks. For the values of D given in the table, we have

$$\sum (D - \bar{D})^2 = (0)^2 + (2)^2 + \cdots + (3)^2 - \frac{(20)^2}{20} = 26.0$$

The standard error of the difference between the two treatment means will then be given by

$$s_{\bar{D}} = \sqrt{\frac{\sum (D - \bar{D})^2}{b(b-1)}} \tag{13.8}$$

Substituting in (13.8), we have

$$s_{\bar{D}} = \sqrt{\frac{26.0}{20(20-1)}} = 0.262$$

For $\bar{X}_{.1}$ we have $180/20 = 9.0$ and for $\bar{X}_{.2}$ we have $160/20 = 8.0$. Testing the null hypothesis $\mu_1 = \mu_2$, we have

$$t = \frac{\bar{X}_{.1} - \bar{X}_{.2}}{s_{\bar{D}}} = \frac{9.0 - 8.0}{0.262} = 3.82$$

with $b - 1 = 19$ d.f., where b is the number of differences or blocks. We note that $t^2 = (3.82)^2 = 14.6$ and t^2 is identical with the value of $F = 14.6$ for the randomized block design with $t = 2$ treatments.

13.13 Grouping Subjects into Blocks

In some cases it will be possible to obtain an initial measure, prior to the experiment proper, on the subjects to be used in the experiment with the same instrument that is to be used to measure the outcome of the experiment. These initial measures may then be used to arrange the subjects into blocks.[4] In other cases, where it is not practical to

[4] If the variable used in establishing the blocks is approximately normally distributed, then difficulties may be encountered in trying to form blocks for the two tails of the distribution, because the frequency of the extreme measures may not be sufficiently great to permit a homogeneous grouping. On the other hand, if we restrict the blocks to those values that are centrally located and for which the frequencies are the largest, then the treatments will be distributed over a less representative sampling of subjects. If we obtain measurements on many more subjects than we intend to use in the experiment, then the problem of forming homogeneous and representative blocks will be considerably simplified.

obtain an initial measure on the same instrument, it may still be possible to group subjects into blocks on the basis of some other available variable that we have reason to believe will tend to result in blocks of subjects that will be relatively homogeneous with respect to the dependent variable of interest.

The variable used in forming the blocks will depend, of course, on the nature of the measurement to be made in the experiment. For example, if we were studying the influence of various diets on gain in weight, we might form blocks on the basis of initial weights. In other cases, the blocks may be formed on the basis of educational level, test performance, intelligence, age, and so forth. In rare cases, a block might consist of a pair of identical twins. In still other cases, a block might consist of rats from the same litter or individuals of the same sex.

The success of the randomized block design depends on the degree to which subjects placed in the same block will, in fact, be relatively homogeneous in their performance on the dependent variable in the absence of treatment effects. Consequently, we should not form blocks on the basis of variables that are irrelevant or unrelated to the measurements to be made in the experiment itself.

13.14 Interactions of Organismic Variables with Treatment Factors

In a randomized block design, organismic variables are often used to group subjects into blocks with the objective of eliminating from the estimate of experimental error some of the systematic differences in performance associated with the different levels of the organismic variable. With a single observation for each treatment in each block, as in the standard randomized block design, there is no way to test the significance of the interaction between the organismic (block) variable and the treatments because the block × treatment interaction itself provides the estimate of experimental error. If the interaction of the organismic variable and the treatments is of experimental interest, that is, if we wish to test the interaction for significance, then we must have more than one observation for each treatment in each block.

Suppose, for example, we are interested in performance under a treatment factor T consisting of $t = 4$ different types of incentives and that 120 subjects are available for the experiment. If we were only interested in differences in the treatment means, we could use a randomized group design and randomly assign $n = 30$ subjects to each of the four treatments. Because the subjects are randomly assigned to the treatment groups, we would expect only random or chance differences in the means of the groups on a test of intelligence or, for that matter, any other organismic variable.

If, however, we are interested in the possible interaction of level of intelligence and treatments, we could group the subjects into blocks of varying levels of intelligence. For example, if intelligence test scores are available for the 120 subjects, they could be rank ordered on this variable and then divided into three blocks of forty subjects each to form $b = 3$ levels of the factor B of intelligence: high, average, and low. Within each block or level of intelligence, we could then assign $n = 10$ subjects at random to each of the four treatments. Note that for each block or level of intelligence, the experimental design is a randomized group design. The blocks themselves simply represent additional replications of the same experiment with varying levels of intelligence. The analysis of variance for this experiment would be the same as that described previously, where we had n subjects for each block-treatment combination, and would result in the following sums of squares with degrees of freedom shown at the right:

Source	d.f.
T: Treatments	3
B: Levels of intelligence	2
$B \times T$	6
Within cell	108
Total	119

The expectations of the mean squares for a random-effect, a fixed-effect, and a mixed model with blocks random and treatments fixed for this experiment would be the same as those given earlier in Table 13.5. For all three models, $F = MS_{BT}/MS_{WC}$ provides a test of significance of the block $\times$ treatment interaction. For any one of the three models, if MS_T is significant and MS_{BT} is not, we have evidence that the differences among the treatments are much the same for the three different levels of intelligence. However, if MS_{BT} is significant, this would indicate that conclusions about the differences in the treatment means need to be qualified by considering the particular level of intelligence involved.

As Table 13.5 shows, a test of significance of the block mean square is possible for all three analysis of variance models. If MS_B is significant, this would indicate that the mean overall performance under the treatment factor is not the same for the three levels of intelligence. When the levels of a factor represent different levels of an organismic variable, as in the present case with respect to the B factor, it is of some importance to recognize that subjects cannot be randomly assigned to the levels of the factor. The intelligence level of a subject, for example, is a property of the subject—it is not something that can be randomly assigned to him.

When subjects can be randomly assigned to the levels of a factor and if the mean square for the factor is significant, we have reason to believe that the differences in the treatment means are a result of the differences in the treatments applied and not the result of systematic differences between the subjects themselves in the various treatment groups. The reason for this is that the random assignment of the subjects to the treatments should result only in random and not systematic differences in the organismic variables associated with the subjects in each of the treatment groups.

When the levels of a factor represent systematic differences in an organismic variable, the main effect of the factor is usually of little experimental interest. In the experiment described, for example, the subjects were grouped into blocks of varying levels of intelligence, presumably because there was reason to believe that there would be some systematic differences in the block means. In this experiment, it is the interaction or lack of interaction of the organismic variable with the treatments that is of primary interest. The nature of the interaction, if significant, can be examined by the methods described previously.

13.15 A Design in which Blocks Represent Random Replications of an Experiment

The design described above is one in which the blocks represent systematic differences with respect to an organismic variable. In this instance, subjects could not be randomly assigned to the blocks. Let us consider a modification of the experimental design such that the subjects are randomly assigned to the blocks. For example, suppose the 120 subjects were divided completely at *random* to form $b = 3$ blocks of forty subjects each. In this case there is no reason to believe that the blocks represent anything other than random differences in organismic variables. Within each block $n = 10$ subjects are then randomly assigned to each of the four treatments. Note that, in this instance, the blocks simply represent *random replications of the same experiment.* Because subjects have been randomly assigned to the blocks and then randomly assigned to the same four treatments within each block, there is no reason to expect the block mean square to be significant.[5] For precisely the same reason, if the experimental technique is reliable and under control, then the interaction of the blocks and treatments should not be significant. If MS_{BT} were to be significant, it would indicate that the differences among the treatment means are not the same in *random replications* of the experiment. Obviously, if an

[5] In this design, we assume that both blocks and subjects are random and that treatments are fixed.

experimenter cannot consistently or reliably obtain comparable results when the same experiment is repeated, there is something wrong with the experimental procedure.

An objection sometimes expressed to the above design is that the test of significance of the treatment mean square is made in terms of MS_{BT} which, in the example described, has only $(b - 1)(t - 1) = 6$ d.f. This objection neglects the fact that if the experimental technique is under control, then σ_{bt}^2 should be equal to zero, and consequently, $E(MS_{BT}) = E(MS_{WC}) = \sigma^2$. In this case, it would be appropriate to pool the sum of squares for blocks $\times$ treatments with the within cell sum of squares and to divide the pooled sum of squares by the pooled degrees of freedom. The result would be an error mean square with 114 d.f. and this error mean square would be used to test the significance of the treatment mean square.

If the experimenter had not randomly assigned subjects to the $b = 3$ blocks but instead had simply assigned $n = 30$ subjects at random to each of the four treatments, the within treatment or error mean square would have had $4(30 - 1) = 116$ d.f. But this design would provide no information regarding the reliability of the experimental technique and procedure. The loss of 2 d.f. for blocks in the design described above is a small price to pay for the additional information provided regarding the replicability of the outcome of the experiment. The outcome of a single experiment, no matter how significant, is of no value if it cannot be demonstrated in repetitions of the experiment.

13.16 Another Variation of a Randomized Block Design

We consider still another variation of a randomized block design. Let A represent a treatment factor with a levels and C another treatment factor with c levels. We regard the levels of A and C as fixed effects which do not involve random selections from larger populations. Let B, however, have b levels each consisting of a senior high school selected at random from a larger population of high schools. In each school students are randomly selected from the student population and these students are then randomly assigned so that we have n students for each of the AC treatment combinations. Note that, in this example, the subjects *cannot* be randomly assigned to blocks but they can be randomly assigned to the treatment combinations within each block. The blocks themselves represent a random sample from a population of schools in which there may be systematic differences.

The analysis of variance model that is appropriate for this experiment is the mixed model with A and C fixed and B random. The expectations of the mean squares for the mixed model are given in Table 13.14. The appropriate

Table 13.14

EXPECTATION OF MEAN SQUARES WITH B (BLOCKS) RANDOM AND A AND C FIXED AND WITH n SUBJECTS IN EACH CELL

Source	Expectation of Mean Square
B: Blocks	$\sigma^2 + nac\sigma_b^2$
A	$\sigma^2 + nc\sigma_{ba}^2 + nbc\theta_a^2$
$B \times A$	$\sigma^2 + nc\sigma_{ba}^2$
C	$\sigma^2 + na\sigma_{bc}^2 + nab\theta_c^2$
$B \times C$	$\sigma^2 + na\sigma_{bc}^2$
$A \times C$	$\sigma^2 + n\sigma_{bac}^2 + nb\theta_{ac}^2$
$B \times A \times C$	$\sigma^2 + n\sigma_{bac}^2$
Within cell	σ^2

error mean squares for the various tests of significance can easily be determined from the expected values of the mean squares.

The appropriate error mean square for A is the $B \times A$ interaction and the appropriate error mean square for C is the $B \times C$ interaction. If the A mean square is significant, we would conclude that the means for the levels of A differ significantly. Because the A effect is tested over a random sample of schools, we may assume that a significant A mean square would also be obtained for the population of schools. Similarly, if the C mean square is significant, when tested with the $B \times C$ interaction, we would conclude that the C mean square would also be significant for the population of schools because it has been tested over a random sample of schools. Similar considerations apply to the test of significance of the $A \times C$ mean square, when tested in terms of MS_{BAC}. Thus, by randomly selecting the levels of B (schools), we are in a position where we can generalize about a significant fixed effect not only with respect to the schools involved in the experiment, but also with respect to the larger population of schools from which we have a random sample.

If MS_B is significant, this would indicate that the overall means for the treatments differ in the random sample of schools and consequently we would expect them to differ for the population of schools. Similar considerations apply to the tests of significance of MS_{BA}, MS_{BC}, and MS_{BAC}.

13.17 Using Blocks to Eliminate a Source of Variation

We have discussed the randomized block design as a means of reducing the error mean square of the analysis of variance, and we have considered a block as being formed on the basis of some prior information about the subjects to be used in the experiment. The randomized block design can also be used, however, to control for certain sources of variation that are not necessarily associated with individual differences between subjects. For example, suppose an experiment involves five treatments and that it requires an experimental period of one hour for each treatment. Thus, it may be possible to test only five subjects on a given day. If there is substantial day-to-day variation, this source of variation could be controlled by regarding the five observations obtained on a single day as a block. In the analysis of variance, the block sum of squares would correspond to the day-to-day variation and would be eliminated from the error sum of squares.

Needless to say, the same treatment should not be applied to the five subjects tested on a given day. If this were done, then differences between the treatments would be completely confounded with differences between the days of testing. To avoid this confounding, all five treatments should be administered on each day.

QUESTIONS AND PROBLEMS

13.1 On the basis of an initial measure, subjects were formed into blocks of three subjects each. Treatments were assigned at random within each block. We have rearranged the observations within the blocks in accordance with the treatments. Find the value of $F = MS_T/MS_{BT}$. To investigate the difference between the various treatment means, it would be possible to use the procedures described previously under multiple comparisons.

	Treatments		
Blocks	1	2	3
1	21	20	22
2	20	19	21
3	20	19	22
4	18	19	20
5	18	17	18
6	18	18	19
7	18	16	19
8	16	15	18
9	16	13	15
10	15	14	16

13.2 Assume we have five treatments and a randomized block design with six blocks. Treatments were assigned at random within each block. We have rearranged the observations within the blocks in accordance with the treatments. Find the value of $F = MS_T/MS_{BT}$.

	Treatments					
Blocks	1	2	3	4	5	Σ
1	25	27	24	28	22	126
2	24	32	29	26	24	135
3	31	35	27	36	26	155
4	40	45	33	42	30	190
5	43	50	38	46	33	210
6	45	48	40	52	36	221
Σ	208	237	191	230	171	1037

13.3 Assume that twenty subjects were pretested and then arranged into blocks with two subjects in each block. Treatments were assigned at random within blocks. We have rearranged the observations within the blocks in accordance with the treatments. (*a*) Find the value of $F = MS_T/MS_{BT}$. (*b*) Analyze the data using the t test of this chapter. You should find that $t^2 = F$.

	Treatments	
Blocks	1	2
1	2.5	3.6
2	4.6	5.7
3	9.3	8.9
4	4.5	6.7
5	1.5	1.9
6	6.4	7.8
7	4.7	4.6
8	5.6	5.9
9	7.3	6.9
10	6.6	7.0

13.4 We have a randomized block design with three treatments and thirty blocks. Treatments were assigned at random within each block. We have rearranged the observations within the blocks in accordance with the treatments. Find the value of $F = MS_T/MS_{BT}$.

Blocks	Treatments 1	2	3
1	18	19	15
2	15	14	13
3	16	18	17
4	18	14	15
5	19	13	13
6	20	16	11
7	19	17	15
8	17	16	14
9	18	17	13
10	20	16	14
11	16	16	19
12	11	15	18
13	11	19	19
14	16	17	21
15	15	15	18
16	14	16	21
17	13	16	20
18	16	15	17
19	14	14	18
20	14	17	19
21	17	17	15
22	13	15	19
23	16	18	18
24	12	15	17
25	13	15	17
26	15	18	18
27	13	16	16
28	13	16	18
29	14	16	15
30	14	14	17

13.5 Discuss the difference between a randomized group design and a randomized block design.

13.6 With three blocks and four treatments, the outcome of a randomized block design is as given below:

Blocks	Treatments 1	2	3	4
1	1	2	3	3
2	2	3	4	2
3	3	4	5	1

13.7 We have a factorial experiment with $a = 3$ levels of A and $c = 2$ levels of C. We use a randomized block design with $b = 5$ blocks. (*a*) Find the sum of squares for the block $\times$ treatment combinations, and the sums of squares for A, C, and $A \times C$. (*b*) Calculate, *independently*, the $B \times A$, $B \times C$, and $B \times A \times C$ sums of squares and show that the sum of these sums of squares is equal to the block $\times$ treatment combinations sum of squares. The outcome of the experiment is given below:

	A_1		A_2		A_3	
Blocks	C_1	C_2	C_1	C_2	C_1	C_2
1	3	4	7	7	6	7
2	2	2	3	6	5	8
3	1	3	4	5	7	10
4	4	5	5	4	8	9
5	5	6	6	3	9	11
Σ	15	20	25	25	35	45

13.8 In the experiment described above, assume that we make the following two orthogonal comparisons on the levels of A:

	A_1	A_2	A_3
D_1:	2	-1	-1
D_2:	0	1	-1

(*a*) Find MS_{D_1} and MS_{D_2}. (*b*) Now show that the $A \times C$ sum of squares is equal to the sum of squares for $D_1 \times C$ and the sum of squares for $D_2 \times C$. (*c*) Show that the $B \times A$ sum of squares is equal to the sum of squares for $B \times D_1$ and the sum of squares for $B \times D_2$. (*d*) Show that the $B \times A \times C$ sum of squares is equal to the sum of squares for $B \times D_1 \times C$ and the sum of squares for $B \times D_2 \times C$.

13.9 One hundred and twenty students were divided into $b = 2$ blocks of sixty students each on the basis of scores on an intelligence test. Block 1 consisted of those students with scores above the median and Block 2 of those students with scores below the median. Within each block $n = 10$ students were assigned at random to each of $t = 6$ treatments. (*a*) Give the analysis of variance summary table showing sources of variation and degrees of freedom associated with each source. (*b*) Write out the expectations of the mean squares for this experiment.

13.10 We have two factors, A and C, each with two levels. Twenty subjects are arranged into five blocks with four subjects in each block. Within blocks, subjects are assigned at random to one of the treatment combinations. (*a*) Test the A, C, and $A \times C$ mean squares for significance using the block $\times$ treatment combination mean square as the error term. (*b*) Find the values of MS_{BA}, MS_{BC}, and MS_{BAC} and use these separate estimates of error to test the A, C, and $A \times C$ mean squares for significance. The outcome of the experiment is given below:

	Treatment Combinations			
Blocks	A_1C_1	A_1C_2	A_2C_1	A_2C_2
1	2	4	5	7
2	3	5	6	8
3	5	6	8	10
4	7	9	11	14
5	6	7	10	11

CHAPTER 14

RANDOMIZED BLOCK DESIGNS: REPEATED MEASURES

14.1 Introduction

In randomized block designs, as used in the behavioral sciences, it is often the case that each block consists of a *single* subject and that each subject is administered all of the treatments involved in the experiment. The order in which the treatments are administered is *independently randomized* for each subject. With $t = 5$ treatments and $s = 4$ subjects, we would need four random permutations of the numbers 1, 2, 3, 4, and 5 to determine the order in which the treatments are to be administered to each of the four subjects. For example, we might obtain the following random permutations: 4, 3, 5, 1, 2; 1, 4, 2, 3, 5; 2, 1, 5, 3, 4; and 5, 4, 3, 1, 2. Then one subject would be assigned to each permutation and administered the treatments in the order specified by that permutation. After the experimental observations are obtained for each subject, they may be rearranged in a table with the columns corresponding to the specific treatments. The analysis of variance would be the same as that for a randomized block design with columns corresponding to treatments and blocks or rows to subjects. We assume that there are no carryover effects from one treatment to the next, so that the differences in the column means reflect treatment effects.[1]

[1] It is important to recognize that if there are carryover effects from one treatment to the next, then we have no way of knowing whether a significant treatment mean square is the result of differences in the treatments or carryover effects. In the randomized block designs discussed in the previous chapter, each subject was tested under a single treatment and there was no basis for believing that carryover effects would be present. If a single subject is administered all of the treatments, then there may be carryover effects. Randomizing the sequence of the treatments independently for each subject is one attempt to minimize systematic carryover effects.

14.2 Sums of Squares

In the analysis of variance of a randomized block design, we have shown that the total sum of squares is equal to the treatment, block, and block $\times$ treatment sums of squares. Similarly, when a single subject represents a block and each subject is administered each of the treatments, then the total sum of squares will be equal to the treatment, subject, and subject $\times$ treatment sums of squares, or

$$\text{Total} = \text{Treatment} + \text{subject} + \text{subject} \times \text{treatment} \qquad (14.1)$$

If we subtract the treatment sum of squares from both sides of the above expression, we have

$$\text{Total} - \text{treatment} = \text{Subject} + \text{subject} \times \text{treatment}$$

and the left side of this expression is what we have called the "within treatment" sum of squares. It differs from the within treatment sum of squares in a randomized group design in that the same group of s subjects is tested under each of the t treatments. For convenience, we shall continue to refer to this sum of squares as the within treatment sum of squares.

We now note that

$$\text{Within treatment} - \text{subject} = \text{Subject} \times \text{treatment} \qquad (14.2)$$

and, in designs of the kind described, the subject $\times$ treatment ($S \times T$) sum of squares is the error sum of squares. It is obvious that the subject $\times$ treatment sum of squares can never be larger than the within treatment sum of squares.

Because each subject is tested under all of the treatments, differences in the block or subject means should not represent treatment effects. However, if there are organismic variables (individual differences) associated with the subjects such that some subjects tend to have consistently higher values on the dependent variable under all treatments and others consistently lower values, these individual differences in the subjects will be reflected in differences in the block or subject means. In the analysis of variance, the variation attributable to differences in the subject means is removed from the estimate of experimental error.

Alternatively, we note that if we subtract the subject sum of squares from the total sum of squares, we have

$$\text{Total} - \text{subject} = \text{Treatment} + \text{subject} \times \text{treatment} \qquad (14.3)$$

and the left side of the above expression is often referred to as the "within subject" sum of squares. This sum of squares reflects variation due to differences in treatments and also the random variation to be expected when the same subject is tested repeatedly. If we subtract the treatment sum of squares from the within subject sum of squares, then we have

$$\text{Within subject} - \text{treatment} = \text{Subject} \times \text{treatment} \qquad (14.4)$$

and the right side of this expression is the same as the right side of (14.2). If the variation attributable to differences in treatments is removed from the within subject sum of squares, it is assumed that the residual or subject × treatment sum of squares will reflect only the random variation to be expected when the same subject is tested repeatedly in the absence of treatment effects. It is anticipated that the error variance based on repeated measures of the same subject, after treatment effects are removed, will be less than the random variance obtained when different subjects are administered the same treatment.

14.3 A Model for Correlated Errors

As a model for a randomized block design in which each subject corresponds to a block, we assume that

$$X_{st} = \mu + t_t + t_s + e_{st} \qquad (14.5)$$

where μ is a common population mean associated with each observation, $t_t = \mu_t - \mu$ represents a treatment effect, $t_s = \mu_s - \mu$ represents a subject effect, and e_{st} is a random error associated with each observation.

We shall derive the expected values of the mean squares using a fixed-effect model, so that $E(t_t) = t_t$ and $E(t_t^2) = t_t^2$, $E(t_s) = t_s$, and $E(t_s^2) = t_s^2$, subject to the restriction that

$$\sum_1^t t_t = \sum_1^s t_s = 0$$

If each subject is administered all of the treatments in an experiment, then the errors associated with the repeated observations of the same subjects may not be independent. Consequently, we shall assume that the errors associated with observations from the same subject are possibly correlated but that errors associated with observations from different subjects are independent. More specifically, we assume that

$$E(e_{st}e_{st}) = \rho\sigma^2 \qquad \text{if } s = s \text{ and } t \neq t$$

$$E(e_{st}e_{st}) = 0 \qquad \text{if } s \neq s$$

$$E(e_{st}e_{st}) = \sigma^2 \qquad \text{if } s = s \text{ and } t = t$$

14.4 Expected Value of the Correction Term

For the expected value of the correction term in the analysis of variance, we have

$$E\left[\frac{(\sum X_{..})^2}{st}\right] = \frac{1}{st} E\left(st\mu + s\sum_1^t t_t + t\sum_1^s t_s + \sum_1^{st} e_{st}\right)^2 \quad (14.6)$$

We note that $(\sum_1^{st} e_{st})^2$ will result in s^2t^2 values of $e_{st}e_{st}$. For st of these values we have $e_{st}e_{st} = e_{st}^2$ and the expected value of each of these is σ^2. For each subject, we have $t(t-1)$ values in which the subscripts s are the same and the t subscripts are different. Summing over all s subjects, we have $st(t-1)$ values for which $E(e_{st}e_{st}) = \rho\sigma^2$. For all other values of $e_{st}e_{st}$ the s subscripts will be different and the expected value of each of these terms is zero. Then

$$E\left[\frac{(\sum X_{..})^2}{st}\right] = \frac{1}{st}\left[s^2t^2\mu^2 + st\sigma^2 + st(t-1)\rho\sigma^2\right]$$

$$= st\mu^2 + \sigma^2 + (t-1)\rho\sigma^2 \quad (14.7)$$

14.5 Expected Value of the Treatment Mean Square

To find the expected value of the treatment sum of squares, we first find

$$E\left[\sum_1^t \frac{(\sum X_{.t})^2}{s}\right] = \frac{1}{s} E\left[\sum_1^t \left(s\mu + st_t + \sum_1^s t_s + \sum_1^s e_{st}\right)^2\right]$$

$$= \frac{1}{s}\left(ts^2\mu^2 + s^2\sum_1^t t_t^2 + ts\sigma^2\right)$$

$$= st\mu^2 + s\sum_1^t t_t^2 + t\sigma^2 \quad (14.8)$$

Subtracting the correction term from (14.8), we have the expected value of the treatment sum of squares, or

$$E(SS_T) = s\sum_1^t t_t^2 + t\sigma^2 - \sigma^2 - (t-1)\rho\sigma^2$$

$$= s\sum_1^t t_t^2 + (t-1)\sigma^2 - (t-1)\rho\sigma^2 \quad (14.9)$$

and for the expected value of the treatment mean square, we have

$$E(MS_T) = \sigma^2(1 - \rho) + s\frac{\sum_1^t t_t^2}{t - 1} \tag{14.10}$$

14.6 Expected Value of the Subject Mean Square

To obtain the expected value of the subject sum of squares, we first find

$$E\left[\sum_1^s \frac{(\sum X_{s\cdot})^2}{t}\right] = \frac{1}{t}E\left[\sum_1^s \left(t\mu + \sum_1^t t_t + tt_s + \sum_1^t e_{st}\right)^2\right]$$

$$= \frac{1}{t}\left[st^2\mu^2 + t^2\sum_1^s t_s^2 + st\sigma^2 + st(t-1)\rho\sigma^2\right]$$

$$= st\mu^2 + t\sum_1^s t_s^2 + s\sigma^2 + s(t-1)\rho\sigma^2 \tag{14.11}$$

Subtracting the correction term from (14.11), we have the expected value of the subject sum of squares, or

$$E(SS_S) = t\sum_1^s t_s^2 + s\sigma^2 + s(t-1)\rho\sigma^2 - \sigma^2 - (t-1)\rho\sigma^2$$

$$= t\sum_1^s t_s^2 + (s-1)\sigma^2 + (s-1)(t-1)\rho\sigma^2 \tag{14.12}$$

Then the expected value of the subject mean square will be

$$E(MS_S) = \sigma^2 + (t-1)\rho\sigma^2 + t\frac{\sum_1^s t_s^2}{s-1}$$

$$= \sigma^2[1 + (t-1)\rho] + t\frac{\sum_1^s t_s^2}{s-1} \tag{14.13}$$

14.7 Expected Value of the Within Treatment Mean Square

We could now find the expected value of the total sum of squares and then subtract the expected values of the subject and treatment sums of squares to obtain the expected value of the subject ×

treatment sum of squares. As an alternative, we find the expected value of the within treatment sum of squares and then subtract the expected value of the subject sum of squares to obtain the expected value of the subject × treatment sum of squares.

For any given observation we have

$$X_{st}^2 = \mu^2 + t_t^2 + t_s^2 + e_{st}^2 + 2\mu(t_t + t_s)$$

and for any given column or treatment

$$E\left(\sum_1^s X_{st}^2\right) = s\mu^2 + st_t^2 + \sum_1^s t_s^2 + s\sigma^2 + 2s\mu t_t \qquad (14.14)$$

The expected value of the correction term for the same treatment will be

$$E\left[\frac{(\sum X_{.t})^2}{s}\right] = \frac{1}{s} E\left(s\mu + st_t + \sum_1^s t_s + \sum_1^s e_{st}\right)^2$$

$$= \frac{1}{s}(s^2\mu^2 + s^2t_t^2 + s\sigma^2 + 2s^2\mu t_t)$$

$$= s\mu^2 + st_t^2 + \sigma^2 + 2s\mu t_t \qquad (14.15)$$

Subtracting (14.15) from (14.14), we have the expected value of the sum of squares within a given treatment. Thus,

$$E\left[\sum_1^s X_{st}^2 - \frac{(\sum X_{.t})^2}{s}\right] = \sum_1^s t_s^2 + (s-1)\sigma^2 \qquad (14.16)$$

The expected value of the sum of squares within each of the other treatments will be exactly the same as that given by (14.16) and, consequently, the expected value of the pooled within treatment sum of squares will be

$$E(SS_W) = t(s-1)\sigma^2 + t\sum_1^s t_s^2 \qquad (14.17)$$

and

$$E(MS_W) = \sigma^2 + t\frac{\sum_1^s t_s}{s-1} \qquad (14.18)$$

Note that the expected value of the within treatment sum of squares includes a component attributable to differences in the subject or block means.

14.8 Expected Value of the Subject × Treatment Mean Square

To obtain the expected value of the subject × treatment sum of squares we subtract the expected value of the subject sum of squares from the expected value of the within treatment sum of squares. Thus,

$$E(SS_{ST}) = E(SS_W) - E(SS_S) \tag{14.19}$$

and subtracting (14.12) from (14.17), we have

$$\begin{aligned} E(SS_{ST}) &= t(s-1)\sigma^2 - (s-1)\sigma^2 - (s-1)(t-1)\rho\sigma^2 \\ &= (s-1)(t-1)\sigma^2 - (s-1)(t-1)\rho\sigma^2 \end{aligned} \tag{14.20}$$

and dividing by $(s-1)(t-1)$, the degrees of freedom associated with the subject × treatment sum of squares, we have

$$E(MS_{ST}) = \sigma^2(1-\rho) \tag{14.21}$$

14.9 Tests of Significance

The expected values of the mean squares for the fixed-effect model are summarized in Table 14.1, where we again let θ^2 represent a fixed effect. We emphasize that for the fixed-effect model we

Table 14.1

EXPECTATION OF MEAN SQUARES FOR A RANDOMIZED BLOCK DESIGN WITH REPEATED MEASURES ON SUBJECTS

Model I: Subjects and Treatments Fixed

Source	Expectation of Mean Square
Subjects	$\sigma^2[1 + (t-1)\rho] + t\theta_s^2$
Treatments	$\sigma^2(1-\rho) + s\theta_t^2$
$S \times T$	$\sigma^2(1-\rho)$

Mixed Model: Subjects Random and Treatments Fixed

Source	Expectation of Mean Square
Subjects	$\sigma^2[1 + (t-1)\rho] + t\sigma_s^2$
Treatments	$\sigma^2(1-\rho) + \sigma_{st}^2 + s\theta_t^2$
$S \times T$	$\sigma^2(1-\rho) + \sigma_{st}^2$

have assumed that θ_{st}^2 is zero. The expected values of the mean squares for a mixed model, with subjects random and treatments fixed, are also shown in Table 14.1. For the mixed model we have *not* assumed that σ_{st}^2 is zero.

With a fixed-effect model, we have $F = MS_T/MS_{ST}$ and

$$\frac{E(MS_T)}{E(MS_{ST})} = \frac{\sigma^2(1 - \rho) + s\theta_t^2}{\sigma^2(1 - \rho)}$$

If the errors associated with the repeated observations of a subject are independent, that is, if $\rho = 0$, then we have

$$\frac{E(MS_T)}{E(MS_{ST})} = \frac{\sigma^2 + s\theta_t^2}{\sigma^2}$$

It is now obvious that if the null hypothesis is false, so that $\theta_t^2 \neq 0$, then for any positive value of ρ we have

$$\frac{\sigma^2(1 - \rho) + s\theta_t^2}{\sigma^2(1 - \rho)} > \frac{\sigma^2 + s\theta_t^2}{\sigma^2}$$

Similarly, for the mixed model with any positive value of ρ, we have

$$\frac{\sigma^2(1 - \rho) + \sigma_{st}^2 + s\theta_t^2}{\sigma^2(1 - \rho) + \sigma_{st}^2} > \frac{\sigma^2 + \sigma_{st}^2 + s\theta_t^2}{\sigma^2 + \sigma_{st}^2}$$

and it is obvious that the efficiency of the repeated measure design depends

Table 14.2

OUTCOME OF A RANDOMIZED BLOCK DESIGN WITH $s = 5$ SUBJECTS AND $t = 3$ TREATMENTS[a]

	Treatments			
Subjects	1	2	3	Σ
1	3	7	3	13
2	1	6	4	11
3	2	3	2	7
4	4	5	5	14
5	5	4	6	15
Σ	15	25	20	60

[a] Each subject was tested under each treatment with the order or sequence of the treatments being independently randomized for each subject. The data have been rearranged so that treatments correspond to columns.

Table 14.3
THE VALUES OF $X_{st} - \bar{X}_{\cdot t}$ FOR THE DATA OF TABLE 14.2

	Treatments			
Subjects	1	2	3	Σ
1	0	2	−1	1
2	−2	1	0	−1
3	−1	−2	−2	−5
4	1	0	1	2
5	2	−1	2	3
Σ	0	0	0	0

on the value of ρ. If ρ is positive, then $E(MS_{ST}) < E(MS_W)$; if ρ is zero, then $E(MS_{ST}) = E(MS_W)$; if ρ is negative, then $E(MS_{ST}) > E(MS_W)$.

For any given set of experimental observations, the efficiency of the repeated measure design depends on $\bar{r}$, the average value of the $t(t-1)$ correlation coefficients, provided that the treatment variances are equal. With equal treatment variances the sample estimate of $E(MS_{ST}) = \sigma^2(1-\rho)$ will be

$$MS_{ST} = MS_W(1 - \bar{r}) \tag{14.22}$$

as we shall see in the next section.

14.10 An Example of a Repeated Measure Design with Equal Variances and Unequal Correlations

In Table 14.2 we show the outcome of an experiment in which we have $t = 3$ treatments and $s = 5$ subjects with each subject tested under each treatment. For the subject sum of squares, we have

$$\text{Subject} = \frac{(13)^2}{3} + \frac{(11)^2}{3} + \cdots + \frac{(15)^2}{3} - \frac{(60)^2}{15} = 13.33$$

The values of the treatment means are: $\bar{X}_{\cdot 1} = 3.0$, $\bar{X}_{\cdot 2} = 5.0$, and $\bar{X}_{\cdot 3} = 4.0$. In Table 14.3 we give the values of $X_{st} - \bar{X}_{\cdot t}$, that is, the deviations of the observations for each treatment from the mean of the treatment. These deviations are free from differences in the treatment means. It is easy to see that within each treatment we have

$$\sum_{1}^{s} (X_{st} - \bar{X}_{\cdot t})^2 = 10.0$$

and, consequently, $s_1^2 = s_2^2 = s_3^2 = 2.5$. Because $\sum_1^s (X_{st} - \bar{X}_{\cdot t})^2 = 10.0$ for each treatment group, we also have

$$SS_W = (3)(10.0) = 30.0$$

and

$$MS_W = \frac{30.0}{3(5-1)} = 2.5$$

The within treatment sum of squares includes the subject sum of squares which we have already found to be equal to 13.33. As a check on our previous calculation of the subject sum of squares, we also have, for the data of Table 14.3,

$$SS_S = \frac{(1)^2}{3} + \frac{(-1)^2}{3} + \frac{(-5)^2}{3} + \frac{(2)^2}{3} + \frac{(3)^2}{3} = \frac{50}{3} = 13.33$$

Then, for the subject $\times$ treatment sum of squares, we have

$$\begin{aligned} SS_{ST} &= SS_W - SS_S \\ &= 30.00 - 13.33 \\ &= 16.67 \end{aligned}$$

and

$$MS_{ST} = \frac{16.67}{(5-1)(3-1)} = 2.08$$

With equal variances within each treatment, we have $MS_{ST} = MS_W(1 - \bar{r})$ and, consequently,

$$\bar{r} = 1 - \frac{MS_{ST}}{MS_W} \tag{14.23}$$

Substituting in the above equation with the values of $MS_{ST} = 2.08$ and $MS_W = 2.50$, we have

$$\bar{r} = 1 - \frac{2.08}{2.50} = 0.17$$

rounded. Note that if $\bar{r} = 0$, then we must also have $MS_{ST} = MS_W$. If $\bar{r}$ is positive, then MS_{ST} must be less than MS_W; if $\bar{r}$ is negative, then MS_{ST} must be greater than MS_W.

In this example, we have $s_1^2 = s_2^2 = s_3^2 = 2.5$, and the treatment variances are equal. The situation is quite different, however, with respect to the values of the correlation coefficients. To determine the values of

r_{12}, r_{13}, and r_{23}, we note that, in general,

$$r_{ij} = \frac{\sum x_i x_j}{\sqrt{(\sum x_i^2)(\sum x_j^2)}} \tag{14.24}$$

where $\Sigma x_i x_j = \sum_1^s (X_{si} - \bar{X}_{.i})(X_{sj} - \bar{X}_{.j})$ is the sum of the products of the paired observations for Treatment i and Treatment j, $\Sigma x_i^2 = \sum_1^s (X_{si} - \bar{X}_{.i})^2$, and $\Sigma x_j^2 = \sum_1^s (X_{sj} - \bar{X}_{.j})^2$. For the data of Table 14.3, we have

$$\begin{aligned}\sum x_1 x_2 &= (0)(2) + (-2)(1) + (-1)(-2) + (1)(0) + (2)(-1)\\ &= -2.0\end{aligned}$$

We also have $\Sigma x_1^2 = 10.0$ and $\Sigma x_2^2 = 10.0$. Then

$$r_{12} = \frac{-2.0}{\sqrt{(10)(10)}} = -0.20$$

In the same manner we find that $r_{13} = 0.70$ and $r_{23} = 0.00$. Then the average value of the correlation coefficients will be

$$\bar{r} = \frac{-0.20 + 0.70 + 0.00}{3} = \frac{0.50}{3} = 0.17$$

rounded, and this is the same value we obtained previously from (14.23).

14.11 An Example of a Repeated Measure Design with Unequal Variances and Unequal Correlations

In Table 14.4 we have an example in which $s_1^2 \neq s_2^2 \neq s_3^2$. The values in this table were obtained by multiplying the values of $X_{st} - \bar{X}_{.t}$ for Treatment 1, Treatment 2, and Treatment 3 in Table 14.3 by the constants 2, 3, and 4, respectively. These linear transformations of $X_{st} - \bar{X}_{.t}$ will not change the values of $r_{12} = -0.20$, $r_{13} = 0.70$, and $r_{23} = 0.00$, but they will change the values of the treatment variances.[2] For the data of Table 14.4 we have $s_1^2 = 10.0$, $s_2^2 = 22.5$, and $s_3^2 = 40.0$.

The within treatment sum of squares for the data of Table 14.4 will be

$$SS_W = 40.0 + 90.0 + 160.0 = 290.0$$

[2] If $Y = aX$, where a is a constant and X is a variable, then $s_Y^2 = a^2 s_X^2$.

and

$$MS_W = \frac{290.0}{3(5-1)} = 24.17$$

rounded. For the subject sum of squares, we have

$$SS_S = \frac{(2)^2}{3} + \frac{(-1)^2}{3} + \frac{(-16)^2}{3} + \frac{(6)^2}{3} + \frac{(9)^2}{3} = \frac{378}{3} = 126.0$$

Then for the subject $\times$ treatment sum of squares, we obtain

$$SS_{ST} = 290.0 - 126.0 = 164.0$$

and

$$MS_{ST} = \frac{164.0}{(5-1)(3-1)} = 20.50$$

For the data of Table 14.4, we not only have unequal correlations but also unequal variances. If the treatment variances are not equal, then MS_{ST} will not be equal to $MS_W(1 - \bar{r})$, but instead

$$MS_{ST} = MS_W - \overline{r_{ij}s_is_j} \qquad (14.25)$$

where $\overline{r_{ij}s_is_j}$ is the average value of the $t(t-1)$ covariances.[3] Substituting in the above equation with the values of $MS_{ST} = 20.50$ and $MS_W = 24.17$ and solving for $\overline{r_{ij}s_is_j}$, we have

$$\overline{r_{ij}s_is_j} = 24.17 - 20.50 = 3.67$$

Table 14.4

THE VALUES OF $2(X_{s1} - \bar{X}_{.1})$, $3(X_{s2} - \bar{X}_{.2})$, AND $4(X_{s3} - \bar{X}_{.3})$ BASED ON THE VALUES OF $X_{st} - \bar{X}_{.t}$ IN TABLE 14.3

	Treatments			
Subjects	1	2	3	Σ
1	0	6	−4	2
2	−4	3	0	−1
3	−2	−6	−8	−16
4	2	0	4	6
5	4	−3	8	9
Σ	0	0	0	0

[3] With equal treatment variances, we have $s_is_j = s^2$ for all possible pairs of treatments and s^2 will be equal to MS_W. Consequently, with equal variances, $MS_{ST} = MS_W - \bar{r}MS_W = MS_W(1 - \bar{r})$.

As a check on the value of $\overline{r_{ij}s_is_j}$ obtained above by subtraction, we calculate

$$r_{12}s_1s_2 = -0.20\sqrt{10.0}\sqrt{22.5} = -3.0$$
$$r_{13}s_1s_3 = 0.70\sqrt{10.0}\sqrt{40.0} = 14.0$$
$$r_{23}s_2s_3 = 0.00\sqrt{22.5}\sqrt{40.0} = 0.0$$

and

$$\overline{r_{ij}s_is_j} = \frac{-3.0 + 14.0 + 0.0}{3} = \frac{11}{3} = 3.67$$

as before.

14.12 Violations of Assumptions of the Repeated Measure Design and Conservative F Tests

In deriving the expected values of MS_T and MS_{ST} we assumed not only homogeneity of variance within treatments but also equal correlations. For example, with $t = 3$ treatments and s subjects, we would have the following variance-covariance matrix:

$$\begin{matrix} s_1^2 & r_{12}s_1s_2 & r_{13}s_1s_3 \\ r_{21}s_2s_1 & s_2^2 & r_{23}s_2s_3 \\ r_{31}s_3s_1 & r_{32}s_3s_2 & s_3^2 \end{matrix}$$

and it is assumed that $E(s^2) = \sigma^2$ for all values of s^2 and $E(r) = \rho$ for all values of r. With equal variances and equal correlations, the expected value of each of the covariance terms would then be $E(r_{ij}s_is_j) = \rho\sigma^2$.

If the assumptions of homogeneity of variance and equal correlations are violated in a repeated measure design, then $F = MS_T/MS_{ST}$ is not distributed as F with $t - 1$ and $(s - 1)(t - 1)$ d.f., but instead is distributed as F with a reduced number of degrees of freedom. The actual reduction in the degrees of freedom depends on the nature of the variance-covariance matrix but, as Geisser and Greenhouse (1958) have been able to show, the *maximum* reduction in the degrees of freedom is obtained when the usual degrees of freedom of the numerator and denominator of the F ratio are multiplied by the reciprocal of the degrees of freedom for the repeated measure variable, or

$$\epsilon = \frac{1}{t - 1} \qquad (14.26)$$

For example, with $t = 3$ treatments and $s = 5$ subjects we would, assuming equal variances and correlations, evaluate the F test of the treatment mean square in terms of the *tabled* value of F with $t - 1 = 2$ and $(s - 1)(t - 1) = 8$ d.f. But, if the assumption of equal variances and correlations is violated, then we might evaluate the obtained value of F in terms of the *tabled* value for

$$\frac{1}{t-1}(t - 1) = 1$$

and

$$\frac{1}{t-1}(s - 1)(t - 1) = s - 1$$

degrees of freedom or, in our example, the tabled value of F with 1 and 4 d.f. This will be a conservative test in the sense that we have made a *maximum* reduction in the number of degrees of freedom.

14.13 An Example of a Factorial Experiment with Repeated Measures on All Treatment Combinations

In an experiment by Schroeder (1945) various factors that influence archery performance were investigated. Eleven women shot at targets at ranges of 30, 40, and 50 yards. Each subject shot at each range on each of six days with a different order of shooting on each day. The six possible orders are as follows:

30	40	50
30	50	40
50	30	40
50	40	30
40	50	30
40	30	50

Nine scores were obtained for each subject, each score corresponding to the sum for a given range in a given position. Thus, a given subject would have a score for the 30-yard range in the first position, in the second position, and in the third position, and similarly for the 40-yard and the 50-yard ranges. We have nine scores for each of eleven subjects and therefore the total number of observations is ninety-nine.

We let $s = 11$ correspond to the number of archers or subjects, $r = 3$ correspond to the number of ranges, and $p = 3$ correspond to the number of positions. It is clear that this experiment is a factorial experiment with $pr = (3)(3) = 9$ treatment combinations in a repeated measure design with $s = 11$ subjects corresponding to blocks. Assuming that subjects are

Table 14.5

EXPECTATION OF MEAN SQUARES FOR A FACTORIAL EXPERIMENT IN WHICH EACH SUBJECT IS TESTED UNDER ALL TREATMENT COMBINATIONS. RANGES AND POSITIONS ARE ASSUMED TO BE FIXED AND SUBJECTS RANDOM

Source	Expectation of Mean Square
Subjects	$\sigma^2[1 + (rp - 1)\rho] + rp\sigma_s^2$
Ranges	$\sigma^2(1 - \rho) + p\sigma_{sr}^2 + sp\theta_r^2$
$S \times R$	$\sigma^2(1 - \rho) + p\sigma_{sr}^2$
Positions	$\sigma^2(1 - \rho) + r\sigma_{sp}^2 + sr\theta_p^2$
$S \times P$	$\sigma^2(1 - \rho) + r\sigma_{sp}^2$
Ranges $\times$ Positions	$\sigma^2(1 - \rho) + \sigma_{srp}^2 + s\theta_{rp}^2$
$S \times R \times P$	$\sigma^2(1 - \rho) + \sigma_{srp}^2$

Source	Error Mean Square for Test of Significance
Ranges	Subject $\times$ Range
Positions	Subject $\times$ Position
Ranges $\times$ Positions	Subject $\times$ Range $\times$ Position

random and that ranges and positions are fixed effects, then the expectations of the mean squares are as shown in Table 14.5.

The appropriate error terms for the various tests of significance are shown at the bottom of Table 14.5. If we have reason to believe that $E(MS_{SRP}) = E(MS_{SP}) = E(MS_{SR})$, we would pool the interaction sums of squares, $S \times R \times P$, $S \times P$, and $S \times R$, and divide this pooled sum of squares by the pooled degrees of freedom to obtain the subject $\times$ treatment combinations mean square. In this case, MS_{ST} would have 80 d.f. and would be used to test the main effects and interaction of ranges and positions for significance. If MS_{SRP}, MS_{SP}, and MS_{SR} are not homogeneous, then the appropriate error mean squares for the tests of significance are those shown at the bottom of Table 14.5.

14.14 A Repeated Measure Design with Two Treatments

In the previous chapter on randomized block designs, we considered a t test for the case of two treatments and showed that t^2 was equal to the value of F when the same data were analyzed by the analysis of variance. For a repeated measure design with two treatments, we assume that for those observations obtained under Treatment 1,

$$X_{s1} = \mu + t_1 + t_s + e_{s1} \tag{14.27}$$

and for those observations obtained under Treatment 2,

$$X_{s2} = \mu + t_2 + t_s + e_{s2} \tag{14.28}$$

where $t_1 = \mu_1 - \mu$ and $t_2 = \mu_2 - \mu$ represent the two treatment effects and $\mu = (\mu_1 + \mu_2)/2$. We also have $t_s = \mu_s - \mu$, where t_s represents a subject effect. For a mixed model, with subjects random and treatments fixed, we have

$$E(t_1) = t_1 \quad \text{and} \quad E(t_1^2) = t_1^2$$

$$E(t_2) = t_2 \quad \text{and} \quad E(t_2^2) = t_2^2$$

$$E(t_s) = 0 \quad \text{and} \quad E(t_s^2) = \sigma_s^2$$

With respect to the random errors, e_{s1} and e_{s2}, we assume that

$$E(e_{s1}) = 0 \quad \text{and} \quad E(e_{s1}^2) = \sigma_1^2$$

$$E(e_{s2}) = 0 \quad \text{and} \quad E(e_{s2}^2) = \sigma_2^2$$

We also assume that

$$E(e_{s1}e_{s2}) = \rho_{12}\sigma_1\sigma_2$$

if $s = s$ and

$$E(e_{s1}e_{s2}) = 0$$

if $s \neq s$.

Now consider the difference between the two observations obtained from the same subject. Subtracting (14.28) from (14.27), we have

$$D = X_{s1} - X_{s2} = (t_1 - t_2) + (t_s - t_s) + (e_{s1} - e_{s2})$$

and

$$E(D) = t_1 - t_2 = \mu_1 - \mu_2 = \mu_D \tag{14.29}$$

Then the variance of D will be

$$\sigma_D^2 = E(D - \mu_D)^2 = E(e_{s1} - e_{s2})^2 = E(e_{s1}^2 + e_{s2}^2 - 2e_{s1}e_{s2})$$

or

$$\sigma_D^2 = \sigma_1^2 + \sigma_2^2 - 2\rho_{12}\sigma_1\sigma_2 \tag{14.30}$$

If $\sigma_1^2 = \sigma_2^2$, then

$$\sigma_D^2 = 2\sigma^2(1 - \rho) \tag{14.31}$$

but it is important to emphasize that (14.30) does *not* assume that $\sigma_1^2 = \sigma_2^2$.

The variance of the mean difference will be $1/s$th the variance of D, and therefore, assuming $\sigma_1^2 = \sigma_2^2 = \sigma^2$,

$$\sigma_{\bar{D}}^2 = \frac{2\sigma^2}{s}(1 - \rho) \tag{14.32}$$

and

$$\sigma_{\bar{D}} = \sqrt{\frac{2\sigma^2}{s}(1 - \rho)} \tag{14.33}$$

Except for the factor $(1 - \rho)$, we recognize (14.33) as the standard error of the difference between the means of two independent random samples drawn from a common population with variance equal to σ^2 and when each sample mean is based on the same number, $s = n$, of observations.

We derived (14.30) for a population model. For the sample estimate of the variance of D, we have

$$E(s_D^2) = E\left[\frac{\sum_1^s (D - \bar{D})^2}{s - 1}\right] = \sigma_D^2 \tag{14.34}$$

and s_D^2 is an unbiased estimate of σ_D^2, both for the case where $\sigma_1^2 = \sigma_2^2$ and for the case where $\sigma_1^2 \neq \sigma_2^2$. For the variance of the difference between the two sample means, we have

$$E(s_{\bar{D}}^2) = E\left(\frac{s_D^2}{s}\right) = E\left[\frac{\sum_1^s (D - \bar{D})^2}{s(s - 1)}\right] = \sigma_{\bar{D}}^2 \tag{14.35}$$

and $s_{\bar{D}}^2$ is an unbiased estimate of $\sigma_{\bar{D}}^2$.

We see that when we have only two treatments with a repeated measure design, we need not be concerned about the assumption of equal correlations because only one value of r is involved. In addition, we need not be concerned about heterogeneity of variance because $s_D^2 = \sum_1^s (D - \bar{D})^2/(s - 1)$ is an unbiased estimate of σ_D^2, regardless of whether $\sigma_1^2 = \sigma_2^2$ or whether $\sigma_1^2 \neq \sigma_2^2$.

With only two treatments, the null hypothesis, $\mu_1 - \mu_2 = 0$, regarding the treatment means can be tested for significance by the t test. In this case, we have

$$t = \frac{\bar{X}_{\cdot 1} - \bar{X}_{\cdot 2}}{s_{\bar{D}}}$$

and t^2 will be equal to $F = MS_T/MS_{ST}$. Note also that when we have only two treatments, then $\epsilon = 1/(t - 1) = 1$ and the degrees of freedom for evaluating F remain unchanged for a "conservative" test.

Table 14.6

A SPLIT-PLOT DESIGN IN WHICH SUBJECTS ARE RANDOMLY ASSIGNED TO THE LEVELS OF A AND EACH SUBJECT IS TESTED UNDER ALL LEVELS OF B

	Subjects	B_1	B_2	B_3
	1	X_{111}	X_{112}	X_{113}
	2	X_{211}	X_{212}	X_{213}
	3	X_{311}	X_{312}	X_{313}
A_1	.	.	.	.
	.	.	.	.
	.	.	.	.
	9	X_{911}	X_{912}	X_{913}
	1	X_{121}	X_{122}	X_{123}
	2	X_{221}	X_{222}	X_{223}
	3	X_{321}	X_{322}	X_{323}
A_2	.	.	.	.
	.	.	.	.
	.	.	.	.
	9	X_{921}	X_{922}	X_{923}

14.15 Split-Plot Designs

Let us suppose that we have an experiment in which two factors are of interest, A and B, and that A has $a = 2$ levels and B has $b = 3$ levels. We have $n = 18$ subjects available and they are divided at random in such a way that $s = 9$ subjects are assigned to A_1 and $s = 9$ are assigned to A_2. Each subject is to be tested under all three levels of B and these are randomized independently for each subject. We assume that the treatments corresponding to the levels of B are such that there are no carryover effects. The layout of the experiment is shown in Table 14.6, in which the levels of B have been rearranged to correspond with columns.

When we have the levels of one factor randomly assigned to blocks (subjects in this instance) and the levels of the other factor are assigned at random within each block, the experimental design is called a *split-plot* design. The experiment we have described is a *repeated measure split-plot* design in which subjects are randomly assigned to the levels of A, or vice versa, and each subject is then tested under all levels of B.

14.16 Expected Values of the Mean Squares in a Split-Plot Design

We let $s = 9$ be the number of subjects assigned to each level of A, and we have $a = 2$ levels of A. We also have $b = 3$ levels of B. To determine the expectations of the mean squares for designs of this kind, the subscript or subscripts that serve to indicate the position in the hierarchy in which a component of variance arises are enclosed in parentheses and those that describe the source are left outside. For example, for the component corresponding to σ_s^2 between subjects in a given level of A, we would write $\sigma_{s(a)}^2$. The subscript s describes the source, subjects, while (a) serves to indicate that the component arises within each level of A. The notation $S(A)$ indicates that subjects are *nested* within the levels of A. When the levels of one factor, subjects in this instance, are nested within the levels of another factor, there can be no interaction between the two factors. In our example, it is easy to see, in Table 14.6, that there can be no $S \times A$ interaction. The subscripts describing the source, those outside the parentheses, are called "essential" by Schultz (1955). For example, for the component $\sigma_{s(a)}^2$, s is an essential subscript, whereas (a) is not.

In determining the expectations of the mean squares for designs of the kind described, we follow the same rules presented earlier except that now we delete a component from the expectation of a mean square, after deleting the subscripts describing the source, only if any of the remaining "essential" subscripts specifies a fixed effect. In the experiment being discussed, we regard the subjects as random and A and B as fixed. Following the rules described, we obtain the expectations of the mean squares shown in Table 14.7. We have again assumed that errors associated with the repeated measures on the subjects are correlated. If $\rho = 0$, then the first component for each of the mean squares would be σ^2.

We note that $MS_{S(A)B}$ is the appropriate error mean square for testing the significance of MS_B and MS_{AB}. For the test of significance of MS_A, the appropriate error mean square is $MS_{S(A)}$. For a given set of experimental data, the average value $\bar{r}$ of the $b(b-1)$ correlations involving the repeated measures is assumed to be an estimate of ρ. If $\bar{r}$ is positive, then $MS_{S(A)B}$ will be smaller than $MS_{S(A)}$. The split-plot design is useful if the test of

Table 14.7

EXPECTATION OF MEAN SQUARES FOR THE SPLIT-PLOT DESIGN OF TABLE 14.6. A AND B ARE ASSUMED TO BE FIXED AND SUBJECTS RANDOM

Source	Expectation of Mean Square
A	$\sigma^2[1 + (b-1)\rho] + b\sigma_{s(a)}^2 + sb\theta_a^2$
$S(A)$	$\sigma^2[1 + (b-1)\rho] + b\sigma_{s(a)}^2$
B	$\sigma^2(1-\rho) + \sigma_{s(a)b}^2 + sa\theta_b^2$
$A \times B$	$\sigma^2(1-\rho) + \sigma_{s(a)b}^2 + s\theta_{ab}^2$
$S(A) \times B$	$\sigma^2(1-\rho) + \sigma_{s(a)b}^2$

significance of the main effect of A is of relatively little interest compared with the test of significance of the main effect of B and the $A \times B$ interaction.

14.17 Another Example of a Split-Plot Design

Let us suppose that we have an experiment in which three factors are of interest, A, B, and C, and that each is at two levels. We have eight subjects and they are divided at random in such a way that we have four subjects assigned to A_1 and four to A_2. The subjects are *nested* within the levels of A. Each subject is tested under all four BC treatment combinations and these are randomized independently for each subject over four periods of testing. We assume that the BC treatment combinations are such that there are no carryover effects. The layout of this experiment is shown in Table 14.8.

Under the assumption that subjects are random and that A, B, and C are fixed, we obtain the expected values of the mean squares shown in Table 14.9. The expectations of the mean squares are obtained following the rules described previously.[4] Note that the $S(A) \times BC$ error sum of squares has been partitioned into three orthogonal components: $S(A) \times B$, $S(A) \times C$, and $S(A) \times B \times C$. The appropriate error mean squares for the various tests of significance can be determined from the expectations of the mean squares given in the table.

If we assume that the mean squares for $S(A) \times B$, $S(A) \times C$, and

Table 14.8

LAYOUT OF A SPLIT-PLOT DESIGN WITH SUBJECTS ASSIGNED AT RANDOM TO LEVELS OF A AND BC COMBINATIONS RANDOMIZED FOR EACH SUBJECT IN EACH LEVEL OF A

Groups	Subjects	Randomization of BC Combinations			
A_1	1	B_2C_1	B_1C_2	B_1C_1	B_2C_2
	2	B_2C_2	B_2C_1	B_1C_2	B_1C_1
	3	B_1C_2	B_1C_1	B_2C_1	B_2C_2
	4	B_2C_2	B_2C_1	B_1C_2	B_1C_1
A_2	1	B_2C_1	B_1C_1	B_1C_2	B_2C_2
	2	B_2C_2	B_1C_2	B_1C_1	B_2C_1
	3	B_2C_2	B_2C_1	B_1C_2	B_1C_1
	4	B_1C_1	B_1C_2	B_2C_1	B_2C_2

[4] The expectations are based on the assumption that all errors are independent or uncorrelated.

Table 14.9

EXPECTATION OF MEAN SQUARES FOR THE SPLIT-PLOT DESIGN OF TABLE 14.7 UNDER THE ASSUMPTION THAT ALL ERRORS ARE INDEPENDENT

Source	d.f.	Expectation of Mean Square
A	1	$\sigma^2 + bc\sigma_{s(a)}{}^2 + sbc\theta_a{}^2$
$S(A)$	6	$\sigma^2 + bc\sigma_{s(a)}{}^2$
B	1	$\sigma^2 + c\sigma_{s(a)b}{}^2 + sac\theta_b{}^2$
$A \times B$	1	$\sigma^2 + c\sigma_{s(a)b}{}^2 + sc\theta_{ab}{}^2$
$S(A) \times B$	6	$\sigma^2 + c\sigma_{s(a)b}{}^2$
C	1	$\sigma^2 + b\sigma_{s(a)c}{}^2 + sab\theta_c{}^2$
$A \times C$	1	$\sigma^2 + b\sigma_{s(a)c}{}^2 + sb\theta_{ac}{}^2$
$S(A) \times C$	6	$\sigma^2 + b\sigma_{s(a)c}{}^2$
$B \times C$	1	$\sigma^2 + \sigma_{s(a)bc}{}^2 + sa\theta_{bc}{}^2$
$A \times B \times C$	1	$\sigma^2 + \sigma_{s(a)bc}{}^2 + s\theta_{abc}{}^2$
$S(A) \times B \times C$	6	$\sigma^2 + \sigma_{s(a)bc}{}^2$

$S(A) \times B \times C$ are homogeneous, then the corresponding sums of squares may be pooled and divided by the pooled degrees of freedom to obtain a single estimate of experimental error. This estimate of experimental error is designated as error (b) in Table 14.10. The $S(A)$ mean square, designated as error (a) in Table 14.10, is the appropriate error mean square for testing the A effect for significance. For all other tests of significance, under the

Table 14.10

SOURCES OF VARIATION AND d.f. WHEN INTERACTIONS OF SUBJECTS AND BC TREATMENT COMBINATIONS ARE POOLED TO OBTAIN ERROR (b)

Source	d.f.
A	1
Error (a)	6
B	1
C	1
$B \times C$	1
$A \times B$	1
$A \times C$	1
$A \times B \times C$	1
Error (b)	18
Total	31

assumption stated, error (*b*) is the appropriate error mean square. For conservative F tests involving error (*b*), the F ratios would be evaluated in terms of the tabled value of F with 1 and 6 d.f.

QUESTIONS AND PROBLEMS

14.1 We have a randomized block design with $s = 5$ subjects and with each subject tested under each of $t = 4$ treatments. The outcome of the experiment is given below:

	Treatments			
Subjects	1	2	3	4
1	5	1	1	1
2	4	2	2	2
3	3	3	3	3
4	2	4	4	4
5	1	5	5	5

(*a*) Show that $MS_W = MS_{ST}$ and explain why this is so. (*b*) Add another treatment, T_5, with values of the observations equal to 5, 4, 3, 2, and 1, respectively, for Subjects 1, 2, 3, 4, and 5. Show that with the values of T_5 included in the experiment, $MS_{ST} > MS_W$ and explain why this is so.

14.2 Assume that $s = 3$ subjects are tested under each of $t = 2$ treatments. The outcome of the experiment is given below:

	Treatments	
Subjects	1	2
1	4	3
2	5	2
3	3	1

(*a*) Find the value of $F = MS_T/MS_{ST}$. (*b*) Find $s_{\bar{D}}^2$ and show that $t^2 = (\bar{X}_{.1} - \bar{X}_{.2})^2/s_{\bar{D}}^2 = F$. (*c*) Show that $MS_{ST} = MS_W(1 - \bar{r})$.

14.3 In a randomized block design with repeated measures, we can calculate the following sums of squares: (1) total, (2) treatment, (3) subject, (4) within treatment, (5) within subject, and (6) subject $\times$ treatment. State which of these sums of squares is given by each of the following:

(*a*) Total − subject − treatment
(*b*) Within treatment − subject
(*c*) Within subject − treatment
(*d*) Subject + subject × treatment
(*e*) Treatment + subject × treatment
(*f*) Total − subject
(*g*) Total − within treatment
(*h*) Total − within subject

14.4 We have $s = 4$ subjects and $t = 3$ treatments. Each subject is tested under each of the treatments. The outcome of the experiment is given below:

	Treatments		
Subjects	1	2	3
1	3	4	2
2	5	2	1
3	4	3	3
4	3	2	1

Find the value of $F = MS_T/MS_{ST}$.

14.5 We have shown that the subject × treatment sum of squares can never be larger than the within treatment sum of squares, but we also stated that if ρ was negative, then $E(MS_{ST}) > E(MS_W)$. Explain why both of these statements can be true. *Hint*: Consider the case of $t = 2$ treatments with $\rho = -1.00$. What is the relationship between the subject × treatment sum of squares and the within treatment sum of squares in this instance? What is the relationship between MS_{ST} and MS_W in this instance?

14.6 We have a factorial experiment with three factors, A, B, and C. Subjects are assigned at random to the levels of A and each subject is then tested under all of the BC treatment combinations. The subjects are nested within the levels of A. The $S(A) \times BC$ error sum of squares is partitioned into the $S(A) \times B$, $S(A) \times C$, and $S(A) \times B \times C$ sums of squares. (*a*) If we have $s = 10$ subjects within each level of A and if we have $a = 3$ levels of A, $b = 2$ levels of B, and $c = 3$ levels of C, show the analysis of variance summary table giving sources of variation and the degrees of freedom associated with each source. (*b*) For the same example give the expectations of the mean squares.

14.7 In the example given above, assume that ten subjects are randomly assigend to each of the levels of A. The subjects within each level of A are then randomly divided into $b = 2$ groups of $s = 5$ subjects each. Each subject in each AB group is then tested under each of the $c = 3$ levels of C. (a) Give the analysis of variance summary table showing sources of variation and the degrees of freedom associated with each source. (b) For the same example give the expectations of the mean squares.

14.8 We have two factors, A and B, that are of interest. We have $a = 2$ levels of A and $b = 6$ levels of B. We have $s = 10$ subjects assigned to each level of A and the subjects within the levels of A are tested under all $b = 6$ levels of B. The order of testing under the levels of B is independently randomized for each subject. The outcome of the experiment is shown below where we have rearranged the levels of B to correspond to columns:

Groups	Subjects	B_1	B_2	B_3	B_4	B_5	B_6	$\sum$
	1	9	3	2	5	10	12	41
	2	10	7	3	11	11	11	53
	3	5	6	10	5	8	15	49
	4	4	11	9	13	5	11	53
	5	8	11	5	13	9	7	53
A_1	6	6	3	4	10	10	12	45
	7	6	2	6	8	9	14	45
	8	4	4	6	10	14	6	44
	9	4	10	4	11	5	15	49
	10	1	6	4	5	11	13	40
	$\sum$	57	63	53	91	92	116	472
	1	8	9	13	9	18	14	71
	2	5	10	9	7	10	12	53
	3	3	4	8	10	11	19	55
	4	5	8	13	11	15	18	70
	5	1	4	13	14	17	15	64
A_2	6	4	8	6	12	9	20	59
	7	10	7	14	13	9	16	69
	8	10	9	9	16	9	11	64
	9	2	9	9	14	12	19	65
	10	9	10	8	12	16	11	66
	$\sum$	57	78	102	118	126	155	636

In this experiment the levels of A represent an organismic variable, intelligence. A_1 consists of subjects *below* average in intelligence and A_2

consists of subjects *above* average in intelligence. The dependent variable is one on which it is expected that students below and above average in intelligence will differ in mean performance. (*a*) Test the A effect for significance. (*b*) Test the B effect for significance. (*c*) Test the $A \times B$ interaction effect for significance. (*d*) Interpret the results of the analysis of variance.

14.9 We have two factors, A and B, each with two levels. Five subjects are tested under each of the treatment combinations. The order of testing is independently randomized for each subject. The outcome of the experiment is given below:

	Treatment Combinations			
Subjects	A_1B_1	A_1B_2	A_2B_1	A_2B_2
1	2	4	5	7
2	3	5	6	8
3	5	6	8	10
4	7	9	11	14
5	6	7	10	11

Complete the analysis of variance.

14.10 A study by Stelmachers and McHugh (1964) investigated seven levels of information on predictive accuracy. The subjects in the experiment consisted of two different groups of judges. One group consisted of forty-two expert judges and the other of seventy naive judges. The forty-two expert judges were divided at random into seven groups of six each and one group was assigned to each level of information. The seventy naive judges were divided at random into seven groups of ten each and one group was assigned to each level of information. Each subject then made predictions about four individuals, in a random order, with different Minnesota Multiphasic Personality Inventory profiles.

Let J correspond to the type of judge and we have $j = 2$ levels of J. Let I correspond to information and we have $i = 7$ levels of I. Let P correspond to the profiles and we have $p = 4$ levels of P. Note that subjects (S) are nested within the levels of JI, that is, $S(JI)$. For the experts we have $s = 6$ subjects and for the naive judges we have $s = 10$ subjects within each combination. (*a*) Show the analysis of variance summary table giving sources of variation and degrees of freedom associated with each source. (*b*) Under the assumption that subjects are random and J, I, and P are fixed, give the expectations of the mean squares.

14.11 In the example below we have $s = 8$ subjects tested under each of $t = 3$ treatments.

	Treatments		
Subjects	1	2	3
1	1	1	0
2	1	1	0
3	1	0	1
4	1	0	1
5	0	0	0
6	0	0	0
7	0	1	1
8	0	1	1

(*a*) What are the values of r_{12}, r_{13}, and r_{23}? (*b*) What are the values of s_1^2, s_2^2, s_3^2, and MS_W? (*c*) What is the value of MS_{ST}?

14.12 Add 8, 7, 6, 5, 4, 3, 2, and 1 to the values of the observations for Subjects 1, 2, 3, 4, 5, 6, 7, and 8, respectively, in the above example. (*a*) What is the value of MS_{ST}? (*b*) What is the value of MS_W? (*c*) Find the value of $\overline{r_{ij}s_is_j}$ and show that $MS_{ST} = MS_W - \overline{r_{ij}s_is_j}$.

CHAPTER 15

LATIN SQUARE DESIGNS

15.1 Introduction

In our discussion of the randomized block design, the blocks were considered primarily as representing a particular group of subjects. The blocks were formed on the basis of prior measurement or prior knowledge of the subjects. In forming blocks it is anticipated that, in the absence of treatment effects, variation within each block will generally be less than would be found if the same number of observations were obtained from subjects selected completely at random. If there is substantial variation among the block means, the analysis of variance effectively eliminates this source of variation from the estimate of experimental error.

Consider now a situation where the laboratory technique is such that only five subjects can be tested on each day and we have five treatments, A, B, C, D, and E. We have no prior measurements or knowledge about the subjects that we can use as a guide in arranging the subjects in blocks, but assume that the variation between the observations made on different days is of some importance. Then a group of five subjects randomly assigned to each day's testing could be considered as forming a block. Within each day, the five subjects could be assigned at random with one subject for each treatment. Table 15.1 shows five random permutations of the five treatments obtained in the manner described previously.

The analysis of variance for this experiment would be the same as that for a randomized block design, with the days corresponding to blocks. If the day-to-day variation is of some importance, then the design will remove this source of variation from the estimate of experimental error.

It may occur to the experimenter, however, that another source of variation may be involved which should also be eliminated from the esti-

Table 15.1
RANDOMIZED BLOCK DESIGN FOR FIVE TREATMENTS WITH DAYS CORRESPONDING TO BLOCKS

Day 1	*A*	*B*	*C*	*E*	*D*
Day 2	*C*	*B*	*E*	*A*	*D*
Day 3	*A*	*C*	*D*	*B*	*E*
Day 4	*B*	*A*	*C*	*D*	*E*
Day 5	*A*	*B*	*C*	*E*	*D*

mate of experimental error. Suppose, for example, that one of the five subjects is tested each day at 1, 2, 3, 4, and 5 o'clock. The variation among the hours may be of importance and might be controlled by restricting the randomization in such a way that each treatment occurs not only once on each day but also only once at each hour. With this restricted randomization we have the arrangement of the treatments shown in Table 15.2.

In Table 15.2, it may be observed that each treatment (letter) occurs once and only once in each row and each column. An arrangement of this kind is called a *Latin square.* To form a Latin square we must have the number of rows equal to the number of columns equal to the number of treatments. We shall consider first the problem of randomization in a Latin square design and then the analysis of variance of the design.

15.2 Randomization Procedures

Table 15.3 lists selected Latin square arrangements for squares of order 3 to 7.[1] Assume that we have four treatments. We first select at random one of the 4×4 squares. We can do this by drawing at

Table 15.2
EXAMPLE OF A 5×5 LATIN SQUARE

	Hours				
Days	1	2	3	4	5
Monday	*B*	*E*	*D*	*C*	*A*
Tuesday	*C*	*A*	*B*	*E*	*D*
Wednesday	*D*	*B*	*C*	*A*	*E*
Thursday	*E*	*C*	*A*	*D*	*B*
Friday	*A*	*D*	*E*	*B*	*C*

[1] Squares larger than 7×7 can be constructed in the same manner as the 7×7 square given in Table 15.3. To construct an 8×8 square, write the first row *A . . . H*, the second row *B . . . HA*, the third row *C . . . HAB*, the fourth row *D . . . HABC*, and so on.

random a number from 1 to 4. We then randomize the rows and columns of the square and assign the treatments at random to the letters. Assume we have randomly selected square (a) from the set of four squares. Using a table of random numbers, we write down three random permutations of the numbers 1, 2, 3, and 4. Suppose our point of entry into Table I in the Appendix is the first block, row 01 and column 10. Reading down, the first permutation we obtain is 1, 3, 4, 2. Continuing, we obtain 4, 1, 2, 3, and 2, 4, 3, 1. We rearrange the rows of square (a) according to the first permutation. This gives us

1	*A*	*B*	*C*	*D*
3	*C*	*D*	*B*	*A*
4	*D*	*C*	*A*	*B*
2	*B*	*A*	*D*	*C*

We now rearrange the columns of the above square according to the second permutation. This gives us

4	1	2	3
D	*A*	*B*	*C*
A	*C*	*D*	*B*
B	*D*	*C*	*A*
C	*B*	*A*	*D*

Then, if the treatments have been numbered 1, 2, 3, and 4, we rearrange them according to the third permutation, which is then used to assign

Table 15.3

EXAMPLES OF SELECTED LATIN SQUARES

3 × 3

A	*B*	*C*
B	*C*	*A*
C	*A*	*B*

4 × 4

(a)

A	*B*	*C*	*D*
B	*A*	*D*	*C*
C	*D*	*B*	*A*
D	*C*	*A*	*B*

(b)

A	*B*	*C*	*D*
B	*C*	*D*	*A*
C	*D*	*A*	*B*
D	*A*	*B*	*C*

(c)

A	*B*	*C*	*D*
B	*D*	*A*	*C*
C	*A*	*D*	*B*
D	*C*	*B*	*A*

(d)

A	*B*	*C*	*D*
B	*A*	*D*	*C*
C	*D*	*A*	*B*
D	*C*	*B*	*A*

5 × 5

A	*B*	*C*	*D*	*E*
B	*A*	*E*	*C*	*D*
C	*D*	*A*	*E*	*B*
D	*E*	*B*	*A*	*C*
E	*C*	*D*	*B*	*A*

6 × 6

A	*B*	*C*	*D*	*E*	*F*
B	*C*	*F*	*A*	*D*	*E*
C	*F*	*B*	*E*	*A*	*D*
D	*E*	*A*	*B*	*F*	*C*
E	*A*	*D*	*F*	*C*	*B*
F	*D*	*E*	*C*	*B*	*A*

7 × 7

A	*B*	*C*	*D*	*E*	*F*	*G*
B	*C*	*D*	*E*	*F*	*G*	*A*
C	*D*	*E*	*F*	*G*	*A*	*B*
D	*E*	*F*	*G*	*A*	*B*	*C*
E	*F*	*G*	*A*	*B*	*C*	*D*
F	*G*	*A*	*B*	*C*	*D*	*E*
G	*A*	*B*	*C*	*D*	*E*	*F*

Table 15.4

OBSERVATIONS OBTAINED WITH A 5 × 5 LATIN SQUARE DESIGN

Days	Hours 1	2	3	4	5	$\sum$	Treatments				
Monday	8	18	5	8	6	45	*B*	*E*	*D*	*C*	*A*
Tuesday	1	6	5	18	9	39	*C*	*A*	*B*	*E*	*D*
Wednesday	5	4	4	8	14	35	*D*	*B*	*C*	*A*	*E*
Thursday	11	4	14	1	7	37	*E*	*C*	*A*	*D*	*B*
Friday	9	9	16	3	2	39	*A*	*D*	*E*	*B*	*C*
$\sum$	34	41	44	38	38	195					

the letters of the Latin square to the treatment. Then we would have

Treatment number:	1	2	3	4
Permutation:	2	4	3	1
Treatment:	*A*	*B*	*C*	*D*

and Treatment 2 will be assigned *A*; Treatment 4, *B*; Treatment 3, *C*; and Treatment 1, *D* in the Latin square.[2]

15.3 Analysis of Variance for a Latin Square Design

Assume that we have the 5 × 5 Latin square shown at the right in Table 15.4 and that the observations corresponding to the cell entries are those shown at the left in the table. The total sum of squares based on the variation of the cell entries of this table may be calculated in the usual manner. Thus, we have

$$\text{Total} = (8)^2 + (1)^2 + \cdots + (2)^2 - \frac{(195)^2}{25} = 590.0$$

The hour (column) sum of squares will be given by

$$\text{Hour} = \frac{(34)^2}{5} + \frac{(41)^2}{5} + \cdots + \frac{(38)^2}{5} - \frac{(195)^2}{25} = 11.2$$

and the day (row) sum of squares will be

$$\text{Day} = \frac{(45)^2}{5} + \frac{(39)^2}{5} + \cdots + \frac{(39)^2}{5} - \frac{(195)^2}{25} = 11.2$$

[2] For a more complete and exact discussion of procedures for selecting a Latin square at random, see Fisher and Yates (1948).

The treatment sums are obtained by adding the cell entries for each treatment. Thus, we have

$$A = 9 + 6 + 14 + 8 + 6 = 43$$

$$B = 8 + 4 + 5 + 3 + 7 = 27$$

$$C = 1 + 4 + 4 + 8 + 2 = 19$$

$$D = 5 + 9 + 5 + 1 + 9 = 29$$

$$E = 11 + 18 + 16 + 18 + 14 = 77$$

Then the treatment sum of squares will be equal to

$$\text{Treatment} = \frac{(43)^2}{5} + \frac{(27)^2}{5} + \cdots + \frac{(77)^2}{5} - \frac{(195)^2}{25} = 420.8$$

If we subtract the sum of squares for days (rows), hours (columns), and treatments, each with 4 d.f., from the total sum of squares, we obtain a residual sum of squares with 12 d.f. Thus,

$$\text{Residual} = \text{Total} - \text{row} - \text{column} - \text{treatment} \qquad (15.1)$$

The residual sum of squares defined by (15.1) is the error sum of squares for the Latin square design. For the data of Table 15.4, we have

$$\text{Residual} = 590.0 - 11.2 - 11.2 - 420.8 = 146.8$$

The summary of the analysis of variance is shown in Table 15.5. To test the treatment mean square for significance, we have $F = MS_T/MS_{res} = 105.20/12.23 = 8.60$ with 4 and 12 d.f. This is a highly significant value and we conclude that the treatment means differ significantly. Specific comparisons on the treatment means may be made by the methods described previously under the heading of multiple comparisons. The estimate of error to be used in making these comparisons is the residual mean square of Table 15.5 with 12 d.f.

Table 15.5
ANALYSIS OF VARIANCE FOR THE DATA OF TABLE 15.4

Source of Variation	Sum of Squares	d.f.	Mean Square	F
Treatments	420.8	4	105.20	8.60
Days	11.2	4	2.80	
Hours	11.2	4	2.80	
Residual error	146.8	12	12.23	
Total	590.0	24		

15.4 General Equation for a Latin Square Design

We obtained the residual or error sum of squares by subtraction; however, it can be calculated directly. For the Latin square, we have r rows, c columns, and t treatments. Let the observation in the rth row, the cth column, with the tth treatment be designated by X_{rct}, with the understanding that when used as subscripts, r, c, and t correspond to variables. Then the deviation of any observation from the overall mean can be expressed as

$$\begin{aligned} X_{rct} - \bar{X}_{...} = & (\bar{X}_{r..} - \bar{X}_{...}) \\ & + (\bar{X}_{.c.} - \bar{X}_{...}) \\ & + (\bar{X}_{..t} - \bar{X}_{...}) \\ & + (X_{rct} - \bar{X}_{r..} - \bar{X}_{.c.} - \bar{X}_{..t} + 2\bar{X}_{...}) \end{aligned}$$

The successive terms on the right correspond to the deviation of the row mean from the overall mean, the deviation of the column mean from the overall mean, the deviation of the treatment mean from the overall mean, and the last term is a residual that results if the first three terms on the right are subtracted from the term on the left. If we square the above expression and sum over all rc observations, we find that all of the products between terms on the right sum to zero. We have $r = c = t$, with a total of rc observations. Then

$$\begin{aligned} \sum_1^{rc} (X_{rct} - \bar{X}_{...})^2 = & \; c \sum_1^{r} (\bar{X}_{r..} - \bar{X}_{...})^2 \\ & + r \sum_1^{c} (\bar{X}_{.c.} - \bar{X}_{...})^2 \\ & + t \sum_1^{t} (\bar{X}_{..t} - \bar{X}_{...})^2 \\ & + \sum_1^{rc} (X_{rct} - \bar{X}_{r..} - \bar{X}_{.c.} - \bar{X}_{..t} + 2\bar{X}_{...})^2 \end{aligned}$$

Table 15.6
A 2 × 2 LATIN SQUARE

	B_1	B_2
A_1	C_1	C_2
A_2	C_2	C_1

The term on the left in the above expression gives the total sum of squares. The successive terms on the right give the sums of squares for rows, columns, treatments, and error. Note that the error sum of squares in a Latin square design is the row × column interaction sum of squares minus the treatment sum of squares.

15.5 Latin Squares and Fractional Replication

Suppose we have a 2 × 2 × 2 factorial experiment and we decide to use a ½ fractional replication. As indicated previously, if only ½ of the treatment combinations are to be replicated, then we choose those that have plus signs in the $A \times B \times C$ comparison. If we take the set with plus signs, then we would replicate each of the following treatment combinations:

$$A_1B_1C_1 \qquad A_1B_2C_2 \qquad A_2B_1C_2 \qquad A_2B_2C_1$$

We may rearrange these treatment combinations in the form shown in Table 15.6, where each cell entry corresponds to one of the treatment combinations.

Examination of Table 15.6 shows that the arrangement is that of a 2 × 2 Latin square for the levels of C, with rows corresponding to the levels of A, and columns to the levels of B. This 2 × 2 Latin square, in essence, corresponds to a ½ fractional replication of the 2 × 2 × 2 factorial experiment. As we have shown previously, in a ½ fractional replication of a 2 × 2 × 2 factorial experiment, the A effect is completely confounded with the $B \times C$ interaction, the B effect with the $A \times C$ interaction, and the C effect with the $A \times B$ interaction. This is also true for the 2 × 2 Latin square. For example, the treatment effect will be completely confounded with the row × column interaction, the row effect with the column × treatment interaction, and the column effect with the row × treatment interaction.

Similarly, consider Table 15.7 where we have a Latin square for five

Table 15.7
A 5 × 5 LATIN SQUARE

	B_1	B_2	B_3	B_4	B_5
A_1	C_1	C_2	C_3	C_4	C_5
A_2	C_2	C_1	C_5	C_3	C_4
A_3	C_3	C_4	C_1	C_5	C_2
A_4	C_4	C_5	C_2	C_1	C_3
A_5	C_5	C_3	C_4	C_2	C_1

levels of C, with rows corresponding to the five levels of A and the columns to five levels of B. For a complete replication of the $5 \times 5 \times 5$ factorial experiment we would have to have 125 observations, whereas, in Table 15.7, we have only twenty-five. Thus, this 5×5 Latin square corresponds to a $\frac{1}{5}$ fractional replication of a $5 \times 5 \times 5$ factorial experiment. If C is the treatment factor, then, as with all fractionally replicated factorial experiments, if we are to have an unbiased test of the treatment effect, we shall have to assume that certain interactions are negligible.

15.6 The 2 × 2 *Latin Square*

For a single Latin square of order 2×2, the error sum of squares is equal to zero because, for all four of the residuals, we have

$$X_{rct} - \bar{X}_{r..} - \bar{X}_{.c.} - \bar{X}_{..t} + 2\bar{X}_{...} = 0$$

It is easy to see why this must be so for the 2×2 Latin square. The total sum of squares will have 3 d.f. The row, column, and treatment sums of squares each will have 1 d.f. Because the row, column, and treatment comparisons are mutually orthogonal, the sum of the sums of squares for these comparisons will be equal to the total sum of squares, and the error sum of squares must therefore be equal to zero. Note also that for the degrees of freedom for the error sum of squares we have $(t - 1)(t - 2)$ and if $t = 2$ then we have no degrees of freedom for the error sum of squares. For Latin square designs involving more than $t = 2$ treatments, the residual sum of squares will not, in general, be equal to zero.

15.7 Expected Values of Mean Squares in a Latin Square Design

In this section we derive the expected values of the mean squares for a Latin square design in which we have $t > 2$ treatments. We shall assume that *all* interactions, $R \times C \times T$, $R \times C$, $R \times T$, and $C \times T$, are zero so that each of the residuals

$$X_{rct} - \bar{X}_{r..} - \bar{X}_{.c.} - \bar{X}_{..t} + 2\bar{X}_{...}$$

is an estimate of e_{rct}, an independent random error associated with each experimental unit. For the population model, assuming all interactions are zero, we have

$$X_{rct} = \mu + t_r + t_c + t_t + e_{rct} \tag{15.2}$$

where μ is a population mean common to all observations, $t_r = \mu_r - \mu$ is a row effect, $t_c = \mu_c - \mu$ is a column effect, and $t_t = \mu_t - \mu$ is a treatment effect. For a fixed-effect model, we have

$$E(t_r) = t_r \quad \text{and} \quad E(t_r^2) = t_r^2$$

$$E(t_c) = t_c \quad \text{and} \quad E(t_c^2) = t_c^2$$

$$E(t_t) = t_t \quad \text{and} \quad E(t_t^2) = t_t^2$$

$$E(e_{rct}) = 0 \quad \text{and} \quad E(e_{rct}^2) = \sigma^2$$

with the restriction that

$$\sum_1^r t_r = \sum_1^c t_c = \sum_1^t t_t = 0$$

For the expected value of the correction term, we have

$$E\left[\frac{(\sum X_{...})^2}{rc}\right] = \frac{1}{rc} E\left(rc\mu + c\sum_1^r t_r + r\sum_1^c t_c + t\sum_1^t t_t + \sum_1^{rc} e_{rct}\right)^2$$

$$= rc\mu^2 + \sigma^2 \tag{15.3}$$

To find the expected value of the row sum of squares, we first find

$$E\left[\sum_1^r \frac{(\sum X_{r..})^2}{c}\right] = \frac{1}{c} E\left[\sum_1^r \left(c\mu + ct_r + \sum_1^c t_c + \sum_1^t t_t + \sum_1^c e_{rct}\right)^2\right]$$

$$= rc\mu^2 + c\sum_1^r t_r^2 + r\sigma^2 \tag{15.4}$$

Subtracting the correction term, as given by (15.3), from (15.4), we have the expected value of the row sum of squares or

$$E(SS_R) = c\sum_1^r t_r^2 + (r-1)\sigma^2 \tag{15.5}$$

and dividing by $r - 1$, the degrees of freedom associated with the row sum of squares, we have the expected value of the row mean square or

$$E(MS_R) = \sigma^2 + c\frac{\sum_1^r t_r^2}{r-1}$$

In the same manner, we would find that the expected values of the column sum of squares and the treatment sum of squares are

$$E(SS_C) = r\sum_1^c t_c^2 + (c-1)\sigma^2 \tag{15.6}$$

and

$$E(SS_T) = t\sum_1^t t_t^2 + (t-1)\sigma^2 \tag{15.7}$$

and dividing these sums of squares by their degrees of freedom, we have

$$E(MS_C) = \sigma^2 + r\frac{\sum_1^c t_c^2}{c-1}$$

and

$$E(MS_T) = \sigma^2 + t\frac{\sum_1^t t_t^2}{t-1}$$

The expected value of the total sum of squares can be obtained by first finding

$$E\left(\sum_1^{rc} X_{rct}^2\right) = E\left[\sum_1^{rc} (\mu + t_r + t_c + t_t + e_{rct})^2\right]$$

$$= rc\mu^2 + c\sum_1^r t_r^2 + r\sum_1^c t_c^2 + t\sum_1^t t_t^2 + rc\sigma^2 \tag{15.8}$$

If we subtract the correction term, as given by (15.3), from (15.8), we have the expected value of the total sum of squares or

$$E(SS_{tot}) = c\sum_1^r t_r^2 + r\sum_1^c t_c^2 + t\sum_1^t t_t^2 + (rc-1)\sigma^2 \tag{15.9}$$

The expected value of the residual or error sum of squares can then be obtained by subtraction. Thus,

$$E(SS_{res}) = E(SS_{tot}) - E(SS_R) - E(SS_C) - E(SS_T)$$

$$= \sigma^2[(rc-1) - (r-1) - (c-1) - (t-1)]$$

But we have $r = c = t$ and, therefore,

$$E(SS_{res}) = \sigma^2(t^2 - 1 - t + 1 - t + 1 - t + 1)$$

$$= \sigma^2(t^2 - 3t + 2)$$

$$= \sigma^2(t-1)(t-2) \tag{15.10}$$

Under the assumptions we have made in deriving the expected values of the mean squares, we note that

$$E(MS_{res}) = \sigma^2$$

Table 15.8

EXPECTATION OF MEAN SQUARES FOR A LATIN SQUARE DESIGN WITH ALL INTERACTIONS EQUAL TO ZERO

Model I: Fixed Effects

Source	Expectation of Mean Square
Rows	$\sigma^2 + c\theta_r^2$
Columns	$\sigma^2 + r\theta_c^2$
Treatments	$\sigma^2 + t\theta_t^2$
Residual	σ^2

Model II: Random Effects

Source	Expectation of Mean Square
Rows	$\sigma^2 + c\sigma_r^2$
Columns	$\sigma^2 + r\sigma_c^2$
Treatments	$\sigma^2 + t\sigma_t^2$
Residual	σ^2

The expected values of the mean squares for a Latin square design with a fixed-effect model, based on the as.umptions stated, are summarized in Table 15.8, where we have again used θ^2 to represent a fixed effect. We could also derive the expected values of the mean squares for a random-effect model from (15.2). For the random-effect model, in addition to assuming that all interactions are equal to zero, we assume that t_r, t_c, and t_t are independent random variables. In this case, $E(t_r) = 0$ and $E(t_r^2) = \sigma_r^2$ and similar considerations apply to t_c and t_t. The expected values of the mean squares for a random-effect model are also summarized in Table 15.8.

15.8 Expected Values of Mean Squares When Interactions Are Present

The expectations of the mean squares given in Table 15.8 are based on the assumption that all interactions are equal to zero. Consider now what happens if the two-factor interactions are not equal to zero; that is, assume that the three-factor interaction, $R \times C \times T$, is equal to zero, but that the two-factor interactions, $R \times C$, $R \times T$, and $C \times T$, are not necessarily equal to zero. If the $R \times C$ interaction is not equal to zero, it will be a component of the expectation of both the error mean square *and* the treatment mean square. If the $R \times T$ interaction is not equal to zero, it will be a component of both the error mean square *and* the column mean square. If the $C \times T$ interaction is not equal to zero, it will be a component of both the error mean square *and* the row mean square. Thus, if these interactions are not equal to zero, we have the expectations of the mean squares shown in Table 15.9 for the fixed-effect model.

It is clear that, with a fixed-effect model, the test of significance of the treatment mean square will be biased unless we can assume that both the $R \times T$ and the $C \times T$ interactions are equal to zero. If either of these interactions is not equal to zero, then we shall be testing the treatment mean square with an inflated error mean square. Consequently, if we fail to obtain a significant treatment mean square, this may be because of the presence of either a $R \times T$ or a $C \times T$ interaction.

Table 15.9 also shows the expectations of the mean squares for a mixed model in which rows are assumed to be random and columns and treatments are assumed to be fixed. If the rows in the Latin square correspond to blocks of subjects, as in a randomized block design, then the assumption that rows are random may not be unreasonable. For the mixed model, we see that if we are to have an unbiased test of significance of the treatment mean square, then the $C \times T$ interaction must be equal to zero. If the $C \times T$ interaction is not equal to zero and we fail to obtain a significant treatment mean square, this may be so because the $C \times T$ interaction has resulted in our obtaining an inflated estimate of experimental error.

Table 15.9

EXPECTATION OF MEAN SQUARES FOR A LATIN SQUARE DESIGN WHEN THE TWO-FACTOR INTERACTIONS ARE NOT EQUAL TO ZERO

Model I: Fixed Effects

Source	Expectation of Mean Square
Rows	$\sigma^2 + \theta_{ct}^2 + c\theta_r^2$
Columns	$\sigma^2 + \theta_{rt}^2 + r\theta_c^2$
Treatments	$\sigma^2 + \theta_{rc}^2 + t\theta_t^2$
Residual	$\sigma^2 + \theta_{rc}^2 + \theta_{rt}^2 + \theta_{ct}^2$

Mixed Model: Rows Random and Columns and Treatments Fixed

Source	Expectation of Mean Square
Rows	$\sigma^2 + \theta_{ct}^2 + c\sigma_r^2$
Columns	$\sigma^2 + \sigma_{rc}^2 + \sigma_{rt}^2 + r\theta_c^2$
Treatments	$\sigma^2 + \sigma_{rc}^2 + \sigma_{rt}^2 + t\theta_t^2$
Residual	$\sigma^2 + \sigma_{rc}^2 + \sigma_{rt}^2 + \theta_{ct}^2$

15.9 Factorial Experiments with a Latin Design

Under the assumptions stated previously, a factorial experiment may also be used with a Latin square design. Suppose, for example, that we have a 2^3 factorial experiment. Then the eight treatment combinations may be arranged in an 8×8 Latin square. The analysis of variance for the Latin square would result in the following sums of squares with degrees of freedom shown at the right:

Source	d.f.
Columns	7
Rows	7
Treatment combinations	7
Error	42
Total	63

The sum of squares for treatment combinations could then be analyzed into the comparisons for the main effects, the two-factor interactions, and the three-factor interactions. Each of these sums of squares would have 1 d.f. and each may be tested for significance using the error mean square with 42 d.f.

15.10 Greco-Latin Squares

Suppose we have two Latin squares each of order $t \times t$. To distinguish between these two squares, let the cell entries of one of the squares be represented by Latin letters and the cell entries of the other by Greek letters. If one of these squares is imposed on the other and if each Greek letter occurs once and only once with each Latin letter, then the two superimposed squares are said to form a *Greco-Latin* square. The following is an example of a 4×4 Greco-Latin square:

$$\begin{array}{cccc} \delta B & \beta D & \gamma A & \alpha C \\ \alpha A & \gamma C & \beta B & \delta D \\ \gamma D & \alpha B & \delta C & \beta A \\ \beta C & \delta A & \alpha D & \gamma B \end{array}$$

We note that each Greek letter occurs once and only once in each row and each column and that the same is true for each Latin letter. Thus, the Greek letters alone meet the requirements of a Latin square as do also the

Latin letters. Furthermore, because each Greek letter occurs once and only once with each Latin letter, the two superimposed squares form a Greco-Latin square.

The Greek letters may represent the levels of one treatment factor and the Latin letters the levels of another treatment factor. The analysis of variance for a Greco-Latin square results in the following sums of squares and degrees of freedom:

Source	d.f.
Columns	$t - 1$
Rows	$t - 1$
Treatments: Greek letters	$t - 1$
Treatments: Latin letters	$t - 1$
Residual error	$(t - 1)(t - 3)$

where t is the order of the square. Thus, for a 4×4 Greco-Latin square, the residual or error sum of squares would have only 3 d.f. and for a 5×5 Greco-Latin square the error sum of squares would have only 8 d.f. In the 6×6 and 7×7 Greco-Latin squares, the error sums of squares would have 15 and 24 d.f., respectively.

Potential applications and limitations of the Greco-Latin square in psychological research are discussed by Archer (1952) and Grant (1948). The analysis of variance for independently replicated Greco-Latin squares is given by Fisher and Yates (1948).

15.11 Replication of a Latin Square Design

If we use a Latin square design with t treatments, then the error mean square will have $(t - 1)(t - 2)$ degrees of freedom. For the 5×5 Latin square, for example, we have 12 d.f. for the residual or error mean square. The error mean square for a 6×6 Latin square will have 20 d.f. and the 7×7 square results in an error mean square with 30 d.f. If the estimate of experimental error based on fewer than 30 d.f. is not very reliable, then it is clear that the smaller Latin squares will not be useful.

If there are treatment effects, that is, if the treatment means differ, then increasing the number of observations on which the treatment means are based will serve to increase the treatment mean square. Thus, if there are treatment effects, they are more apt to be detected and declared significant as the number of observations on which the treatment means are based increases. With Latin squares of order 7 × 7 or less, the treatment means will be based on seven or fewer observations. Even though there may be treatment effects, the number of observations on which the treatment means are based may be so small that the test of significance of the treatment mean square will have relatively little power.

By replicating an experiment involving a Latin square design with additional Latin squares, we may expect to obtain not only a more reliable estimate of experimental error but also a more sensitive test of significance of the treatment mean square. In the next chapter we discuss the analysis of variance of replicated Latin square designs.

QUESTIONS AND PROBLEMS

15.1 Complete the analysis of variance for the Latin square design given below. The cell entries at the right are the measures corresponding to the treatments given at the left.

B	*A*	*C*	*D*	1	1	1	2
C	*D*	*A*	*B*	2	3	2	2
A	*B*	*D*	*C*	2	2	1	1
D	*C*	*B*	*A*	4	0	1	3

15.2 For the example given above find the values of $X_{rct} - \bar{X}_{r..} - \bar{X}_{.c.} - \bar{X}_{..t} + 2\bar{X}_{...}$ for each observation and show that the sum of the squared values is equal to the residual or error sum of squares obtained by subtraction.

15.3 For a 2 × 2 Latin square, show that the row × column interaction sum of squares is exactly equal to the treatment sum of squares. Explain why this is so.

15.4 Sixteen subjects were randomly assigned to the treatments in a 4 × 4 Latin square. The outcome of the experiment is given below:

A	*B*	*C*	*D*	1	5	5	6
C	*D*	*A*	*B*	5	4	2	3
B	*A*	*D*	*C*	3	1	7	4
D	*C*	*B*	*A*	5	4	3	1

The cell entries at the right are the measures corresponding to the treatments given at the left. Complete the analysis of variance.

15.5 The experiment described in 15.4 was replicated. In the replication the experimenter obtained the outcome shown below:

C	*D*	*A*	*B*	8	9	4	5
D	*C*	*B*	*A*	7	6	6	3
B	*A*	*D*	*C*	5	3	7	7
A	*B*	*C*	*D*	1	4	5	6

Complete the analysis of variance. Are the results obtained in the replication comparable to those obtained in the first experiment?

15.6 What is meant by a Latin square design? How many degrees of freedom will the error or residual sum of squares have?

15.7 What is meant by a Greco-Latin square? How many degrees of freedom will the error or residual sum of squares have?

CHAPTER 16

LATIN SQUARE DESIGNS: REPEATED MEASURES

16.1 Introduction

Replications of experiments involving a Latin square design may take one of two forms: the complete experiment may be replicated by using additional independently randomized Latin squares or the same Latin square may be replicated by assigning additional subjects to each row of the square.

In the examples considered in this chapter, each subject is tested under each of the treatments and the design is a *repeated measure design*. The arithmetic of the analysis of variance, however, is exactly the same regardless of whether we have repeated measures on subjects or whether each subject is tested under only one of the treatments. If each subject is tested under only one treatment, then we may assume that the errors associated with the observations are independent. However, if each subject is tested under each treatment, then we assume that the errors associated with the repeated measures are not independent.

16.2 Replications of a Latin Square Design with Independently Randomized Squares

In an experiment by Bliss and Rose (1940), twenty dogs were divided at random into five sets of four dogs each. For each set of four dogs an independently randomized 4×4 Latin square was used. The treatments consisted of a dosage of extract of parathyroid and four different

Table 16.1

MILLIGRAMS PERCENT CALCIUM SECRETION OF DOGS GIVEN FOUR DIFFERENT PREPARATIONS OF PARATHYROID EXTRACT

Latin Squares				Dogs	Days 3/15	3/25	4/5	4/15	$\sum$
S_1	S_2	U_2	U_1	1	13.8	17.0	16.0	16.0	62.8
U_2	U_1	S_1	S_2	2	15.8	14.3	14.8	15.4	60.3
S_2	S_1	U_1	U_2	3	15.0	14.5	14.0	15.0	58.5
U_1	U_2	S_2	S_1	4	14.7	15.4	14.8	14.0	58.9
				$\sum$	59.3	61.2	59.6	60.4	240.5
U_2	U_1	S_1	S_2	5	17.0	16.5	15.0	15.4	63.9
U_1	U_2	S_2	S_1	6	15.1	15.0	15.8	13.4	59.3
S_2	S_1	U_1	U_2	7	15.0	14.0	14.6	15.6	59.2
S_1	S_2	U_2	U_1	8	12.0	13.8	14.0	13.8	53.6
				$\sum$	59.1	59.3	59.4	58.2	236.0
S_2	U_2	S_1	U_1	9	14.6	15.4	14.0	14.8	58.8
U_1	S_1	U_2	S_2	10	13.6	15.3	17.2	15.3	61.4
U_2	S_2	U_1	S_1	11	14.4	13.8	14.4	15.0	57.6
S_1	U_1	S_2	U_2	12	15.8	15.0	15.2	15.8	61.8
				$\sum$	58.4	59.5	60.8	60.9	239.6
U_1	U_2	S_1	S_2	13	14.0	13.8	14.0	14.0	55.8
U_2	U_1	S_2	S_1	14	16.2	14.0	13.0	13.0	56.2
S_2	S_1	U_2	U_1	15	13.0	14.0	14.0	13.0	54.0
S_1	S_2	U_1	U_2	16	13.2	16.0	14.9	16.4	60.5
				$\sum$	56.4	57.8	55.9	56.4	226.5
S_1	U_1	S_2	U_2	17	14.2	14.1	15.0	14.4	57.7
U_1	S_1	U_2	S_2	18	13.0	13.4	13.8	14.0	54.2
U_2	S_2	U_1	S_1	19	15.8	16.0	15.0	15.4	62.2
S_2	U_2	S_1	U_1	20	15.2	16.2	15.0	15.3	61.7
				$\sum$	58.2	59.7	58.8	59.1	235.8

preparations were investigated. The experiment was one in biological assay in which two dosages, U_1 and U_2, of an unknown preparation of parathyroid were administered along with two dosages, S_1 and S_2, of a standard preparation. On the first day all twenty dogs were given the treatment prescribed by the separate Latin square entries corresponding to the first column of all five Latin squares. At the time of the second test all animals were given the treatment prescribed by the separate Latin square entries corresponding to the second column, and so on. In this instance, each row of the five Latin squares corresponds to a single subject with the columns corresponding to successive tests or periods. The time interval between the successive tests was sufficiently long that it was believed that all effects of the previous treatment would be dissipated, or, in other words, there would be no carryover effects from one treatment to the next. The observations recorded in Table 16.1 consist of the milligrams percent calcium secretion of the dogs for each treatment.

16.3 Analysis of Variance for a Single Square

If we consider only the first Latin square of Table 16.1, we obtain the following sums of squares:

$$\text{Total} = (13.8)^2 + (15.8)^2 + \cdots + (14.0)^2 - \frac{(240.5)^2}{16} = 11.294$$

$$\text{Day} = \frac{(59.3)^2}{4} + \frac{(61.2)^2}{4} + \cdots + \frac{(60.4)^2}{4} - \frac{(240.5)^2}{16} = 0.546$$

$$\text{Dog} = \frac{(62.8)^2}{4} + \frac{(60.3)^2}{4} + \cdots + \frac{(58.9)^2}{4} - \frac{(240.5)^2}{16} = 2.832$$

Summing the entries for each preparation, we obtain $S_1 = 57.1$, $S_2 = 62.2$, $U_1 = 59.0$, and $U_2 = 62.2$. Then the sum of squares for drugs will be

$$\text{Drug} = \frac{(57.1)^2}{4} + \frac{(62.2)^2}{4} + \cdots + \frac{(62.2)^2}{4} - \frac{(240.5)^2}{16} = 4.756$$

and for the residual sum of squares, we have

$$\text{Residual} = 11.294 - 0.546 - 2.832 - 4.756 = 3.160$$

The analysis of variance for the first square is shown in Table 16.2.

Table 16.2

ANALYSIS OF VARIANCE FOR THE FIRST LATIN SQUARE OF TABLE 16.1

Source of Variation	Sum of Squares	d.f.	Mean Square	F
Treatments	4.756	3	1.585	3.01
Dogs	2.832	3	0.944	
Days	0.546	3	0.182	
Residual error	3.160	6	0.527	
Total	11.294	15		

Table 16.3

POOLED SUMS OF SQUARES AND DEGREES OF FREEDOM FOR THE FIVE LATIN SQUARES OF TABLE 16.1

Source of Variation	Sum of Squares	d.f.
Treatments	22.57	15
Dogs	35.50	15
Days	2.62	15
Residual error	18.21	30
Latin Squares	7.69	4
Total	86.59	79

Testing the treatment mean square for significance, we have $F = 1.585/0.527 = 3.01$ with 3 and 6 d.f. From the table of F we find that $F = 3.01$ is a nonsignificant value.

16.4 Analysis of Variance for Replications with Independent Squares

Now a similar analysis of variance can be made for each of the other four Latin squares. If we then add together the corresponding sums of squares and degrees of freedom, we obtain the analysis shown in Table 16.3. The next to last row in the table is a sum of squares between the five Latin squares with 4 d.f. and can be obtained by calculating

$$\text{Square} = \frac{(240.5)^2}{16} + \frac{(236.0)^2}{16} + \cdots + \frac{(235.8)^2}{16} - \frac{(1178.4)^2}{80} = 7.69$$

Table 16.4

TWO-WAY TABLE FOR SQUARES AND DAYS

	Days				
	1	2	3	4	Σ
Square 1	59.3	61.2	59.6	60.4	240.5
Square 2	59.1	59.3	59.4	58.2	236.0
Square 3	58.4	59.5	60.8	60.9	239.6
Square 4	56.4	57.8	55.9	56.4	226.5
Square 5	58.2	59.7	58.8	59.1	235.8
Σ	291.4	297.5	294.5	295.0	1178.4

The sum of squares for days in Table 16.3 is composed of two component parts. For example, consider Table 16.4. If we calculate the sum of squares between cells of this table and subtract the row and column sums of squares from the cell sum of squares, we would have the row × column interaction sum of squares. In the present instance, the row × column sum of squares would be the square × day sum of squares and would have $(5 - 1)(4 - 1) = 12$ d.f. The column sum of squares is the overall day sum of squares with 3 d.f. Making the necessary calculations, we find that the square × day sum of squares is equal to 1.68 and the day sum of squares is equal to 0.94. The sum of these two sums of squares is 2.62 and this is the value shown in Table 16.3 for days with 15 d.f.

Just as we did in the case of the two-way table for squares and days, we can set up a two-way table for squares and drugs, as shown in Table 16.5. The cell entries of Table 16.5 represent the sums for each drug in each square. The overall column sum of squares would be the sum of squares for drugs with 3 d.f. and the row × column interaction sum of squares would be the square × drug sum of squares with $(5 - 1)(4 - 1) = 12$ d.f. In Table 16.3 the sum of squares for drugs (treatments) consists of both of these components. By making the necessary calculations, we find that the square × drug sum of squares is equal to 7.42 and the sum of squares for drugs is equal to 15.15.

16.5 Summary of the Analysis of Variance for Independent Squares

Table 16.6 summarizes the calculations we have made. It is the drug or treatment mean square that is of primary interest, but before we can test this mean square for significance we must first determine the appropriate error term for the test. In the next section we describe an analysis of variance model for the experiment and obtain the expected values of the mean squares in Table 16.6 for this model.

Table 16.5
TWO-WAY TABLE FOR SQUARES AND DRUGS

	Drugs			
Square 1	S_1	S_2	U_1	U_2
Square 2	S_1	S_2	U_1	U_2
Square 3	S_1	S_2	U_1	U_2
Square 4	S_1	S_2	U_1	U_2
Square 5	S_1	S_2	U_1	U_2

Table 16.6
ANALYSIS OF VARIANCE FOR THE DATA OF TABLE 16.1

Source of Variation	Sum of Squares	d.f.	Mean Square
Squares	7.69	4	1.922
Between dogs in same square	35.50	15	2.367
Treatments	15.15	3	5.050
Days	0.94	3	0.313
Squares × treatments	7.42	12	0.618
Squares × days	1.68	12	0.140
Residual error	18.21	30	0.607
Total	86.59	79	

16.6 Expectation of Mean Squares When Replications of an Experiment Consist of Independently Randomized Latin Squares

As a model for experiments of the kind described above, we assume that

$$X_{rcts} = \mu + t_r + t_c + t_t + t_s + t_{sc} + t_{st} + e_{rcts}$$

where t_r, t_c, and t_t correspond to a row (subject), a column, and a treatment effect, respectively, and t_s represents an effect due to differences in the Latin squares. The only interaction effects assumed in the model are those involving squares and columns (t_{sc}) and squares and treatments (t_{st}). We note that each group of r subjects is *nested* within squares, so that there is no interaction of rows and squares. Furthermore, we have c repeated measures, corresponding to the c columns, for each of the r subjects or rows. We assume that the errors associated with the repeated measures of the same subjects are *not* independent.

It seems reasonable to regard the columns (C) and the treatments (T) as fixed effects. If the squares (S) are selected at random and if r subjects are assigned at random to the rows of each square, then we may regard both S and R as random. The expected values of the mean squares for this mixed model are shown in Table 16.7.[1]

For the model described, we see that the square × treatment mean

[1] The expectations of the mean squares are obtained following the rules cited previously with respect to split-plot designs. Note that the rows (subjects) are nested within squares.

In determining the coefficients of the components of variance, we observe that although a given observation, X_{rcts}, is identified by the subscripts *rcts*, the total number of observations is *not* equal to *rcts*, but instead is only equal to $rcs = rts = cts$. Consequently, the more general rule, given in Chapter 11, must be used in obtaining the coefficients of the components of variance. For example, the coefficients of σ_s^2 are rc because the sample estimate, $\bar{X}_{...s}$, of μ_s is based on only rc observations.

square (MS_{ST}) is the appropriate mean square for testing the significance of the treatment mean square (MS_T). For this test we have $F = MS_T/MS_{ST} = 5.050/0.618 = 8.17$, with 3 and 12 d.f.

Suppose, however, that the square × treatment interaction effect, σ_{st}^2, is equal to zero. Then we would have $E(MS_T) = \sigma^2(1 - \rho) + sc\theta_t^2$ and $E(MS_{ST}) = E(MS_{res}) = \sigma^2(1 - \rho)$ and both MS_{ST} and MS_{res} would be appropriate error terms for testing the significance of MS_T. Similarly, if the square × column interaction effect, σ_{sc}^2, is equal to zero, then $E(MS_{SC}) = E(MS_{ST}) = E(MS_{res}) = \sigma^2(1 - \rho)$ and, in this case, it would be reasonable to combine these three estimates to obtain a single estimate of $\sigma^2(1 - \rho)$.

It is apparent in Table 16.6 that the square × treatment mean square (0.618) and the residual mean square (0.607) are comparable. Furthermore, the square × column mean square (0.140) is smaller than the residual mean square and, in terms of the model described, can be so only as a result of chance or random variation. Thus, in this experiment, it seems reasonable to combine the three estimates to obtain a single estimate of experimental error.

Assuming $\sigma_{st}^2 = \sigma_{sc}^2 = 0$, let the combined error sum of squares be

$$\text{Error }(b) = SS_{ST} + SS_{SC} + SS_{res}$$

and this combined error sum of squares will have

$$(s - 1)(t - 1) + (s - 1)(c - 1) + s(t - 1)(t - 2)$$

degrees of freedom. Because $c = t$, the above expression reduces to $(t - 1)(st - 2)$ d.f., where s is the number of independent Latin squares.

For the data of Table 16.6, we have

$$\text{Error }(b) = 7.42 + 1.68 + 18.21 = 27.31$$

with $(4 - 1)(20 - 2) = 54$ d.f. Then the error mean square (MS_b) will

Table 16.7

EXPECTATION OF MEAN SQUARES FOR AN EXPERIMENT INVOLVING INDEPENDENTLY RANDOMIZED LATIN SQUARES

Source	d.f.	Expectation of Mean Square
Squares	$s - 1$	$\sigma^2[1 + (c - 1)\rho] + c\sigma_{r(s)}^2 + rc\sigma_s^2$
$R(S)$	$s(r - 1)$	$\sigma^2[1 + (c - 1)\rho] + c\sigma_{r(s)}^2$
Treatments	$t - 1$	$\sigma^2(1 - \rho) + c\sigma_{st}^2 + sc\theta_t^2$
Columns	$c - 1$	$\sigma^2(1 - \rho) + r\sigma_{sc}^2 + sr\theta_c^2$
$S \times T$	$(s - 1)(t - 1)$	$\sigma^2(1 - \rho) + c\sigma_{st}^2$
$S \times C$	$(s - 1)(c - 1)$	$\sigma^2(1 - \rho) + r\sigma_{sc}^2$
Residual error	$s(t - 1)(t - 2)$	$\sigma^2(1 - \rho)$

be $27.31/54 = 0.506$ and this is the error mean square that was used by Bliss and Rose (1940) to test the significance of the treatment or drug mean square in their published analysis of variance of the data presented. In this case, we have $F = MS_T/MS_b = 5.050/0.506 = 9.98$ with 3 and 54 d.f.

In general, with a reliable experimental technique one might anticipate that differences in the treatment means will be comparable from square to square because each of the Latin squares is nothing more than a replication or repetition of the same experiment. Similarly, if the columns represent successive test periods, these test periods will be the same for each of the Latin squares and we would expect the differences among the column means to be comparable from square to square. Under these conditions it would be reasonable to assume that σ_{st}^2 and σ_{sc}^2 are both equal to zero. If this is the case, then MS_b can be used to test the significance of MS_T.

In some experiments, one may find that $F = MS_{ST}/MS_{res}$ is a significant value. This would indicate that the differences among the treatment means are not the same for the various Latin squares or, in other words, that there is an interaction between squares and treatments. We note, however, that if the Latin squares involved in the experiment have been randomly selected from a population of Latin squares of the same order, then the obtained overall treatment means will have been averaged over a *random sample* of Latin squares. Thus if, in the same experiment, $F = MS_T/MS_{ST}$ is also significant, we would expect comparable treatment differences to be found for the population of Latin squares.

Note that a test of significance of the mean square for Latin squares (MS_S) is possible. In this case, the appropriate error mean square is $MS_{(R)S}$. If MS_S is significant, it would indicate that the overall means, as given by each of the replications of the experiment, that is, the means for the different Latin squares, differ significantly. Ordinarily, we would not expect MS_S to be significant and in the Bliss and Rose experiment we find that $MS_S = 1.922$ is less than $MS_{(R)S} = 2.367$. Note also that $MS_{(R)S}$ is considerably larger than $MS_{res} = 0.607$ and this will, in general, be the case when the errors associated with the repeated measures are positively intercorrelated.

16.7 Replication of an Experiment with the Same Square

Let us now consider the data of Table 16.8. The experiment was concerned with the ability of subjects to locate targets when they appeared on circular screens of varying size. Screen sizes of 3, 4, 5, 6, and 7 inches were used in the experiment. Radii were marked on the screens at twenty-degree intervals. Each screen size was also marked by a

Table 16.8
NUMBER OF CORRECT JUDGMENTS FOR EACH SUBJECT FOR EACH SCREEN SIZE

		Periods					
Latin Square	Subjects	1	2	3	4	5	$\sum$
	1	11	15	10	13	10	59
	2	7	18	12	17	16	70
3 6 4 7 5	3	12	17	9	19	18	75
	4	13	16	17	15	15	76
	5	14	17	18	18	15	82
	$\sum$	57	83	66	82	74	362
	6	8	11	11	15	14	59
	7	13	13	13	5	10	54
4 7 5 3 6	8	19	16	18	14	21	88
	9	8	17	12	14	17	68
	10	9	9	16	10	11	55
	$\sum$	57	66	70	58	73	324
	11	10	8	13	11	17	59
	12	16	16	18	14	19	83
5 3 6 4 7	13	15	15	16	14	14	74
	14	12	14	15	13	16	70
	15	12	10	14	12	16	64
	$\sum$	65	63	76	64	82	350
	16	13	12	16	14	9	64
	17	16	10	19	16	11	72
6 4 7 5 3	18	13	15	13	13	9	63
	19	11	10	13	11	7	52
	20	6	16	13	17	10	62
	$\sum$	59	63	74	71	46	313
	21	11	8	9	12	15	55
	22	20	18	10	18	17	83
7 5 3 6 4	23	16	14	10	12	13	65
	24	14	15	16	16	17	78
	25	11	12	10	15	15	63
	$\sum$	72	67	55	73	77	344

series of expanding circles representing intervals of ten miles from the center of the target. The screens were exposed to the subjects by means of an automatic timer at the rate of one screen every fifteen seconds. The screens had been photographed on a film strip and were projected to a round glass plate. There were thirty-six projections for each of the five screen sizes and the subject was asked to locate the position of a target that appeared on the screen in terms of both degrees and miles. The data given in Table 16.8 are the number of correct judgments for degrees.

With five subjects a single Latin square may be formed with the rows corresponding to subjects, the columns to successive test periods, and the cell entries to the screen sizes. In accordance with a Latin square design, each screen size would appear once and only once in each column and each row. Instead of replicating the experiment with additional and independently randomized Latin squares, the experimenter, in this instance, chose to replicate the *same* Latin square five times with five subjects assigned at random to each row of the square. Each subject was tested under each screen size in the order prescribed by the row of the Latin square to which he was assigned. This particular design makes possible the isolation of a sum of squares corresponding to the five sequences or orders of presentation of the screen sizes, although we must keep in mind that only five of the 120 possible different orders were investigated. Note also that subjects are *nested* within the various orders.

16.8 Analysis of Variance When the Same Square Is Replicated

For the data of Table 16.8 we have the following sums of squares:

$$\text{Total} = (11)^2 + (7)^2 + \cdots + (15)^2 - \frac{(1693)^2}{125} = 1327.01$$

$$\text{Row} = \frac{(59)^2}{5} + \frac{(70)^2}{5} + \cdots + \frac{(63)^2}{5} - \frac{(1693)^2}{125} = 495.41$$

Table 16.9
TWO-WAY TABLE FOR ORDERS AND PERIODS

	Periods					
Order	1	2	3	4	5	$\sum$
3–6–4–7–5	57	83	66	82	74	362
4–7–5–3–6	57	66	70	58	73	324
5–3–6–4–7	65	63	76	64	82	350
6–4–7–5–3	59	63	74	71	46	313
7–5–3–6–4	72	67	55	73	77	344
$\sum$	310	342	341	348	352	1693

Table 16.10
ANALYSIS OF VARIANCE FOR THE DATA OF TABLE 16.8

Source of Variation	Sum of Squares	d.f.	Mean Square	F
Orders	63.01	4	15.75	
Between subjects: Error (a)	432.40	20	21.62	
Screen size	232.05	4	58.01	9.67
Periods	44.13	4	11.03	1.84
Residual error	75.02	12	6.25	1.04
Subjects $\times$ periods: Error (b)	480.40	80	6.00	
Total	1327.01	124		

Then, from Table 16.8, we form Table 16.9, where each cell entry is the sum of five observations. For this table we find the sum of squares for rows corresponding to the five orders or sequences of screen sizes. Thus,

$$\text{Order} = \frac{(362)^2}{25} + \frac{(324)^2}{25} + \cdots + \frac{(344)^2}{25} - \frac{(1693)^2}{125} = 63.01$$

This sum of squares is part of the row sum of squares, 495.41. Subtracting the order sum of squares from the sum of squares for rows, we have a residual, $495.41 - 63.01 = 432.40$, with $24 - 4 = 20$ d.f. This residual sum of squares is the pooled sum of squares between subjects tested with the same order. For example, the sum of squares between subjects tested with order 3-6-4-7-5 would have $5 - 1 = 4$ d.f. Calculating this sum of squares for each order, we would have five such sums of squares and the pooled sum of squares would have $5(5 - 1) = 20$ d.f. This is the sum of squares on which error mean square (a) of Table 16.10 is based.

From Table 16.9 we also calculate the sum of squares for periods and this is equal to

$$\text{Period} = \frac{(310)^2}{25} + \frac{(342)^2}{25} + \cdots + \frac{(352)^2}{25} - \frac{(1693)^2}{125} = 44.13$$

From the same table we also find the sums for each screen size. For example, the sum for the 3-inch screen size is equal to $57 + 58 + 63 + 46 + 55 = 279$. The sums for the 4-, 5-, 6-, and 7-inch screens are 327, 347, 364, and 376, respectively. Then the sum of squares for screen size will be

$$\text{Screen} = \frac{(279)^2}{25} + \frac{(327)^2}{25} + \cdots + \frac{(376)^2}{25} - \frac{(1693)^2}{125} = 232.05$$

If we now find the sum of squares between the cells of Table 16.9, we have

$$\text{Cell} = \frac{(57)^2}{5} + \frac{(57)^2}{5} + \cdots + \frac{(77)^2}{5} - \frac{(1693)^2}{125} = 414.21$$

Then, by subtraction,

$$\begin{aligned}\text{Order} \times \text{period} &= \text{Cell} - \text{order} - \text{period}\\ &= 414.21 - 63.01 - 44.13\\ &= 307.07\end{aligned}$$

and for the Latin square error or residual sum of squares, we have

$$\begin{aligned}\text{Residual} &= \text{Order} \times \text{period} - \text{treatment}\\ &= 307.07 - 232.05\\ &= 75.02\end{aligned}$$

For each of the five orders we have five subjects tested at five different periods. It is therefore possible to find a sum of squares for each order that is the row $\times$ column interaction for that order. Because subjects correspond to rows and periods to columns, this interaction sum of squares is the subject $\times$ period sum of squares. To calculate the sum of squares for the first order we would find the sum of squares between the twenty-five observations for that order. From this sum of squares we would then subtract the subject (row) sum of squares and the period (column) sum of squares. The residual is the subject $\times$ period sum of squares for the first order with $(5 - 1)(5 - 1) = 16$ d.f. Repeating these calculations for each order, we would have five different subject $\times$ period sums of squares. The sum of these sums of squares is equal to 480.40 with $5(5 - 1)(5 - 1) = 80$ d.f. This is the sum of squares on which error mean square (b) of Table 16.10 is based. If we are only interested in the pooled subject $\times$ period sum of squares with 80 d.f., it should be clear that this sum of squares can be obtained by subtraction. We point out the fact that the subject $\times$ period sum of squares is exactly the same as the subject $\times$ treatment sum of squares and we shall show why in a subsequent section.

16.9 Summary of the Analysis of Variance When the Same Square Is Replicated

The summary of the analysis of variance is shown in Table 16.10. Using the subject $\times$ period mean square, error (b), as an error mean square, we have for the test of significance of the Latin square error or residual mean square, $F = MS_{res}/MS_{(b)} = 6.25/6.00 = 1.04$ with 12 and 80 d.f., a nonsignificant value. For reasons to be discussed later, this test of significance presumably provides information regarding the possible interaction of periods and treatments and, in the absence of a significant interaction, the Latin square error or residual sum of squares is often pooled with the subject $\times$ treatment sum of squares to provide a single estimate of experimental error. In the present example, this pooled error sum of squares is equal to 75.02 + 480.40 = 555.42 with 12 + 80 = 92 d.f., and the mean square is equal to 555.42/92 = 6.04. No conclusions regarding significance would be changed using this value as the error mean square instead of $MS_{(b)} = 6.00$.

Error mean square (b) is the appropriate error mean square for testing the significance of the screen size and period mean squares. The error mean square designated (a) is the appropriate error term for testing the order mean square for significance. We note, in this analysis, that error (a) is considerably larger than error (b) and this will usually be the case.

16.10 Test of Significance of the Linear Component of Screen Size

The only significant effect, as shown in Table 16.10, is the screen size. Dividing each of the screen sums by the number of observations, twenty-five, we have the following means for each screen size:

Screen size:	3	4	5	6	7
Mean:	11.16	13.08	13.88	14.56	15.04

The means are plotted against the screen size in Figure 16.1. To determine

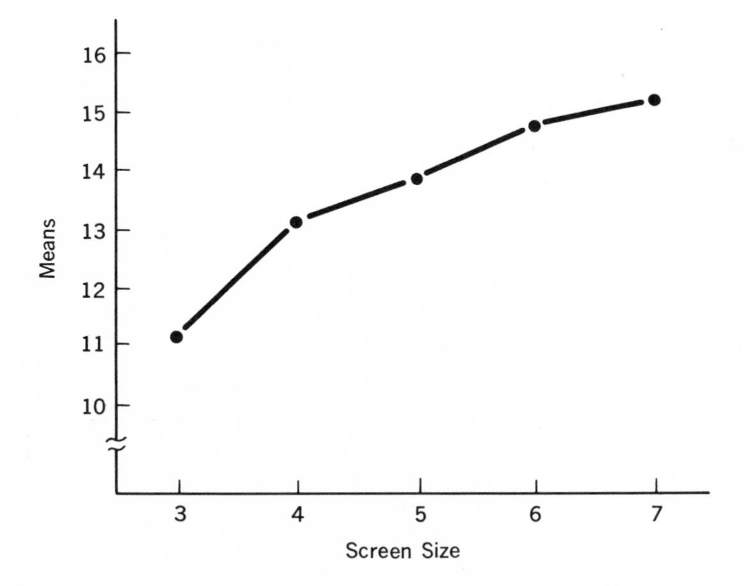

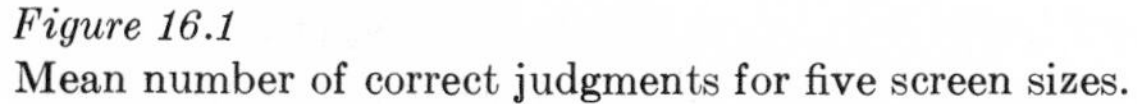
Figure 16.1
Mean number of correct judgments for five screen sizes.

whether the linear component of the trend is significant, we use the orthogonal coefficients for $k = 5$, obtained from Table XI in the Appendix. Then, using the treatment sums rather than the treatment means, we have

$$D_1 = (-2)(279) + (-1)(327) + (0)(347) + (1)(364) + (2)(376) = 231$$

The sum of squares for the coefficients is equal to 10 and we have $n = 25$ observations for each screen size. Then the mean square for the linear component will be

$$MS_{D_1} = \frac{(231)^2}{(25)(10)} = 213.44$$

with 1 d.f. For the test of significance, we have

$$F = \frac{213.44}{6.00} = 35.57$$

with 1 and 80 d.f. This is a significant value and we conclude that the linear component of the trend is significant.

16.11 Test of Significance of the Quadratic Component of Screen Size

It would appear from Figure 16.1 that there is a slight curvature in the trend of the means. To test for the significance of the curvature, we use the orthogonal coefficients for the quadratic component, as obtained from Table XI in the Appendix, with $k = 5$. Then, multiplying the sums for the screen sizes by these coefficients, we have

$$D_2 = (2)(279) + (-1)(327) + (-2)(347) + (-1)(364) + (2)(376) = -75$$

The sum of the squared coefficients is equal to 14 and we have $n = 25$ observations for each screen size. Then the mean square for the quadratic comparison will be

$$MS_{D_2} = \frac{(-75)^2}{(25)(14)} = 16.07$$

with 1 d.f. For the test of significance of curvature, we have

$$F = \frac{16.07}{6.00} = 2.68$$

with 1 and 80 d.f., a nonsignificant value. We conclude, therefore, that the trend of the means is essentially linear and without significant curvature.

16.12 Expectation of Mean Squares When the Same Latin Square Is Replicated

Let G represent the rows of a given Latin square, P represent the columns or periods, and T the cell entries or treatments. Then, for the levels of G, P, and T, we have $g = p = t$. We now divide subjects at random into g groups with s subjects in each group. It is not necessary that the number of subjects in each group be equal to the number of groups, as happened to be the case in the experiment described above. Preferably, s will be larger than g and particularly so when g is small. Each group of s subjects is assigned at random to one of the rows of the Latin square and tested under all treatments in the order or sequence prescribed by that row of the Latin square. The s subjects are *nested* within the levels of G.

For any given group of subjects, the subject $\times$ period sum of squares will be identical with the subject $\times$ treatment sum of squares. For example, in Table 16.8, note that for the group of subjects tested with the treatments in the sequence 3-6-4-7-5, the differences in the treatment means are completely confounded with differences in the column or period means. For this group of subjects, the subject $\times$ treatment sum of squares is exactly the same as the subject $\times$ period sum of squares, and this is true for all other groups. Thus, the $S(G) \times P$ sum of squares is equal to the $S(G) \times T$ sum of squares. It follows that the variance component for this interaction can be written either as $\sigma_{s(g)p}^2$ or as $\sigma_{s(g)t}^2$.

Assuming the Latin square has been selected at random, we shall regard the rows or levels of G as random. Because the s subjects were assigned at random to the levels of G, we shall also regard S as random. In addition, because each group of subjects receives exactly the same treatments but in a different order or sequence, a significant group mean square should be obtained only if the different orders of presentation of the treatments have an influence on the overall mean of the treatments.

In the case of a single Latin square, we have pointed out previously that the Latin square residual or error mean square is an estimate of σ^2 only if all interactions of rows, columns, and treatments can be assumed to be equal to zero, or, in the design under discussion, only if the interactions of G, P, and T are equal to zero. If G is random so that the levels of G simply represent random replications of the same experiment with the treatments administered in a different sequence or order and if there are no carryover effects, then it seems reasonable to assume that the component for the $G \times T$ interaction, σ_{gt}^2, is equal to zero. For similar reasons, we shall assume that the component for the $G \times P$ interaction, σ_{gp}^2, is equal to zero. These two assumptions are, in essence, the same as those made when we considered the analysis of variance for a design in which different Latin squares provided replications of an experiment. In that case, we assumed that the square × treatment and the square × period interaction components were zero. With respect to the $P \times T$ interaction component, we regard both P and T as fixed effects and it is possible that θ_{pt}^2 may not be equal to zero.

It is of some consequence to note that from the experimental data given in Table 16.8 we can obtain any one of the three two-way tables shown in Table 16.11. In our analysis of the data we considered only Table 16.9,

Table 16.11
TWO-WAY TABLES FOR $G \times P$, $G \times T$, AND $P \times T$

Table (1): Two-Way Table for $G \times P$

	Periods				
Group	P_1	P_2	P_3	P_4	P_5
G_1	T_1	T_4	T_2	T_5	T_3
G_2	T_2	T_5	T_3	T_1	T_4
G_3	T_3	T_1	T_4	T_2	T_5
G_4	T_4	T_2	T_5	T_3	T_1
G_5	T_5	T_3	T_1	T_4	T_2

Table (2): Two-Way Table for $G \times T$

	Treatments				
Group	T_1	T_2	T_3	T_4	T_5
G_1	P_1	P_3	P_5	P_2	P_4
G_2	P_4	P_1	P_3	P_5	P_2
G_3	P_2	P_4	P_1	P_3	P_5
G_4	P_5	P_2	P_4	P_1	P_3
G_5	P_3	P_5	P_2	P_4	P_1

Table (3): Two-Way Table for $P \times T$

	Treatments				
Period	T_1	T_2	T_3	T_4	T_5
P_1	G_1	G_2	G_3	G_4	G_5
P_2	G_3	G_4	G_5	G_1	G_2
P_3	G_5	G_1	G_2	G_3	G_4
P_4	G_2	G_3	G_4	G_5	G_1
P_5	G_4	G_5	G_1	G_2	G_3

which is the same as Table (1), the two-way table with groups (orders) corresponding to rows and columns to period. The cell entries of Table (1) are the treatment sums and it is obvious that they form a Latin square.

In Table (2) we have a two-way table with groups corresponding to rows and treatments corresponding to columns. The cell entries in this case are the period sums and it is obvious that they form a Latin square. In Table (3) the data are rearranged with periods corresponding to rows and treatments to columns. The cell entries are the group sums and again we note that they form a Latin square. Note that the cell sums for any one of the three two-way tables are exactly the same, and the cell sum of squares for each of the three tables is equal to 414.21.

For each of the three tables, we can find the sum of squares for the row $\times$ column interaction. Thus,

$$\begin{aligned}
\text{Table (1)}: G \times P &= \text{Cell} - \text{group} - \text{period} \\
&= 414.21 - 63.01 - 44.13 = 307.07 \\
\text{Table (2)}: G \times T &= \text{Cell} - \text{group} - \text{treatment} \\
&= 414.21 - 63.01 - 232.05 = 119.15 \\
\text{Table (3)}: P \times T &= \text{Cell} - \text{period} - \text{treatment} \\
&= 414.21 - 44.13 - 232.05 = 138.03
\end{aligned}$$

Unfortunately, each of these two-factor interaction sums of squares includes the sum of squares for the main effect of the third factor; that is, $G \times P$ includes the treatment sum of squares, $G \times T$ includes the period sum of squares, and $P \times T$ includes the group sum of squares. Consequently, we have no way of interpreting the differences in these two-factor interactions. For example, if we subtract the sum of squares for the main effect included in each of the two-factor interactions, then the Latin square error or residual sum of squares will be

$$\begin{aligned}
\text{Residual} &= G \times P - T = 307.07 - 232.05 = 75.02 \\
\text{Residual} &= G \times T - P = 119.15 - 44.13 = 75.02 \\
\text{Residual} &= P \times T - G = 138.03 - 63.01 = 75.02
\end{aligned}$$

and we have no way of knowing which, if any, of the two-factor interactions may be contributing to the Latin square error or residual sum of squares.

With G random, we have assumed that σ_{gp}^2 and σ_{gt}^2 are both equal to zero, but that θ_{pt}^2 may not necessarily be equal to zero. With these assumptions, the expectations of the mean squares with groups and subjects random and periods and treatments fixed are shown in Table 16.12. Because

Table 16.12

EXPECTATION OF MEAN SQUARES IN AN EXPERIMENT IN WHICH THE SAME LATIN SQUARE IS REPLICATED

Source	d.f.	Expectation of Mean Square
Groups	$g - 1$	$\sigma^2[1 + (p - 1)\rho] + p\sigma_{s(g)}^2 + sp\sigma_g^2$
$S(G)$	$g(s - 1)$	$\sigma^2[1 + (p - 1)\rho] + p\sigma_{s(g)}^2$
Treatments	$t - 1$	$\sigma^2(1 - \rho) + \sigma_{s(g)p}^2 + sp\theta_t^2$
Periods	$p - 1$	$\sigma^2(1 - \rho) + \sigma_{s(g)p}^2 + st\theta_p^2$
Residual	$(t - 1)(t - 2)$	$\sigma^2(1 - \rho) + \sigma_{s(g)p}^2 + s\theta_{pt}^2$
$S(G) \times P$ [a]	$g(s - 1)(p - 1)$	$\sigma^2(1 - \rho) + \sigma_{s(g)p}^2$

[a] *Note:* $S(G) \times P = S(G) \times T$ and $\sigma_{s(g)p}^2 = \sigma_{s(g)t}^2$.

$\sigma_{s(g)p}^2 = \sigma_{s(g)t}^2$, this component appears as part of the expectation of *both* the period and the treatment mean squares. Note also that the component $\sigma_{s(g)p}^2 = \sigma_{s(g)t}^2$ includes all of the necessary subscripts to identify either $G \times P$ or $G \times T$, sources that *may* be included in the Latin square error or residual mean square. When the subscripts gp and gt are deleted from $\sigma_{s(g)p}^2 = \sigma_{s(g)t}^2$, the remaining subscript is s and because s is random, $\sigma_{s(g)p}^2 = \sigma_{s(g)t}^2$ remains as part of the expectation of the Latin square error or residual mean square.

If we are reasonably confident that σ_{gt}^2 and σ_{gp}^2 are both equal to zero, then $F = MS_{res}/MS_{S(G)P}$ provides a test of significance of θ_{pt}^2. For this test of significance in the experiment considered previously, we had $F = 6.25/6.00 = 1.04$, a nonsignificant value. For reasons discussed earlier, a significant F for this test *may* also be obtained if σ_{gt}^2 or σ_{gp}^2 are not equal to zero. However, because the levels of G represent replications of the experiment with the treatments presented in a different order or sequence, we should expect both σ_{gt}^2 and σ_{gp}^2 to be negligible, unless there are systematic differences in carryover effects for the different sequences. With a nonsignificant value of F, we have some assurance that all of the two-factor interactions, $G \times T$, $G \times P$, and $P \times T$, are negligible.

16.13 Conservative F Tests

Because Latin square designs with repeated measures involve the same assumptions regarding homogeneity of variance and equal correlations as randomized block designs involving repeated measures, conservative F tests can also be made for Latin square designs with repeated measures. For a conservative test, we evaluate those F ratios involving repeated measures in terms of the tabled values of F with reduced degrees of freedom. The reduced degrees of freedom are obtained by dividing the degrees of freedom associated with the numerator and denominator of the F ratio by $p - 1$, where p is the number of repeated measures for each subject.

16.14 Replication of the 2 × 2 Latin Square

In Table 16.13, we have five replications of a 2 × 2 Latin square. It should be obvious, in the case of the 2 × 2 square, that we have only two possible orders, AB and BA, and it does not make any difference whether we replicate the same square or a series of independently randomized squares. Regardless of which procedure we use, each order or sequence will occur an equal number of times.

Table 16.13
FIVE REPLICATIONS OF THE 2 × 2 LATIN SQUARE

Squares

	Periods	
Subjects	1	2
1	*A*	*B*
2	*B*	*A*
3	*A*	*B*
4	*B*	*A*
5	*B*	*A*
6	*A*	*B*
7	*A*	*B*
8	*B*	*A*
9	*A*	*B*
10	*B*	*A*

Observations

	Periods		
Subjects	1	2	Σ
1	9	6	15
2	5	8	13
3	10	9	19
4	7	7	14
5	4	7	11
6	8	9	17
7	8	4	12
8	5	11	16
9	5	1	6
10	6	4	10
Σ	67	66	133

	Periods		
	1	2	Σ
Square 1	14	14	28
Square 2	17	16	33
Square 3	12	16	28
Square 4	13	15	28
Square 5	11	5	16
Σ	67	66	133

	Treatments		
	A	*B*	Σ
Square 1	17	11	28
Square 2	17	16	33
Square 3	15	13	28
Square 4	19	9	28
Square 5	9	7	16
Σ	77	56	133

For the analysis of variance of Table 16.13, we obtain the following sums of squares:

$$\text{Total} = (9)^2 + (5)^2 + \cdots + (4)^2 - \frac{(133)^2}{20} = 114.55$$

$$\text{Row} = \frac{(15)^2}{2} + \frac{(13)^2}{2} + \cdots + \frac{(10)^2}{2} - \frac{(133)^2}{20} = 64.05$$

$$\text{Column} = \frac{(67)^2}{10} + \frac{(66)^2}{10} - \frac{(133)^2}{20} = 0.05$$

$$\text{Treatment} = \frac{(77)^2}{10} + \frac{(56)^2}{10} - \frac{(133)^2}{20} = 22.05$$

As we have pointed out previously, we have no Latin square or residual error for the 2 × 2 square. To obtain an estimate of experimental error, we can calculate the square × period and square × treatment interactions from the two tables shown at the bottom of Table 16.13. Making these calculations, we find

$$\text{Square} \times \text{period} = 14.20$$

Table 16.14
THE OBSERVATIONS OF TABLE 16.13 REARRANGED ACCORDING TO THE ORDER, *AB* OR *BA*, OF THE TREATMENTS

Order	Subjects	Period 1	Period 2	Σ
	1	9	6	15
	3	10	9	19
AB	6	8	9	17
	7	8	4	12
	9	5	1	6
	Σ	40	29	69
	2	5	8	13
	4	7	7	14
BA	5	4	7	11
	8	5	11	16
	10	6	4	10
	Σ	27	37	64

Table 16.15
ANALYSIS OF VARIANCE FOR THE DATA OF TABLE 16.13

Source of Variation	Sum of Squares	d.f.	Mean Square	F
Orders	1.25	1	1.25	
Error (*a*)	62.80	8	7.85	
Treatments	22.05	1	22.05	6.21
Periods	0.05	1	0.05	
Error (*b*)	28.40	8	3.55	
Total	114.55	19		

with 4 d.f., and

$$\text{Square} \times \text{treatment} = 14.20$$

with 4 d.f. These two sums of squares are pooled in Table 16.15 to provide the error (*b*) sum of squares, $14.20 + 14.20 = 28.40$, with 8 d.f. It is not necessary, except as a check on our arithmetic, to calculate the square × period and square × treatment interactions. We can obtain the pooled sum of squares by subtraction. Thus,

$$\text{Error } (b) = \text{Total} - \text{row} - \text{column} - \text{treatment}$$

$$= 114.55 - 64.05 - 0.05 - 22.05 = 28.40$$

It is also possible to rearrange the observations of Table 16.13 in such a way as to obtain Table 16.14. If we calculate the subject × period sum of squares for the first order, *AB*, and also for the second order, *BA*, each of these sums of squares will have 4 d.f. The sum of these two sums of squares is equal to 28.40 and is identical with the error (*b*) sum of squares of Table 16.15. It is also easy to see, in this example, that the subject × period sum of squares is identical with the subject × treatment sum of squares.

The row sum of squares, which is equal to 64.05, with 9 d.f., can be analyzed into a sum of squares for orders with 1 d.f. and a pooled sum of squares between subjects tested with the same order with 8 d.f. We can obtain these sums of squares from the data as arranged in Table 16.14. Thus,

$$\text{Order} = \frac{(69)^2}{10} + \frac{(64)^2}{10} - \frac{(133)^2}{20} = 1.25$$

and the pooled sum of squares between subjects with the same order will be equal to $64.05 - 1.25 = 62.80$.

By direct calculation, we obtain as the sum of squares between subjects with order AB

$$\frac{(15)^2}{2} + \frac{(19)^2}{2} + \cdots + \frac{(6)^2}{2} - \frac{(69)^2}{10} = 51.4$$

and between subjects with order BA

$$\frac{(13)^2}{2} + \frac{(14)^2}{2} + \cdots + \frac{(10)^2}{2} - \frac{(64)^2}{10} = 11.4$$

and the sum of these two sums of squares is 62.80, the same value we obtained by subtraction. The pooled sum of squares between subjects tested with the same order with 8 d.f. provides the error mean square (a) for testing the significance of the order mean square.

The analysis of variance is summarized in Table 16.15. For the test of significance of the treatment mean square, we have $F = 22.05/3.55 = 6.21$ with 1 and 8 d.f. With $\alpha = 0.05$, this is a significant value of F.

16.15 Carryover Effects When Subjects Are Administered All Treatments

A basic condition in the analysis of variance for the Latin square designs involving repeated measures is that the value of an observation for one treatment is not influenced by the effects of treatments applied during earlier periods. When this condition is not met, we refer to the treatments as having *carryover* effects.[2]

With a 5×5 Latin square and with twenty-five subjects, so that each subject is tested under only one treatment, there is no reason to believe that carryover effects would be operative. However, with a 5×5 Latin square and with only five subjects, so that each subject is tested under all treatments, there *may* be carryover effects.

One way in which the experimenter may hope to eliminate the possibility of carryover effects, when the same subject is tested under all treatments, is to increase the time interval between the various periods in which the treatments are applied. For example, in the Bliss and Rose experiment described earlier, it may be noted that a ten-day interval separated the periods in which the drugs were administered. Obviously, one might expect carryover effects, in drug research, if the various drugs being investigated are administered in close succession before the effects of the previous drugs have had an opportunity to wear off.

[2] In many experiments carryover effects resulting from practice or learning are of experimental interest, but these effects should be investigated in experimental designs where they are not confounded with differences in treatments. Designs for investigating practice and learning effects are discussed in Chapter 17.

In psychological research, where each subject is given a series of *different* treatments, the possibility of carryover effects must always be considered. For example, early treatments may produce fatigue effects that carry over to later treatments, if the treatments are administered in close succession. Furthermore, if an early treatment is such as to produce a feeling of anxiety, failure, or fear in a subject, then it seems likely that such feelings would influence the subject's motivation and behavior during subsequent treatments.

16.16 Balanced Latin Square Designs for Repeated Measures

Williams (1949) has suggested a variation of the Latin square arrangement in which each treatment follows every other treatment the same number of times.[3] Latin squares of this kind are often called *balanced* squares. If the number of treatments is even, then for the first row of the balanced square we take

$$1, 2, t, 3, t-1, 4, t-2, 5, t-3, \cdots$$

in which the sequence $1, t, t-1, t-2, t-3, \ldots$ alternates with the sequence $2, 3, 4, 5, \ldots$. For example, with $t = 4$ treatments, the first row of the square would be

$$1 \quad 2 \quad 4 \quad 3$$

Then the remaining rows of the square are obtained by adding 1 to each previous row, with the understanding that if the number obtained is greater than t we subtract t. Thus, for the 4×4 square, we obtain

$$\begin{array}{cccc} 1 & 2 & 4 & 3 \\ 2 & 3 & 1 & 4 \\ 3 & 4 & 2 & 1 \\ 4 & 1 & 3 & 2 \end{array}$$

We see that, in this Latin square, Treatment 1 follows immediately after Treatments 2, 3, and 4 one time each. Similarly, Treatment 2 follows immediately after Treatments 1, 3, and 4 one time each; Treatment 3 follows immediately after Treatments 1, 2, and 4 one time each; and Treatment 4 follows immediately after Treatments 1, 2, and 3 one time each.

If the number of treatments is odd, then two Latin squares are required to have each treatment follow immediately after every other treatment an

[3] See also Wagenaar (1969).

equal number of times. For one of the two squares we follow the same sequence for the first row as when the number of treatments is even. For the other square we reverse the sequence of the first row. For five treatments, for example, the first row of the first square would be 1, 2, 5, 3, 4, and for the first row of the second square we would have 4, 3, 5, 2, 1. Then, following the addition rule for each square, we obtain

Square 1						Square 2				
1	2	5	3	4		4	3	5	2	1
2	3	1	4	5		5	4	1	3	2
3	4	2	5	1		1	5	2	4	3
4	5	3	1	2		2	1	3	5	4
5	1	4	2	3		3	2	4	1	5

In these two squares each treatment follows immediately after every other treatment exactly two times.

Suppose, for example, that Treatment 4 functions as a general depressant, lowering the value of the observation by a constant for any treatment immediately following it. Because every treatment follows immediately after Treatment 4 exactly twice, the mean for Treatments 1, 2, 3, and 5 will be influenced in the same way and the differences among them will not be changed. On the other hand, the means for Treatments 1, 2, 3, and 5 relative to the mean for Treatment 4 will not be the same as they would be if Treatment 4 had no influence on the other treatments.

If Treatment 4 acts nonadditively or differentially on the treatments immediately following it, the treatment means will not be influenced in the same manner and the differences among them will not be the same as we might expect to find if each treatment mean were based on a set of independent observations as, for example, in a randomized group design.

16.17 Analysis of Variance for Balanced Designs

The method of analysis for Latin square designs that we have presented involves the assumption that treatment effects are additive and constant, and that there is no carryover effect from one treatment to the next. The analysis of variance for the balanced Latin square provides an estimate of both the treatment effects and the carryover or residual effects of the immediately preceding treatment.[4] A somewhat

[4] For another solution to the problem of residual effects, see Pearce (1957).

better estimate of the residual effects can be obtained if an additional period or column is added to the Latin square that duplicates exactly the treatments of what would ordinarily be the last column or period. In this way, each treatment will be preceded equally often by every other treatment including itself. The analysis of variance for the balanced Latin square designs can be found in Cochran and Cox (1957).

QUESTIONS AND PROBLEMS

16.1 Sleight (1948) used a Latin square design to study the influence of the shape of instrument dials and exposure time on legibility. Five dial shapes were used: (*H*)orizontal, (*O*)pen window, (*R*)ound, (*V*)ertical, and (*S*)emicircular. The exposure times used were 0.28, 0.20, 0.17, 0.14, and 0.12 seconds. In a preliminary experiment, five subjects were tested. The measurements reported are the number of errors made by the subjects in reading the various dials under the various exposure times. The data are given below:

	Exposure Speed in Seconds						Exposure Speed in Seconds				
Subjects	0.28	0.20	0.17	0.14	0.12	Subjects	0.28	0.20	0.17	0.14	0.12
1	*H*	*O*	*S*	*R*	*V*	1	3	0	4	2	6
2	*S*	*R*	*V*	*H*	*O*	2	2	2	6	1	0
3	*V*	*H*	*O*	*S*	*R*	3	10	6	1	6	0
4	*O*	*S*	*R*	*V*	*H*	4	0	4	4	12	2
5	*R*	*V*	*H*	*O*	*S*	5	3	6	8	0	7

(*a*) Find the value of $F = MS_T/MS_{res}$. Note that the means and variances for three of the dial shapes are much the same, suggesting a square root transformation. (*b*) Add 0.5 to each cell entry and then take the square root. Find the value of $F = MS_T/MS_{res}$ for the transformed data. (*c*) Is there reason to believe that carryover effects might be present? (*d*) Are there good arguments for or against having the longest exposure time on the first trial?

16.2 De Lury (1946) reports on a Latin square design concerned with the investigation of the reactions of rabbits to four doses of a drug. The observations reported below are in terms of milligrams of glucose per 100 cc of blood. Four independently randomized Latin squares were used. The data reported below by De Lury were made available through the courtesy of Dr. D. M. Young and the Connaught Laboratories of the University of Toronto. Find the value of $F = MS_T/MS_{res}$ for each Latin square.

	Days					Days			
Rabbits	1	2	3	4	Rabbits	1	2	3	4
1	*C*	*A*	*B*	*D*	1	59	56	41	54
2	*B*	*D*	*C*	*A*	2	56	58	73	69
3	*A*	*C*	*D*	*B*	3	45	41	30	28
4	*D*	*B*	*A*	*C*	4	62	49	63	84
5	*A*	*B*	*D*	*C*	5	42	39	44	61
6	*C*	*A*	*B*	*D*	6	49	61	38	43
7	*B*	*D*	*C*	*A*	7	83	81	101	96
8	*D*	*C*	*A*	*B*	8	56	54	65	58
9	*B*	*D*	*A*	*C*	9	47	46	62	76
10	*A*	*C*	*B*	*D*	10	90	74	61	63
11	*C*	*B*	*D*	*A*	11	79	63	58	87
12	*D*	*A*	*C*	*B*	12	50	69	66	59
13	*D*	*A*	*C*	*B*	13	45	61	45	71
14	*C*	*B*	*D*	*A*	14	52	31	35	81
15	*A*	*D*	*B*	*C*	15	57	30	57	50
16	*B*	*C*	*A*	*D*	16	64	83	74	67

16.3 The experiment on screen size described in the chapter was concerned not only with the ability of the subjects to locate the position of the target in terms of degrees, but also with the ability of the subjects to judge the distance of the target from the center of the screen. The latter judgments were made in terms of miles. (*a*) Complete the analysis of variance. (*b*) Calculate the subject × period sum of squares for each order. The sum of these sums of squares should be equal to the value you obtained by subtraction.

16.4 A 4 × 4 Latin square is selected at random and this same square is replicated six times by using twenty-four subjects with each subject tested under each of the four treatments. (*a*) Give the analysis of variance summary table showing sources of variation and degrees of freedom. (*b*) Give the expectations of the mean squares.

16.5 An experiment involves five treatments. Four independently randomized Latin squares are used with a single subject assigned to each of the rows of each square. (*a*) Give the analysis of variance summary table showing sources of variation and degrees of freedom. (*b*) Give the expectations of the mean squares.

16.6 An experimenter used five independently randomized 4 × 4 Latin squares. In his analysis of variance table there is a line labeled "error" with 54 d.f. What are the components of this error sum of squares and how many degrees of freedom are associated with each component?

16.7 The same 6 × 6 Latin square is replicated five times. In the published analysis of variance table there is a line labeled "error" with

Scores of Subjects in Locating Targets on Screens of Five Different Sizes

Latin Square	Subjects	Periods 1	2	3	4	5	Σ
	1	19	21	25	27	22	114
	2	22	20	23	31	24	120
3 6 4 7 5	3	26	28	26	31	32	143
	4	17	20	17	14	18	86
	5	28	30	30	31	28	147
	Σ	112	119	121	134	124	610
	6	23	30	29	24	28	134
	7	24	33	28	19	32	136
4 7 5 3 6	8	29	30	31	29	28	147
	9	24	26	27	25	31	133
	10	11	18	27	18	24	98
	Σ	111	137	142	115	143	648
	11	18	16	24	24	18	100
	12	29	26	29	29	27	140
5 3 6 4 7	13	30	27	30	30	31	148
	14	25	22	28	26	30	131
	15	15	17	15	16	15	78
	Σ	117	108	126	125	121	597
	16	27	24	27	29	26	133
	17	22	14	12	18	15	81
6 4 7 5 3	18	28	30	34	31	28	151
	19	23	22	25	23	16	109
	20	29	26	28	28	26	137
	Σ	129	116	126	129	111	611
	21	29	28	21	30	24	132
	22	30	31	30	33	30	154
7 5 3 6 4	23	19	19	17	25	24	104
	24	28	20	16	20	21	105
	25	24	27	25	28	27	131
	Σ	130	125	109	136	126	626

140 d.f. What are the components of this error sum of squares and how many degrees of freedom are associated with each component?

16.8 An experimenter has two treatments, A and B, of interest. Each subject in the experiment receives both treatments. The experimenter uses a 2×2 Latin square and has a total of twenty subjects. (a) How many degrees of freedom will be associated with the error sum of squares? (b) What are the components of the error sum of squares and how many degrees of freedom are associated with each? (c) No matter what your answer is to (b), provided it is correct, there is an alternative way of partitioning the error sum of squares. What is this alternative and how many degrees of freedom will be associated with each component?

16.9 Twelve subjects were divided at random into $g = 3$ groups of $s = 4$ subjects each. Each subject was tested under $t = 3$ treatments. Group 1 received the treatments in the order ACB, Group 2 received the treatments in the order CBA, and Group 3 received the treatments in the order BAC. The design is a Latin square design with $s = 4$ subjects assigned to each row of a 3×3 Latin square. Find the various sums of squares involved in the analysis of variance.

		Periods		
Group	Subjects	1	2	3
	1	0	2	6
ACB	2	3	1	4
	3	5	9	7
	4	8	1	4
	5	0	7	6
CBA	6	1	10	5
	7	6	7	8
	8	1	3	4
	9	2	0	8
BAC	10	13	10	13
	11	14	11	14
	12	9	6	12

(a) What is the value of $MS_{S(G)}$? (b) What is the value of the Latin square error or residual mean square, MS_{res}? (c) What is the value of the subject $\times$ period mean square, $MS_{S(G)P}$? (d) How would you account for the fact that $MS_{S(G)P}$ is considerably smaller than MS_{res}?

16.10 An experiment involved two independently randomized 4×4 Latin squares. Subjects were assigned at random to the rows of each square.

The outcome of the experiment is given below:

Subjects	Periods 1	2	3	4	Latin Squares
1	1	5	5	6	*ABCD*
2	5	4	2	3	*CDAB*
3	3	1	7	4	*BADC*
4	5	4	3	1	*DCBA*
5	8	9	4	5	*CDAB*
6	7	6	6	3	*DCBA*
7	5	3	7	7	*BADC*
8	1	4	5	6	*ABCD*

Find the various sums of squares involved in the analysis of variance. (*a*) What is the value of $MS_{R(S)}$, that is, rows within squares? (*b*) What is the value of the square × treatment, MS_{ST}, mean square? (*c*) What is the value of the square × column, MS_{SC}, mean square? (*d*) What is the value of the mean square based on the pooled Latin square error or residual sums of squares, that is, MS_{res}? (*e*) Is it reasonable to conclude that the experimental technique is reliable?

16.11 Twelve subjects were divided at random into three groups of four subjects each. One group was assigned at random to each row of a 3 × 3 Latin square and each subject was tested under all three treatments. The first group received the treatments in the order *CBA*, the second in the order *BAC*, and the third in the order *ACB*. The outcome of the experiment is shown below:

Group	Subjects	Periods 1	2	3
CBA	1	8	9	11
	2	8	4	3
	3	13	10	11
	4	12	4	7
BAC	1	10	7	11
	2	13	7	6
	3	8	7	15
	4	9	11	10
ACB	1	9	7	4
	2	5	6	6
	3	7	8	6
	4	10	15	11

Complete the analysis of variance. (*a*) What is the value of $MS_{S(G)}$? (*b*) What is the value of MS_{res}? (*c*) What is the value of $MS_{S(G)P}$?

CHAPTER 17

TREND ANALYSIS DESIGNS

17.1 Introduction

In studies of learning, our interest is centered on improvement or change in performance as a result of practice. In a sense, practice can be considered a factor with the successive periods of practice or trials as levels. If, for example, we were interested in change in performance over five trials, and if fifty subjects were available, we might randomly divide the subjects into five groups of ten subjects each. One group of subjects would then be given a single trial, another two trials, a third three trials, and so on. For each subject in each group, we would use only the final measure, under the assumption that it provides an estimate of performance for a specified number of trials. The analysis of variance of the experimental data would be the same as for a randomized group design.

If we used the measure obtained on Trial 1 for the subjects, it would also be possible to arrange the subjects in blocks of five such that within each block the subjects are relatively homogeneous with respect to performance on Trial 1. Then using random methods, one subject in each block could be assigned to each of the five levels or trials. In this instance, the analysis of variance would correspond to that of a randomized block design.

Most experimenters, however, would feel that both of the above procedures are inefficient in the sense that four observations (all but the last) would be discarded for the subjects with five trials. Similarly, we would discard the first three observations for those subjects receiving four trials, the first two observations for those subjects receiving three trials, and the first observation for those subjects receiving two trials. The argument would be that the same amount of information could be obtained by giving a single group of ten subjects five trials. Thus, each subject would have a score or measure for each trial. In this instance, each subject would correspond to a block in a randomized block design with repeated measures.

However, we should note that the trials (treatments) would not be independently randomized for each subject, as they would be in a randomized block design with repeated measures, but rather would occur in exactly the same sequence or order for each subject.

In this chapter we shall be concerned with the analysis of variance for experiments concerned with the trend of a series of means in which we have repeated measures on subjects. The treatments will consist of successive trials and the trials are the same for all subjects. In these experiments, the primary objective is to study the trend of the means over the successive trials. The observations for each trial are obtained under a standard condition and it is assumed that any differences found among the trial means are the result of differing amounts of *practice* or *carryover effects* from preceding trials. An extension of this experimental design involves the introduction of one or more factors. These factors may be treatments that can be randomized or organismic factors that cannot be randomized.

An examination of the means for a series of trials may reveal that the trend is either upward or downward; that is, the means may either increase or decrease. Now such a trend can, of course, occur as a result of random variation. From the experimenter's point of view, the important question is whether the upward or downward trend can be regarded as meeting the requirements of statistical significance or as a random or chance affair. Similarly, the trend of the means, in addition to being downward or upward, as the case may be, may also show a bend or degree of curvature. Again, if there is a bend or curvature in the trend, we wish to be able to determine whether the curvature is such as to meet the requirements of statistical significance.

If trial means are available for two or more different treatment groups, then we may wish to determine whether there are significant differences between certain characteristics of the trends of the means for the various treatment groups. For example, if we have trial means for two different treatment groups, we may have reason to believe that the trend of the means for one treatment should be sharply downward, whereas the trend for the other treatment should be only slightly downward. An examination of the trial means for each treatment group may indicate that the trends are in accord with expectation. A corresponding test of significance provides a basis for determining whether the difference in the trends for the two treatments is significant.

It should be emphasized that the methods of analysis described in this chapter are not concerned with the problem of finding an equation that will describe the trend of the trial means. The problem of curve fitting is, of course, of importance. Our concern in this chapter, however, is in providing the experimenter with methods for determining whether certain characteristics of the trend of the trial means are statistically significant or whether they can be attributed to random variation.

It may also be emphasized that before undertaking the analyses and tests of significance described, it is always advisable to plot the means for the successive trials for each treatment or experimental condition. Examination of these plots prior to the data analysis is of value in that the plots suggest what the data analysis may confirm. Furthermore, it is advisable to present either the means or the plots in reporting the data analysis because they assist others in understanding the results of the analysis.

17.2 Trial Means: One Standard Condition

Table 17.1 gives measures of performance for each of five subjects on each of three trials. The subjects were tested under the same conditions and the only effect of interest is the differences in the trial means. We find the total sum of squares, the sum of squares for subjects (blocks or rows), and the sum of squares for trials (columns) in the usual way. For the data of Table 17.1, we have

$$\text{Total} = (3)^2 + (7)^2 + \cdots + (12)^2 - \frac{(105)^2}{15} = 144.00$$

$$\text{Trial} = \frac{(20)^2}{5} + \frac{(35)^2}{5} + \frac{(50)^2}{5} - \frac{(105)^2}{15} = 90.00$$

$$\text{Subject} = \frac{(20)^2}{3} + \frac{(27)^2}{3} + \cdots + \frac{(27)^2}{3} - \frac{(105)^2}{15} = 48.67$$

Table 17.1

OBSERVATIONS OBTAINED FOR FIVE SUBJECTS ON THREE TRIALS

	Trials			
Subjects	1	2	3	Σ
1	3	7	10	20
2	7	9	11	27
3	2	4	7	13
4	2	6	10	18
5	6	9	12	27
Σ	20	35	50	105

Table 17.2
ANALYSIS OF VARIANCE OF THE DATA OF TABLE 17.1

Source of Variation	Sum of Squares	d.f.	Mean Square	F
Subjects	48.67	4	12.17	
Trials	90.00	2	45.00	67.2
$S \times T$	5.33	8	0.67	
Total	144.00	14		

If we subtract the trial and subject sums of squares from the total, we have the subject $\times$ trial ($S \times T$) sum of squares. Thus,

$$S \times T = \text{Total} - \text{subject} - \text{trial} \tag{17.1}$$

or, for the present example,

$$S \times T = 144.00 - 48.67 - 90.00 = 5.33$$

17.3 Expectation of Mean Squares and Test of Significance

Table 17.2 summarizes the analysis of variance and in Table 17.3 we give the expected values of the mean squares for a mixed model with subjects random and trials fixed. It is obvious that, for the mixed model, MS_{ST} is the appropriate error mean square for testing the significance of the trial mean square. Then, for the experimental data, we have $F = MS_T/MS_{ST} = 45.00/0.67 = 67.2$ with 2 and 8 d.f., and this is a highly significant value.

The fact that the error mean square, MS_{ST}, is not very large indicates that the learning curves for the subjects have comparable forms. This could be examined graphically by plotting each subject's learning curve for the three trials.

Table 17.3
EXPECTATION OF MEAN SQUARES IN TABLE 17.2

Source	Expectation of Mean Square
Subjects	$\sigma^2[1 + (t - 1)\rho] + t\sigma_s^2$
Trials	$\sigma^2(1 - \rho) + \sigma_{st}^2 + s\theta_t^2$
$S \times T$	$\sigma^2(1 - \rho) + \sigma_{st}^2$

17.4 Linear and Quadratic Components of the Trial Sum of Squares

The sum of squares for trials, with 2 d.f., may be partitioned into a sum of squares for linear regression, the linear component of the trend, with 1 d.f., and a sum of squares for curvature, the quadratic component of the trend, also with 1 d.f. From Table XI in the Appendix, with $k = 3$ trials, we have as the orthogonal coefficients for the linear component of the trend of the trial means -1, 0, and 1. Then the sum of squares for the linear component of the trend of the trial means will be

$$\text{Linear} = \frac{[(-1)(20) + (0)(35) + (1)(50)]^2}{(5)(2)} = 90.00$$

and we note that this sum of squares is exactly equal to the sum of squares for trials. Therefore, the sum of squares for curvature, the quadratic component, must be equal to zero. The trend of the trial means can thus be accurately represented by a straight line. All of this, of course, is obvious in this simple example, and could easily be shown by plotting the trial means. In actual experiments involving actual data, results are seldom so obvious or so simple.

17.5 Trial Means with Different Treatments

In learning experiments we may be interested not only in the change in performance over a series of trials under a standard condition, but under different experimental treatments. Various other factors, in addition to practice, may influence the shape of the learning curve. For example, we may be interested in the progress of learning under different dosages of a drug or under different drugs at a standard dosage. Frequency of reinforcement may be varied in several ways and we may wish to know whether the different levels of reinforcement influence the shape of the learning curve. For treatment factors such as those described, randomization in the assignment of the subjects to the levels of the factor is possible and should be used.

Let us suppose we are interested in the influence of three drugs, each at a standard dosage, on learning. We designate this treatment factor as A and let the three drugs be represented by A_1, A_2, and A_3. We have fifteen subjects available and the subjects are assigned at random in such a way that we have $s = 5$ subjects for each drug. Let us also assume that we have decided to test each subject on $t = 3$ trials; that is, we shall have three observations for each subject. We designate the trials by T and the successive trials by T_1, T_2, and T_3. Each subject corresponds to a block and the levels of A have been randomized over the blocks. Other than the fact that the levels of T

have not been independently randomized for each subject or block, this design is similar to the split-plot design with repeated measures discussed in an earlier chapter.

The outcome of the experiment is shown in Table 17.4. We find the total sum of squares, the sum of squares for subjects (blocks or rows), and the sum of squares for trials (columns) in the usual manner. Thus,

$$\text{Total} = (2)^2 + (2)^2 + \cdots + (10)^2 - \frac{(340)^2}{5} = 369.11$$

$$\text{Subject} = \frac{(13)^2}{3} + \frac{(18)^2}{3} + \cdots + \frac{(27)^2}{3} - \frac{(340)^2}{45} = 175.78$$

$$\text{Trial} = \frac{(80)^2}{15} + \frac{(110)^2}{15} + \frac{(150)^2}{15} - \frac{(340)^2}{45} = 164.44$$

If we subtract the subject and trial sums of squares from the total, we will

Table 17.4

OBSERVATIONS FOR THREE GROUPS WITH EACH GROUP TESTED UNDER A DIFFERENT DRUG AND WITH THREE TRIALS FOR EACH SUBJECT

Drugs	Subjects	Trials B_1	B_2	B_3	Σ
	1	2	4	7	13
	2	2	6	10	18
A_1	3	3	7	10	20
	4	7	9	11	27
	5	6	9	12	27
	1	5	6	10	21
	2	4	5	10	19
A_2	3	7	8	11	26
	4	8	9	11	28
	5	11	12	13	36
	1	3	4	7	14
	2	3	6	9	18
A_3	3	4	7	9	20
	4	8	8	10	26
	5	7	10	10	27
	Σ	80	110	150	340

have the $S \times T$ interaction sum of squares or

$$S \times T = 369.11 - 175.78 - 164.44 = 28.89$$

The number of subjects tested with each drug is $s = 5$. We have $a = 3$ drugs and $t = 3$ trials. Then the total sum of squares will have $sat - 1 = (5)(3)(3) - 1 = 44$ d.f. The sum of squares for subjects will have $sa - 1 = (5)(3) - 1 = 14$ d.f., and the sum of squares for trials will have $t - 1 = 2$ d.f. The $S \times T$ sum of squares will have $(sa - 1)(t - 1) = 28$ d.f.

17.6 Partitioning the Subject Sum of Squares

The sum of squares between subjects with 14 d.f. can be partitioned into two component parts. One of these sums of squares will be the sum of squares for A (drugs) with $a - 1 = 2$ d.f. The sum of squares for A can be obtained from the marginal entries of Table 17.5. Thus,

$$A = \frac{(105)^2}{15} + \frac{(130)^2}{15} + \frac{(105)^2}{15} - \frac{(340)^2}{45} = 27.78$$

If we subtract the A or drug sum of squares from the sum of squares for subjects, we will have a residual sum of squares equal to $175.78 - 27.78 = 148.00$ with $14 - 2 = 12$ d.f. This residual sum of squares is the *pooled* sum of squares between subjects in each drug group. As we have done previously, we use the notation $S(A)$ to indicate that this sum of squares is based on the variation between subjects *nested* within each level of A. For example, for subjects tested with A_1, we have

$$S(A_1) = \frac{(13)^2}{3} + \frac{(18)^2}{3} + \cdots + \frac{(27)^2}{3} - \frac{(105)^2}{15} = 48.67$$

Similarly, for subjects tested with A_2 and A_3, we have

$$S(A_2) = \frac{(21)^2}{3} + \frac{(19)^2}{3} + \cdots + \frac{(36)^2}{3} - \frac{(130)^2}{15} = 59.33$$

and

$$S(A_3) = \frac{(14)^2}{3} + \frac{(18)^2}{3} + \cdots + \frac{(27)^2}{3} - \frac{(105)^2}{15} = 40.00$$

Then

$$\begin{aligned} S(A) &= S(A_1) + S(A_2) + S(A_3) \\ &= 48.67 + 59.33 + 40.00 \\ &= 148.00 \end{aligned}$$

and this is exactly equal to the value we obtained above by subtraction. Each of the separate sums of squares has $s - 1 = 4$ d.f. and because we have a of these sums of squares, the pooled $S(A)$ sum of squares has $a(s - 1) = 3(5 - 1) = 12$ d.f.

17.7 Partitioning the Subject $\times$ Trial Sum of Squares

The $S \times T$ sum of squares, with 28 d.f., can also be partitioned into two component parts. One of these sums of squares will be the drug $\times$ trial ($A \times T$) sum of squares with $(a - 1)(t - 1) = (3 - 1)(3 - 1) = 4$ d.f. This sum of squares can be obtained by first calculating the cell sum of squares for Table 17.5. Thus,

$$\text{Cell} = \frac{(20)^2}{5} + \frac{(35)^2}{5} + \cdots + \frac{(45)^2}{5} - \frac{(340)^2}{45} = 201.11$$

We have already calculated the row or drug (A) sum of squares and the column or trial (T) sum of squares for Table 17.5. Then, by subtraction, we obtain

$$A \times T = 201.11 - 27.78 - 164.44 = 8.89$$

If we now subtract the $A \times T$ sum of squares from the $S \times T$ sum of squares, we obtain a residual that is equal to $28.89 - 8.89 = 20.00$ and this residual sum of squares will have $28 - 4 = 24$ d.f. This residual sum of squares is the *pooled* $S(A) \times T$ interaction and could be calculated directly by finding the $S(A_1) \times T$, $S(A_2) \times T$, and $S(A_3) \times T$ sums of squares. If we were to make these calculations, using the data of Table 17.4, we

Table 17.5

SUMS FOR DRUGS AND TRIALS FOR THE DATA OF TABLE 17.4

	Trials			
Drugs	1	2	3	Σ
A_1	20	35	50	105
A_2	35	40	55	130
A_3	25	35	45	105
Σ	80	110	150	340

would find that

$$S(A_1) \times T = 5.33$$
$$S(A_2) \times T = 6.67$$
$$S(A_3) \times T = 8.00$$

and

$$\begin{aligned} S(A) \times T &= S(A_1) \times T + S(A_2) \times T + S(A_3) \times T \\ &= 5.33 + 6.67 + 8.00 \\ &= 20.00 \end{aligned}$$

which is exactly equal to the value we obtained by subtraction. The $S(A) \times T$ sum of squares will have $a(s - 1)(t - 1) = (3)(5 - 1) \times (3 - 1) = 24$ d.f.

17.8 *Trial Means with Different Treatments: Expectation of Mean Squares and Tests of Significance*

Table 17.6 summarizes the calculations and Table 17.7 gives the expectations of the mean squares under the assumption that subjects (S) are random and drugs (A) and trials (T) are fixed. We see that $MS_{S(A)}$ is the appropriate error mean square for testing the significance of the A effect and that $MS_{S(A)T}$ is the appropriate error mean square for testing the significance of the trial and the $A \times T$ interaction effects.

For the A effect we have $F = 13.89/12.33 = 1.13$ with 2 and 12 d.f., and this is not a significant value. Because the means for A_1, A_2, and A_3 have been averaged over three trials, they correspond to a general measure of performance for each drug. The test of significance indicates that the means for the three drugs do not differ significantly.

Table 17.6
ANALYSIS OF VARIANCE OF THE DATA OF TABLE 17.4

Source of Variation	Sum of Squares	d.f.	Mean Square	F
A: Drugs	27.78	2	13.89	1.13
$S(A)$: Error (a)	148.00	12	12.33	
T: Trials	164.44	2	82.22	99.06
$A \times T$	8.89	4	2.22	2.67
$S(A) \times T$: Error (b)	20.00	24	0.83	
Total	369.11	44		

Table 17.7
EXPECTATION OF MEAN SQUARES IN TABLE 17.6

Source	Expectation of Mean Square
A: Drugs	$\sigma^2[1 + (t - 1)\rho] + t\sigma_{s(a)}^2 + st\theta_a^2$
$S(A)$: Error (a)	$\sigma^2[1 + (t - 1)\rho] + t\sigma_{s(a)}^2$
T: Trials	$\sigma^2(1 - \rho) + \sigma_{s(a)t}^2 + sa\theta_t^2$
$A \times T$	$\sigma^2(1 - \rho) + \sigma_{s(a)t}^2 + s\theta_{at}^2$
$S(A) \times T$: Error (b)	$\sigma^2(1 - \rho) + \sigma_{s(a)t}^2$

For the T effect, we have $F = 82.22/0.83 = 99.06$ with 2 and 24 d.f., and this is a highly significant value. The three trial means, T_1, T_2, and T_3, have been averaged over the three drugs and we conclude that these means differ significantly.

Testing the $A \times T$ mean square for significance, we have $F = 2.22/0.83 = 2.67$ with 4 and 24 d.f. The tabled value of F for 4 and 24 d.f., with $\alpha = 0.05$, is 2.78 and our obtained value just misses being declared significant at this level. A significant $A \times T$ mean square would indicate that the learning curves for the three levels of A are not of the same form. To examine this interaction, we plot the curves for each drug group in Figure 17.1. If these curves were exactly parallel, the $A \times T$ interaction sum of squares would have to be equal to zero. As it is, there is some tendency for the curves to have somewhat different forms, but, with $\alpha = 0.05$, we cannot say that the forms differ significantly.

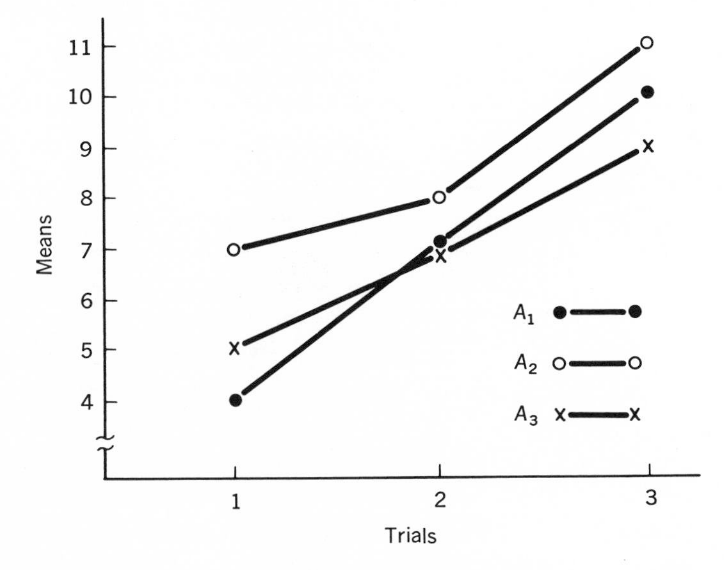

Figure 17.1
Trial means for each of three drug groups.

In experiments of the kind described, where the levels of the A factor are randomized over blocks (subjects), we will, as we have pointed out previously, generally find that $MS_{S(A)}$ is larger than $MS_{S(A)T}$. Our primary interest is usually in the T (trial) effect and the $A \times T$ interaction, and both of these are tested for significance with $MS_{S(A)T}$. The significance or lack of significance of the $A \times T$ interaction tells us whether the trend of the trial means is of the same form for the various levels of A. Because the levels of A have been randomized over subjects, a significant $A \times T$ interaction should not occur as a result of systematic differences between subjects in the various A groups.

In the next section, we illustrate some additional analyses that can be made with respect to the trend of the trial means and of the differences in the trends of the trial means for the separate levels of A.

17.9 Trial Means with Different Treatments: Linear and Quadratic Components of the Trial Sum of Squares

In Table 17.8 we give the sums for each drug group for each trial. At the top of the table we show the orthogonal coefficients for the linear comparison (L) and also for the quadratic comparison (Q). These coefficients were obtained from Table XI in the Appendix. We multiply the trial sums for A_1 by the linear coefficients to obtain the first entry in column L or

$$(-1)(20) + (0)(35) + (1)(50) = 30$$

Similarly, multiplying the trial sums for A_1 by the quadratic coefficients

Table 17.8
TREND ANALYSIS FOR THE TRIAL SUMS OF TABLE 17.5

	Orthogonal Coefficients				
Linear:	-1	0	1		
Quadratic:	1	-2	1	Comparison	
Group	T_1	T_2	T_3	L	Q
A_1	20	35	50	30	0
A_2	35	40	55	20	10
A_3	25	35	45	20	0
$\sum$	80	110	150	70	10

we obtain the first entry in column Q or

$$(1)(20) + (-2)(35) + (1)(50) = 0$$

The linear and quadratic comparisons for the trial sums for A_2 and A_3 are obtained in the same manner. The last entries in columns L and Q are obtained by multiplying the overall trial sums by the linear and quadratic coefficients and these values must be equal to ΣL and ΣQ, respectively.

We consider first the trend of the trial means. The sum of the squared coefficients for the linear component of the trend is $\Sigma a._l^2 = 2$. Each of the trial sums is based on as observations, where a is the number of levels of A and s is the number of subjects in each level of A. We have $a = 3$ and $s = 5$, so that each of the trial sums is based on $(3)(5) = 15$ observations. Then, for the sum of squares for the linear component of the trend, we have

$$\text{Linear} = \frac{(70)^2}{(3)(5)(2)} = 163.33$$

with 1 d.f.

To test the linear component for significance we use $MS_{S(A)T} = 0.83$ as the error mean square. Thus, we have $F = 163.33/0.83 = 196.78$, a highly significant value for 1 and 24 d.f.

For the quadratic component of the trend we have $\Sigma a._q^2 = 6$ and, because each trial sum is based on $as = 15$ observations, the sum of squares for the quadratic component of the trend will be

$$\text{Quadratic} = \frac{(10)^2}{(3)(5)(6)} = 1.11$$

with 1 d.f. To test the curvature of the trend of the trial means for significance, we again use $MS_{S(A)T}$ as the error term. Then we have $F = 1.11/0.83 = 1.34$ and this is a nonsignificant value. We conclude that the trend of the trial means is essentially linear and that there is no significant curvature.

We note also that the sum of the two sums of squares, 163.33 and 1.11, each with 1 d.f., is equal to 164.44 or the sum of squares for trials with 2 d.f. Thus, we have made two orthogonal comparisons regarding the trend of the trial means. The linear comparison was significant, whereas the quadratic was not.

17.10 Trial Means with Different Treatments: Linear and Quadratic Components of the Group × Trial Sum of Squares

We now examine the differences among the linear components of the trends for the three levels of A or drug groups. The *linear* component ($G \times L$) of a group × trial ($G \times T$) interaction sum of squares will be given by a general formula. Thus,

$$G \times L = \frac{\sum_1^k L^2}{s \sum a._l^2} - \frac{\left(\sum_1^k L\right)^2}{ks \sum a._l^2} \tag{17.2}$$

where k is the number of groups, s is the number of subjects in each group, and $\Sigma a._l^2$ is the sum of the squares of the coefficients for the linear comparison. The first term on the right is the sum of k sums of squares of linear comparisons, one for each group, and each of these comparisons has 1 d.f. Consequently, the first term on the right has k d.f. The second term on the right is the linear comparison for the overall trial sum of squares and has 1 d.f. Thus, the sum of squares defined by (17.2) will have $k - 1$ d.f.[1] and provides a measure of the variation among the k linear components.

In our example, the sum of squares defined by (17.2) is the linear component, $A \times L$, of the $A \times T$ interaction sum of squares. We have $k = a = 3$ groups with $s = 5$ subjects in each group and $\Sigma a._l^2 = 2$. Then, substituting in (17.2), we have

$$A \times L = \frac{(30)^2 + (20)^2 + (20)^2}{(5)(2)} - \frac{(70)^2}{(3)(5)(2)} = 6.67$$

with $3 - 1 = 2$ d.f. Therefore, the mean square is $6.67/2 = 3.33$. For the test of significance, we have $F = 3.33/0.83 = 4.01$ with 2 and 24 d.f. This value of F is significant with $\alpha = 0.05$ and indicates that the linear components of the trends of the trial means for the three drug groups differ significantly.

It is apparent from Figure 17.1 that the trends for A_1 and A_3 are exactly linear. For each of these two groups the linear component will be equal to the corresponding trial sum of squares. Thus, the quadratic components of the trends for both groups must, of necessity, be equal to zero, as shown in Table 17.8.

The general equation for the quadratic component ($G \times Q$) of a group $\times$ trial ($G \times T$) interaction sum of squares will be given by

$$G \times Q = \frac{\sum_1^k Q^2}{s \sum a._q^2} - \frac{\left(\sum_1^k Q\right)^2}{ks \sum a._q^2} \tag{17.3}$$

where k is the number of groups, s is the number of subjects in each group, and $\Sigma a._q^2$ is the sum of the squares of the coefficients for the quadratic comparison. The sum of squares defined by (17.3) will have $k - 1$ d.f., and provides a measure of the differences among the k comparisons, quadratic components.

[1] The degrees of freedom for the $G \times L$ sum of squares can be obtained following the usual multiplication rule for the degrees of freedom for an interaction sum of squares.

In our example, the sum of squares defined by (17.3) is the quadratic component, $A \times Q$, of the $A \times T$ interaction sum of squares. We have $k = a = 3$, $s = 5$, and $\Sigma a_{\cdot q}^2 = 6$. Then, substituting in (17.3), we have

$$A \times Q = \frac{(0)^2 + (10)^2 + (0)^2}{(5)(6)} - \frac{(10)^2}{(3)(5)(6)} = 2.22$$

and the mean square will be equal to $2.22/2 = 1.11$. We have $F = 1.11/0.83 = 1.34$, a nonsignificant value for 2 and 24 d.f. We conclude that the trends of the trial means for the different levels of A, although differing significantly in their linear components, do not show any significant differences in curvature.

We observe that the sum of squares for the linear component of the $A \times T$ interaction is 6.67 and the sum of squares for the quadratic component is 2.22. Each of these sums of squares has 2 d.f. and their sum is equal to 8.89 with 4 d.f. This, of course, is the sum of squares for the $A \times T$ interaction. In essence, we have analyzed the $A \times T$ interaction sum of squares into two components, $A \times L$ and $A \times Q$. The first corresponds to differences in the linear components of the group trends and the second to differences in the quadratic components.

Note that a test of significance of the $A \times T$ interaction, if significant, simply tells us that the trend of the means is not comparable for the different levels of A. The tests of significance of the linear and quadratic components of the $A \times T$ interaction provide additional and more precise information regarding the nature of the differences in the trends.

17.11 *Trial Means with Different Treatments: Linear and Quadratic Components of the* $S(A) \times T$ *Sum of Squares*

Suppose, in the experiment described, we also multiply the scores of each subject in each group by the linear coefficients. For the subjects tested with Drug A_1, we would have

$$(-1)(2) + (0)(4) + (1)(7) = 5$$
$$(-1)(2) + (0)(6) + (1)(10) = 8$$
$$(-1)(3) + (0)(7) + (1)(10) = 7$$
$$(-1)(7) + (0)(9) + (1)(11) = 4$$
$$(-1)(6) + (0)(9) + (1)(12) = 6$$

Then, using (17.2), we have

$$S(A_1) \times L = \frac{(5)^2 + (8)^2 + (7)^2 + (4)^2 + (6)^2}{(1)(2)} - \frac{(30)^2}{(5)(1)(2)} = 5.0$$

as the linear component of the $S(A_1) \times T$ interaction sum of squares with 4 d.f. Similarly, for the group tested with Drug A_2, if we multiply each subject's scores by the linear coefficients, we would have

$$S(A_2) \times L = \frac{(5)^2 + (6)^2 + (4)^2 + (3)^2 + (2)^2}{(1)(2)} - \frac{(20)^2}{(5)(1)(2)} = 5.0$$

as the linear component of the $S(A_2) \times T$ interaction sum of squares. And, for the subjects tested with Drug A_3, we would have

$$S(A_3) \times L = \frac{(4)^2 + (6)^2 + (5)^2 + (2)^2 + (3)^2}{(1)(2)} - \frac{(20)^2}{(5)(1)(2)} = 5.0$$

Each of the above sums of squares has 4 d.f. and the pooled sum of squares, which is equal to 15.0 with 12 d.f., is the linear component, $S(A) \times L$, of the $S(A) \times T$ sum of squares.

To obtain the quadratic component of the $S(A) \times T$ sum of squares we would proceed in the same way using the quadratic coefficients. Doing this for each of the three groups separately, we have

$$S(A_1) \times Q = \frac{(1)^2 + (0)^2 + (-1)^2 + (0)^2 + (0)^2}{(1)(6)} - \frac{(0)^2}{(5)(1)(6)} = 0.33$$

$$S(A_2) \times Q = \frac{(3)^2 + (4)^2 + (2)^2 + (1)^2 + (0)^2}{(1)(6)} - \frac{(10)^2}{(5)(1)(6)} = 1.67$$

$$S(A_3) \times Q = \frac{(2)^2 + (0)^2 + (-1)^2 + (2)^2 + (-3)^2}{(1)(6)} - \frac{(0)^2}{(5)(1)(6)} = 3.00$$

Each of the above sums of squares has 4 d.f. and the pooled sum of squares has 12 d.f. The pooled sum of squares, which is equal to 5.0, is the quadratic component, $S(A) \times Q$, of the $S(A) \times T$ sum of squares.

We note that the sum of the linear and quadratic components of the $S(A) \times T$ sum of squares or $S(A) \times L + S(A) \times Q = 15.0 + 5.0 = 20.0$ and this is equal to the $S(A) \times T$ sum of squares.

17.12 Expectation of Mean Squares for Linear and Quadratic Components

In our tests of significance we have assumed that the linear and quadratic components of the $S(A) \times T$ sum of squares are estimates of the same common error variance; that is, we have assumed that

$$E(MS_{S(A)L}) = E(MS_{S(A)Q}) = E(MS_{S(A)T})$$

Table 17.9

SUMMARY OF THE ANALYSIS OF VARIANCE WITH LINEAR AND QUADRATIC COMPONENTS

Source of Variation	Sum of Squares	d.f.	Mean Square	F
A	27.78	2	13.89	1.13
$S(A)$: Error (a)	148.00	12	12.33	
Linear	163.33	1	163.33	130.66
$A \times L$	6.67	2	3.33	2.66
$S(A) \times L$: Error (l)	15.00	12	1.25	
Quadratic	1.11	1	1.11	2.64
$A \times Q$	2.22	2	1.11	2.64
$S(A) \times Q$: Error (q)	5.00	12	0.42	

However, the calculations above show that it is possible to partition the $S(A) \times T$, or error sum of squares, in the same way that we partition the other sums of squares into linear and quadratic components, if this assumption is not reasonable. In Table 17.9, we summarize the analysis in which the error sum of squares, $S(A) \times T$, has been partitioned into the two estimates of experimental error, $S(A) \times L$ and $S(A) \times Q$.

Table 17.10 shows the expectations of the mean squares of Table 17.9. We have assumed that subjects (S) are random and drugs (A) and trials (T) are fixed. We have also assumed that the linear (L) and quadratic (Q) comparisons on trials represent fixed effects. Under these assumptions the expectations of the mean squares are easily determined from the rules previously given for obtaining the expectations of mean squares. The rules applicable, in this instance, are essentially the same as those presented earlier in connection with a split-plot design in which subjects are nested within the levels of A. We have $t = 3$ repeated measures on each subject

Table 17.10

EXPECTATION OF MEAN SQUARES IN TABLE 17.9

Source	Expectation of Mean Square
A	$\sigma^2[1 + (t-1)\rho] + t\sigma_{s(a)}^2 + st\theta_a^2$
$S(A)$: Error (a)	$\sigma^2[1 + (t-1)\rho] + t\sigma_{s(a)}^2$
Linear	$\sigma^2(1-\rho) + \sigma_{s(a)l}^2 + sa\theta_l^2$
$A \times L$	$\sigma^2(1-\rho) + \sigma_{s(a)l}^2 + s\theta_{al}^2$
$S(A) \times L$: Error (l)	$\sigma^2(1-\rho) + \sigma_{s(a)l}^2$
Quadratic	$\sigma^2(1-\rho) + \sigma_{s(a)q}^2 + sa\theta_q^2$
$A \times Q$	$\sigma^2(1-\rho) + \sigma_{s(a)q}^2 + s\theta_{aq}^2$
$S(A) \times Q$: Error (q)	$\sigma^2(1-\rho) + \sigma_{s(a)q}^2$

and we have assumed that errors associated with the repeated measures are correlated. If $\rho = 0$, then the first component in the expectation of each mean square would be σ^2.

The appropriate error terms for the various tests of significance can easily be determined from Table 17.10. Under the assumption that $E(MS_{S(A)L}) = E(MS_{S(A)Q})$, the sums of squares for these two estimates of experimental error would be pooled and divided by the pooled degrees of freedom to obtain a single estimate. The resulting mean square would, of course, be $MS_{S(A)T}$.

Table 17.11

PERSEVERATIVE ERROR SCORES AT DIFFERENT STAGES (C) FOR ANXIETY GROUPS (A) AT THREE LEVELS OF SHOCK (B)[a]

Anxiety-Shock Combination	Subjects	Stages					$\sum$
		C_1	C_2	C_3	C_4	C_5	
A_1B_1	1	1.4	1.0	0.0	1.4	1.0	4.8
A_1B_1	2	1.7	2.4	2.0	3.7	2.2	12.0
A_1B_1	3	1.7	4.1	1.7	4.0	3.2	14.7
A_1B_1	4	2.8	1.0	2.4	3.7	3.2	13.1
A_1B_2	1	2.0	1.7	2.4	1.0	1.4	8.5
A_1B_2	2	1.7	1.4	0.0	0.0	0.0	3.1
A_1B_2	3	1.4	0.0	1.0	0.0	0.0	2.4
A_1B_2	4	1.4	1.4	1.0	1.0	0.0	4.8
A_1B_3	1	1.4	1.0	2.0	2.4	1.0	7.8
A_1B_3	2	1.4	2.4	1.0	1.4	2.8	9.0
A_1B_3	3	1.0	2.0	1.4	1.0	1.4	6.8
A_1B_3	4	4.2	1.7	2.2	4.9	4.7	17.7
A_2B_1	1	1.7	1.0	1.7	1.0	1.0	6.4
A_2B_1	2	3.0	1.7	1.0	1.0	1.0	7.7
A_2B_1	3	2.4	1.4	3.0	2.4	0.0	9.2
A_2B_1	4	1.4	1.0	1.0	1.0	0.0	4.4
A_2B_2	1	2.0	2.4	1.0	0.0	0.0	5.4
A_2B_2	2	2.8	1.0	1.0	1.0	1.0	6.8
A_2B_2	3	5.3	2.0	0.0	0.0	1.0	8.3
A_2B_2	4	2.0	4.1	1.4	1.4	1.4	10.3
A_2B_3	1	2.2	1.7	1.7	1.4	1.0	8.0
A_2B_3	2	4.4	0.0	1.7	1.0	0.0	7.1
A_2B_3	3	2.0	4.7	1.4	1.0	1.0	10.1
A_2B_3	4	2.8	1.7	2.0	0.0	0.0	6.5
$\sum$		54.1	42.8	34.0	35.7	28.3	194.9

[a] From Grant (1956).

17.13 Trial Means: A Treatment Factor and an Organismic Factor

Using data from an experiment by Grant and Patel (1957), Grant (1956) has provided an example of a trend analysis design in which an organismic and a treatment factor were investigated. The organismic factor of interest was anxiety. On the basis of a test of anxiety, two groups, a group of twelve high-anxious and a group of twelve low-anxious subjects were selected. We designate the anxiety factor by A and the two levels by A_1 and A_2, with A_1 corresponding to the high-anxious group and A_2 to the low-anxious group. The treatment factor involved in the experiment was shock and this factor had three levels. We let the shock factor be B and let B_1, B_2, and B_3 correspond to the three levels. The dependent variable consisted of perseverative error scores at five different stages on the Wisconsin Card Sorting Test. We let the stage factor be C and let the successive stages be C_1, C_2, C_3, C_4, and C_5.

The twelve high-anxious subjects were divided at random into three groups of $s = 4$ subjects each. One of these groups was then tested under B_1, another under B_2, and the third under B_3. Each subject in each group was tested at all five stages of C. The procedure was exactly the same for the twelve low-anxious subjects. We note that subjects are nested within the levels of AB. The layout of the experiment is shown in Table 17.11, which also gives the observations for each subject at each stage. The values given in the table are based on a square root transformation of the original perseverative error scores.

From Table 17.11 we find the total sum of squares, the subject sum of squares, and the stage sum of squares. Thus,

$$\text{Total} = (1.4)^2 + (1.7)^2 + \cdots + (0.0)^2 - \frac{(194.9)^2}{120} = 165.9799$$

$$\text{Subject} = \frac{(4.8)^2}{5} + \frac{(12.0)^2}{5} + \cdots + \frac{(6.5)^2}{5} - \frac{(194.9)^2}{120} = 59.4639$$

$$\text{Stage} = \frac{(54.1)^2}{24} + \frac{(42.8)^2}{24} + \cdots + \frac{(28.3)^2}{24} - \frac{(194.9)^2}{120} = 16.3678$$

Subtracting the stage and subject sums of squares from the total, we obtain

$$\text{Subject} \times \text{stage} = 165.9799 - 59.4639 - 16.3678 = 90.1482$$

The degrees of freedom for the sums of squares we have calculated above can be obtained in the usual way. The total sum of squares will have $120 - 1 = 119$ d.f., the subject sum of squares will have $24 - 1 = 23$ d.f., the stage sum of squares will have $5 - 1 = 4$ d.f., and the subject $\times$ stage sum of squares will have $(24 - 1)(5 - 1) = 92$ d.f.

17.14 Partitioning the Subject Sum of Squares

The sum of squares for subjects, with 23 d.f., can be partitioned into the sum of squares for anxiety (A) with 1 d.f., the sum of squares for shock (B) with 2 d.f., the sum of squares for anxiety $\times$ shock ($A \times B$) with 2 d.f., and the *pooled* sum of squares between subjects in each AB combination, $S(AB)$, with $(2)(3)(4-1) = 18$ d.f.

The sum of squares for anxiety (A), shock (B), and the anxiety $\times$ shock ($A \times B$) interaction can be obtained from Table 17.12. For the cell sum of squares for this table, we have

$$\text{Cell} = \frac{(44.6)^2}{20} + \frac{(27.7)^2}{20} + \cdots + \frac{(31.7)^2}{20} - \frac{(194.9)^2}{120} = 21.9054$$

For the row or anxiety (A) sum of squares, we have

$$A = \frac{(104.7)^2}{60} + \frac{(90.2)^2}{60} - \frac{(194.9)^2}{120} = 1.7521$$

and for the column or shock (B) sum of squares, we have

$$B = \frac{(72.3)^2}{40} + \frac{(49.6)^2}{40} + \frac{(73.0)^2}{40} - \frac{(194.9)^2}{120} = 8.8611$$

Then the anxiety $\times$ shock sum of squares can be obtained by subtraction and is equal to

$$A \times B = 21.9054 - 1.7521 - 8.8611 = 11.2922$$

The pooled sum of squares between subjects in the various AB groups can be obtained by direct calculation in the manner described previously,

Table 17.12

TWO-WAY TABLE FOR ANXIETY AND SHOCK

Each cell entry is the sum of twenty observations.

	Shock			
Anxiety	B_1	B_2	B_3	$\sum$
A_1	44.6	18.8	41.3	104.7
A_2	27.7	30.8	31.7	90.2
$\sum$	72.3	49.6	73.0	194.9

Table 17.13
TWO-WAY TABLE FOR ANXIETY AND STAGES

Each cell entry is the sum of twelve observations.

	Stages					
Anxiety	C_1	C_2	C_3	C_4	C_5	$\sum$
A_1	22.1	20.1	17.1	24.5	20.9	104.7
A_2	32.0	22.7	16.9	11.2	7.4	90.2
$\sum$	54.1	42.8	34.0	35.7	28.3	194.9

but it can also be obtained by subtraction. Thus,

$$\begin{aligned} S(AB) &= \text{Subject} - \text{anxiety} - \text{shock} - \text{anxiety} \times \text{shock} \\ &= 59.4639 - 1.7521 - 8.8611 - 11.2922 \\ &= 37.5585 \end{aligned}$$

17.15 Partitioning the Subject × Stage Sum of Squares

The subject × stage sum of squares can also be partitioned in the manner described previously. To obtain the anxiety × stage sum of squares, we first find the sum of squares for the cells of Table 17.13 and then subtract the row (anxiety) and column (stage) sums of squares, which we have already calculated, from the cell sum of squares. For the cell sum of squares, we have

$$\text{Cell} = \frac{(22.1)^2}{12} + \frac{(32.0)^2}{12} + \cdots + \frac{(7.4)^2}{12} - \frac{(194.9)^2}{120} = 35.6991$$

and, by subtraction, we obtain the anxiety × stage $(A \times C)$ sum of squares or

$$A \times C = 35.6991 - 1.7521 - 16.3678 = 17.5792$$

and this sum of squares will have $(2 - 1)(5 - 1) = 4$ d.f.

Similarly, to find the shock × stage sum of squares, we first calculate the cell sum of squares for Table 17.14 and then subtract the row (shock) and column (stage) sums of squares, which we have already calculated, from the cell sum of squares. For the cell sum of squares, we have

$$\text{Cell} = \frac{(16.1)^2}{8} + \frac{(18.6)^2}{8} + \cdots + \frac{(11.9)^2}{8} - \frac{(194.9)^2}{120} = 35.8483$$

Table 17.14
TWO-WAY TABLE FOR SHOCK AND STAGES

Each cell entry is the sum of eight observations.

	Stages					
Shock	C_1	C_2	C_3	C_4	C_5	Σ
B_1	16.1	13.6	12.8	18.2	11.6	72.3
B_2	18.6	14.0	7.8	4.4	4.8	49.6
B_3	19.4	15.2	13.4	13.1	11.9	73.0
Σ	54.1	42.8	34.0	35.7	28.3	194.9

and, by subtraction, we obtain the shock $\times$ stage ($B \times C$) sum of squares or

$$B \times C = 35.8483 - 8.8611 - 16.3678 = 10.6194$$

and the shock $\times$ stage sum of squares will have $(3 - 1)(5 - 1) = 8$ d.f.

The anxiety $\times$ shock $\times$ stage interaction sum of squares can be obtained by subtraction or calculated directly in the manner we have previously described for finding a three-factor interaction sum of squares. We obtain this sum of squares by subtraction. Consider, for example, Table 17.15, where we show the sums for each anxiety-shock group for each stage. The cell sum of squares for this table is equal to

$$\text{Cell} = \frac{(7.6)^2}{4} + \frac{(6.5)^2}{4} + \cdots + \frac{(2.0)^2}{4} - \frac{(194.9)^2}{120} = 73.5324$$

with 29 d.f. The cell sum of squares for this table is equal to the sum of the sums of squares shown below:

	Source	d.f.	Sum of Squares
A:	Anxiety	1	1.7521
B:	Shock	2	8.8611
C:	Stage	4	16.3678
$A \times B$:	Anxiety $\times$ shock	2	11.2922
$A \times C$:	Anxiety $\times$ stage	4	17.5792
$B \times C$:	Shock $\times$ stage	8	10.6194
$A \times B \times C$:	Anxiety $\times$ shock $\times$ stage	8	
	Σ = Cell	29	73.5324

We have the sum of squares for cells and because we have obtained all of the other sums of squares except that for anxiety × shock × stage, this sum of squares can be obtained by subtraction. The sum of the first six sums of squares is equal to 66.4718 and, by subtraction,

$$A \times B \times C = 73.5324 - 66.4718 = 7.0606$$

The anxiety × stage ($A \times C$) sum of squares (17.5792), the shock × stage ($B \times C$) sum of squares (10.6194), and the anxiety × shock × stage ($A \times B \times C$) sum of squares (7.0606) are all *components* of the subject × stage ($S \times C$) sum of squares (90.1482). Subtracting these sums of squares from the subject × stage sum of squares, we have

$$\begin{aligned} S(AB) \times C &= S \times C - A \times C - B \times C - A \times B \times C \\ &= 90.1482 - 17.5792 - 10.6194 - 7.0606 \\ &= 54.8890 \end{aligned}$$

which is the pooled $S(AB)$ × stage or $S(AB) \times C$ sum of squares. We could, of course, calculate $S(A_1B_1) \times C$ for the subjects in A_1B_1 and repeat these calculations for each of the other AB groups. Adding these sums of squares for each group, we would also obtain the $S(AB) \times C$ sum of squares. Because we have $ab = (2)(3) = 6$ different groups and because the interaction sum of squares for each group has $(s - 1)(c - 1) = (4 - 1)(5 - 1) = 12$ d.f., the $S(AB) \times C$ sum of squares will have $(6)(12) = 72$ d.f.

Table 17.15
SUMS FOR ANXIETY (A), SHOCK (B), AND STAGES (C)

Each cell entry is the sum of four observations.

		C_1	C_2	C_3	C_4	C_5	Σ
High: A_1	B_1	7.6	8.5	6.1	12.8	9.6	44.6
	B_2	6.5	4.5	4.4	2.0	1.4	18.8
	B_3	8.0	7.1	6.6	9.7	9.9	41.3
	Σ	22.1	20.1	17.1	24.5	20.9	104.7
Low: A_2	B_1	8.5	5.1	6.7	5.4	2.0	27.7
	B_2	12.1	9.5	3.4	2.4	3.4	30.8
	B_3	11.4	8.1	6.8	3.4	2.0	31.7
	Σ	32.0	22.7	16.9	11.2	7.4	90.2

Table 17.16

ANALYSIS OF VARIANCE OF THE DATA OF TABLE 17.11

Source of Variation	Sum of Squares	d.f.	Mean Square	F
A: Anxiety	1.7521	1	1.7521	
B: Shock	8.8611	2	4.4306	2.12
$A \times B$	11.2922	2	5.6461	2.71
$S(AB)$: Error (a)	37.5585	18	2.0866	
C: Stages	16.3678	4	4.0920	5.37*
$A \times C$	17.5792	4	4.3948	5.77*
$B \times C$	10.6194	8	1.3274	1.74
$A \times B \times C$	7.0606	8	0.8823	1.16
$S(AB) \times C$: Error (b)	54.8890	72	0.7623	
Total	165.9799	119		

17.16 *Expectation of Mean Squares and Tests of Significance*

Table 17.16 summarizes the analysis of variance. To determine the appropriate error terms for tests of significance, we give the expectations of the mean squares in Table 17.17. We have assumed that subjects (S) are random and that A, B, and C are fixed. Errors associated with the repeated measures on the subjects are assumed to be correlated.

We note that $MS_{S(AB)}$ is the appropriate error term for testing the significance of the A effect, the B effect, and the $A \times B$ interaction. $MS_{S(AB)C}$ is the appropriate error term for all other tests of significance. We also note, in this example as in previous ones, that $MS_{S(AB)C}$ is smaller than $MS_{S(AB)}$ and, as we have stated before, this will usually be the case.

Table 17.17

EXPECTATION OF MEAN SQUARES OF TABLE 17.16

Source	Expectation of Mean Square
A	$\sigma^2[1 + (c-1)\rho] + c\sigma_{s(ab)}^2 + sbc\theta_a^2$
B	$\sigma^2[1 + (c-1)\rho] + c\sigma_{s(ab)}^2 + sac\theta_b^2$
$A \times B$	$\sigma^2[1 + (c-1)\rho] + c\sigma_{s(ab)}^2 + sc\theta_{ab}^2$
$S(AB)$: Error (a)	$\sigma^2[1 + (c-1)\rho] + c\sigma_{s(ab)}^2$
C	$\sigma^2(1-\rho) + \sigma_{s(ab)c}^2 + sab\theta_c^2$
$A \times C$	$\sigma^2(1-\rho) + \sigma_{s(ab)c}^2 + sb\theta_{ac}^2$
$B \times C$	$\sigma^2(1-\rho) + \sigma_{s(ab)c}^2 + sa\theta_{bc}^2$
$A \times B \times C$	$\sigma^2(1-\rho) + \sigma_{s(ab)c}^2 + s\theta_{abc}^2$
$S(AB) \times C$: Error (b)	$\sigma^2(1-\rho) + \sigma_{s(ab)c}^2$

Only the two values of F marked with an asterisk in Table 17.16 are significant. The stage means, averaged over anxiety and shock, differ significantly. The significant anxiety $\times$ stage mean square tells us that the trend of the stage means does not have the same form for the two anxiety groups. In Figure 17.2 the stage sums are shown for each anxiety group. We observe that the trend for the low-anxiety group (A_2) is downward and appears to be approximately linear. For the high-anxiety group (A_1), the trend is downward for the first three stages, but then it tends slightly upward and then again downward.

17.17 Linear and Quadratic Components of the Stage Sum of Squares

The trend of the stage sums is shown in Figure 17.3. The sum of squares for stages is equal to 16.3678 with 4 d.f. To obtain the sum of squares for the linear component of the trend, we make use of the orthogonal coefficients of Table XI in the Appendix. With five stages, the orthogonal coefficients are -2, -1, 0, 1, and 2. Multiplying each of the

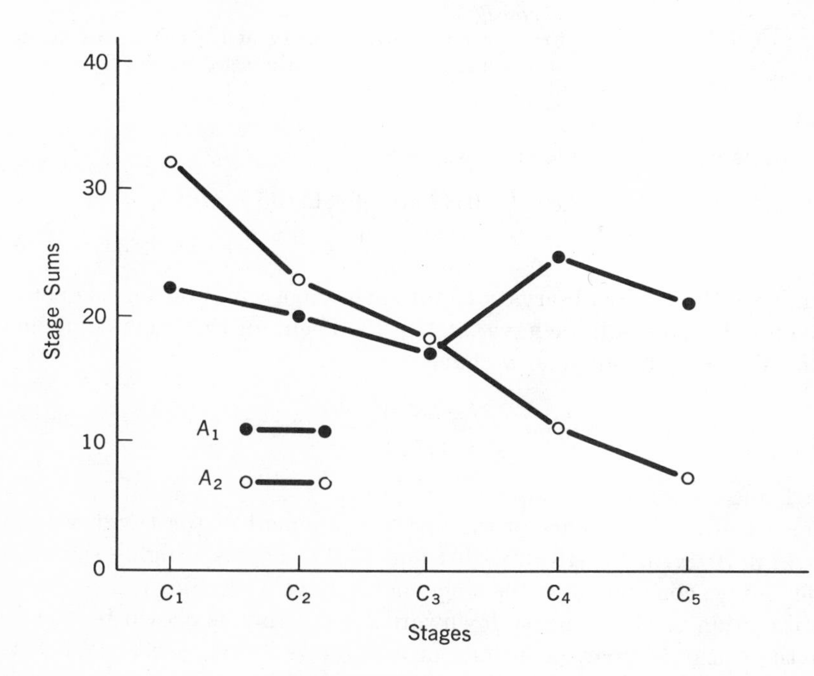

Figure 17.2
Sums of errors at each of five stages on the Wisconsin Card Sorting Test for a high-anxiety (A_1) and a low-anxiety (A_2) group.

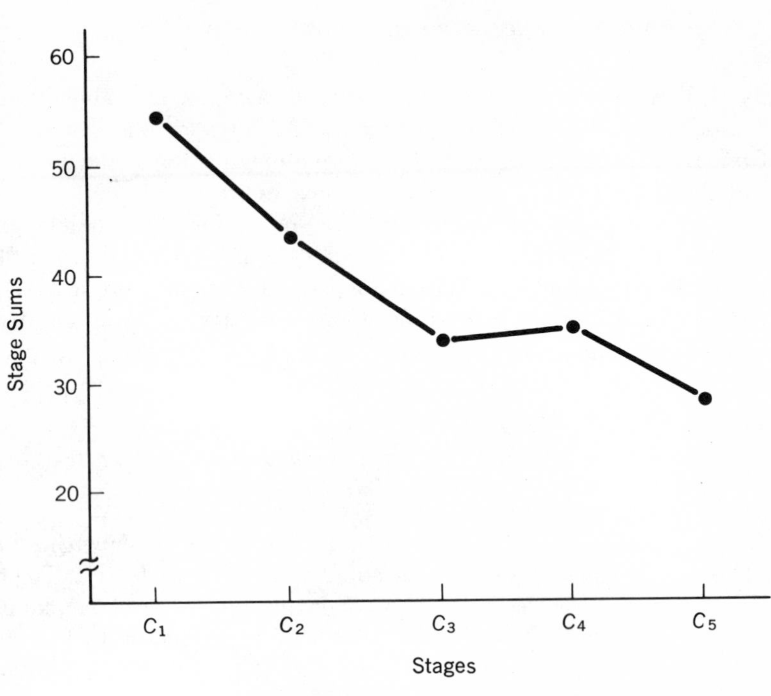

Figure 17.3
Errors summed over anxiety and shock groups at each of five stages on the Wisconsin Card Sorting Test.

stage sums by these coefficients, we have

$$(-2)(54.1) + (-1)(42.8) + (0)(34.0) + (1)(35.7) + (2)(28.3) = -58.7$$

We have twenty-four observations for each stage sum and for the sum of the squared coefficients we have $\Sigma a_{.i}^2 = 10$. Then, for the linear component of the stage sum of squares, we have

$$\text{Linear} = \frac{(-58.7)^2}{(24)(10)} = 14.3570$$

with 1 d.f.

For a test of significance of the linear component of the trend, we have $F = 14.3570/0.7623 = 18.83$ with 1 and 72 d.f. This is a highly significant value and we conclude that the stage means have a significant linear trend. The direction of the trend is downward or negative as shown by the fact that the linear comparison results in a negative value, -58.7. When the linear comparison results in a positive value, this indicates an upward trend of the means.

To determine whether there is a significant curvature in the trend of the stage means, we use the orthogonal coefficients for the quadratic component.

These coefficients, obtained from Table XI, are 2, −1, −2, −1, and 2 and for the sum of the squared coefficients we have $\Sigma a._q^2 = 14$. Multiplying each of the stage sums by the coefficients, we have

$$(2)(54.1) + (-1)(42.8) + (-2)(34.0) + (-1)(35.7) + (2)(28.3) = 18.3$$

and the quadratic component of the stage sum of squares will be

$$\text{Quadratic} = \frac{(18.3)^2}{(24)(14)} = 0.9967$$

with 1 d.f.

To test the significance of the quadratic component, we have $F = 0.9967/0.7623 = 1.31$ with 1 and 72 d.f. and this is not a significant value. It is obvious that almost all of the variation in the stage means can be accounted for by the linear comparison.

17.18 Linear Component of the Group × Stage Sum of Squares

In Table 17.18 we give the stage sums for each of the $k = ab = (2)(3) = 6$ anxiety-shock (AB) groups. Multiplying the stage sums for each group by the linear coefficients we obtain the entries in column L. For example, the first entry is

$$(-2)(7.6) + (-1)(8.5) + (0)(6.1) + (1)(12.8) + (2)(9.6) = 8.3$$

Table 17.18

TREND ANALYSIS FOR THE STAGE SUMS OF TABLE 17.15

		Orthogonal Coefficients					Comparison	
Linear:		−2	−1	0	1	2		
Quadratic:		2	−1	−2	−1	2		
		C_1	C_2	C_3	C_4	C_5	L	Q
	B_1	7.6	8.5	6.1	12.8	9.6	8.3	0.9
A_1	B_2	6.5	4.5	4.4	2.0	1.4	−12.7	0.5
	B_3	8.0	7.1	6.6	9.7	9.9	6.4	5.8
	B_1	8.5	5.1	6.7	5.4	2.0	−12.7	−2.9
A_2	B_2	12.1	9.5	3.4	2.4	3.4	−24.5	12.3
	B_3	11.4	8.1	6.8	3.4	2.0	−23.5	1.7
	Σ	54.1	42.8	34.0	35.7	28.3	−58.7	18.3

The last entry in this column, -58.7, is obtained by multiplying the overall stage sums by the orthogonal coefficients and this must be equal to ΣL. The sum of the squared coefficients is $\Sigma a._l^2 = 10$ and the stage sums for each group are based on $s = 4$ observations. Then, for the linear component of the group $\times$ stage sum of squares, we have

$$G \times L = \frac{(8.3)^2 + (-12.7)^2 + \cdots + (-23.5)^2}{(4)(10)} - \frac{(-58.7)^2}{(6)(4)(10)} = 25.2662$$

17.19 Partitioning the Linear Component of the Group $\times$ Stage Sum of Squares

We have previously partitioned the sum of squares for the $k = ab = 6$ groups into the anxiety (A), shock (B), and anxiety $\times$ shock ($A \times B$) sums of squares. Then, as we have shown previously, if we partition a row (G) sum of squares into orthogonal components, we can also partition the row $\times$ column ($G \times L$) sum of squares into orthogonal components. Thus, for the $G \times L$ sum of squares, we have

$$G \times L = A \times L + B \times L + A \times B \times L$$

We now proceed to partition the $G \times L$ sum of squares into the components shown on the right in the above expression.

To obtain the linear component for the high-anxiety (A_1) group, we multiply the stage sums for this group by the orthogonal coefficients. It should be obvious from Table 17.18 that the results of this multiplication would give us the same value as

$$(8.3) + (-12.7) + (6.4) = 2.0$$

Similarly, if we multiply the stage sums for the low-anxiety (A_2) group by the orthogonal coefficients, we would obtain the same value as

$$(-12.7) + (-24.5) + (-23.5) = -60.7$$

We have $s = 12$ subjects in each of the $k = a = 2$ anxiety groups. Then, substituting in (17.2), we have as the linear component of the anxiety stage sum of squares

$$A \times L = \frac{(2.0)^2 + (-60.7)^2}{(12)(10)} - \frac{(-58.7)^2}{(2)(12)(10)} = 16.3804$$

with 1 d.f. The mean square for this component provides a measure of the difference between the linear components of the trend of the means for the two anxiety groups.

We now find the linear component of the shock $\times$ stage sum of squares. In Table 17.18, it is easy to see that if we multiply the stage sums for B_1 by the orthogonal coefficients, we would obtain $L = (8.3) + (-12.7) =$

-4.4. Similarly, for the second level of shock, B_2, we would have $L = (-12.7) + (-24.5) = -37.2$, and for the third level, B_3, we would have $L = (6.4) + (-23.5) = -17.1$. For each level of B, the stage sums are based on $s = 8$ observations and we have $k = b = 3$ groups or levels. Then, substituting in (17.2), we have

$$B \times L = \frac{(-4.4)^2 + (-37.2)^2 + (-17.1)^2}{(8)(10)} - \frac{(-58.7)^2}{(3)(8)(10)} = 6.8381$$

with 2 d.f. The mean square provides a measure of the differences among the linear trends for the three shock groups.

The linear component of the three-factor interaction, anxiety × shock × stage, can be obtained by subtraction. Thus,

$$\begin{aligned} A \times B \times L &= G \times L - A \times L - B \times L \\ &= 25.2662 - 16.3804 - 6.8381 \\ &= 2.0477 \end{aligned}$$

According to the usual multiplication rule for degrees of freedom for an interaction sum of squares, we have $(a - 1)(b - 1)(1) = (2 - 1) \times (3 - 1)(1) = 2$ d.f. for this interaction sum of squares.

17.20 Partitioning the Quadratic Component of the Group × Stage Sum of Squares

Multiplying each of the row entries in Table 17.18 by the orthogonal coefficients for the quadratic comparison, we obtain the entries in column Q of the table. For example, the first entry is

$$(2)(7.6) + (-1)(8.5) + (-2)(6.1) + (-1)(12.8) + (2)(9.6) = 0.9$$

The last entry in column Q is obtained by multiplying the overall stage sums by the quadratic coefficients and this must be equal to ΣQ.

The calculations of the quadratic components for the interactions are shown below. The calculations are based on the same procedures we followed in calculating the linear components. We have $\Sigma a._q^2 = 14$. Then, for the quadratic component of the group × stage sum of squares, we have

$$G \times Q = \frac{(0.9)^2 + (0.5)^2 + \cdots + (1.7)^2}{(4)(14)} - \frac{(18.3)^2}{(6)(4)(14)} = 2.5263$$

The quadratic component of the anxiety × stage sum of squares is

$$\begin{aligned} A \times Q &= \frac{(0.9 + 0.5 + 5.8)^2 + (-2.9 + 12.3 + 1.7)^2}{(12)(14)} - \frac{(18.3)^2}{(2)(12)(14)} \\ &= 0.0453 \end{aligned}$$

and the quadratic component of the shock $\times$ stage sum of squares is

$$B \times Q = \frac{(0.9 - 2.9)^2 + (0.5 + 12.3)^2 + (5.8 + 1.7)^2}{(8)(14)} - \frac{(18.3)^2}{(3)(8)(14)}$$

$$= 1.0041$$

Then, by subtraction, we obtain the quadratic component of the three-factor interaction or

$$\begin{aligned} A \times B \times Q &= G \times Q - A \times Q - B \times Q \\ &= 2.5263 - 0.0453 - 1.0041 \\ &= 1.4769 \end{aligned}$$

The degrees of freedom for each of the quadratic components can easily be obtained by the multiplication rule for determining the degrees of freedom for an interaction sum of squares. The degrees of freedom for each of the quadratic components will be exactly the same as those for the corresponding linear components.

17.21 Tests of Significance of Linear and Quadratic Components of the Interaction Sums of Squares

Table 17.19 summarizes the analysis of the linear and quadratic components of the interaction sums of squares. The significance of the $A \times L$ mean square with $F = 21.49$ confirms our impression that the linear component of the trend of the stage means for the low-anxiety and the linear component for the high-anxiety group differ significantly. We note, for example, that the linear component for the low-anxiety group is $(-60.7)^2/(12)(10) = 30.7041$, whereas for the high-anxiety group the linear component is $(2.0)^2/(12)(10) = 0.0333$. It is the marked discrepancy between these two components that results in a significant $A \times L$ mean square.

Suppose, for example, that the linear components for the two anxiety groups were the same. Because ΣL must be equal to -58.7, then L for each anxiety group would have to be equal to $-58.7/2 = -29.35$. Then, for each anxiety group, we would have as the linear component $(-29.35)^2/(12)(10) = 7.1785$. The linear component for the overall trend, both anxiety groups combined, is $(-58.7)^2/(12)(10) = 14.3570$. By substitution in (17.2) we see that, under this condition, the linear component of the anxiety $\times$ stage sum of squares would be

$$A \times L = 7.1785 + 7.1785 - 14.3570 = 0$$

In the present analysis, we also find that the linear components of the

Table 17.19

ANALYSIS OF VARIANCE OF LINEAR AND QUADRATIC COMPONENTS OF INTERACTIONS WITH STAGES

Source of Variation	Sum of Squares	d.f.	Mean Square	F
Linear components:				
$A \times L$	16.3804	1	16.3804	21.49
$B \times L$	6.8381	2	3.4190	4.49
$A \times B \times L$	2.0477	2	1.0238	1.34
$G \times L$	25.2662	5		
Quadratic components:				
$A \times Q$	0.0453	1	0.0453	
$B \times Q$	1.0041	2	0.5020	
$A \times B \times Q$	1.4769	2	0.7384	
$G \times Q$	2.5263	5		
$S(AB) \times C$: Error (b)[a]	54.8890	72	0.7623	

[a] From Table 17.16.

trends for the three shock groups differ significantly. The graphs of the stage sums for the three shock groups are shown in Figure 17.4. For the three levels of B, we have L equal to -4.4, -37.2, and -17.1, respectively, and it is because these values are not comparable that the $B \times L$ mean square is significant.

If the $A \times B \times L$ mean square had been significant, we would qualify our interpretations of the $A \times L$ and $B \times L$ mean squares accordingly. A significant $A \times B \times L$ mean square, for example, would indicate that the difference between the linear components of the stage means for the two anxiety groups is not independent of the level of shock.

None of the mean squares for the quadratic components given in Table 17.19 is significant, indicating that the difference in curvature of trends for the two anxiety groups and the differences in curvature of the trends for the three shock groups are not significant.

17.22 Expectation of Mean Squares for Linear and Quadratic Components

Earlier in the chapter, we have shown how the linear, $S(AB) \times L$, and quadratic, $S(AB) \times Q$, components of the $S(AB) \times C$ sum of squares may be calculated. The expected values of the $S(AB) \times L$ and $S(AB) \times Q$ mean squares are given in Table 17.20, where

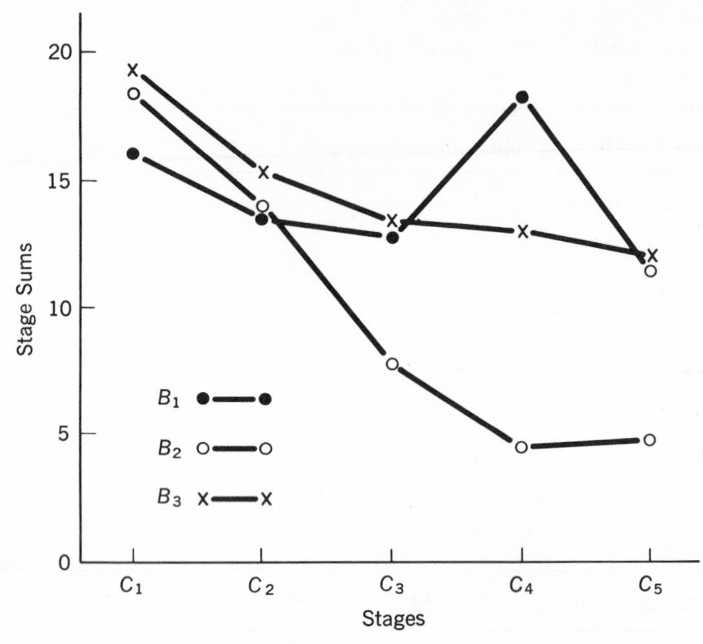

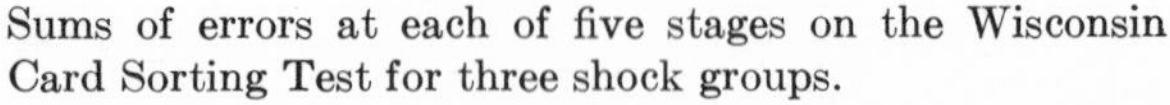

Figure 17.4
Sums of errors at each of five stages on the Wisconsin Card Sorting Test for three shock groups.

we also give the expectations of the mean squares for the other linear and quadratic components.[2] We have assumed that subjects are random and that A, B, and C are fixed. We have also assumed that L and Q represent fixed effects.

It is obvious that $S(AB) \times L$ is the appropriate error mean square for testing all of the other linear components for significance. Similarly, $S(AB) \times Q$ is the appropriate error mean square for testing all of the other quadratic components for significance. In our tests of significance we have used a single estimate, $MS_{S(AB)C}$ or error (b), as the error mean square for these tests. We have done so under the assumption that the expected values of $MS_{S(AB)L}$, $MS_{S(AB)Q}$, and $MS_{S(AB)C}$ are all equal. If we have reason to doubt this assumption, then we should calculate the separate error estimates, $MS_{S(AB)L}$ and $MS_{S(AB)Q}$. In the present example, no conclusions concerning significance would be changed by using $MS_{S(AB)L}$ and $MS_{S(AB)Q}$ as error estimates rather than $MS_{S(AB)C}$.

[2] We have examined only the linear and quadratic components of the stage sum of squares and of the interactions with stages because ordinarily these are the components of major interest. Because the stage sum of squares has 4 d.f., it would also be possible to make two additional orthogonal comparisons on the stage sums, the cubic and quartic. The sum of the linear, quadratic, cubic, and quartic sums of squares will be equal to the stage sum of squares. Similar analyses may be made of the interaction sum of squares. Coefficients for the cubic, quartic, and higher degree polynomials can be found in Fisher and Yates (1948).

17.23 Conservative F Tests in Trend Analysis Experiments

In trend analysis experiments we have t trials. For all tests of significance involving error mean square (b), the error mean square based on the subject $\times$ trial sum of squares, we may make a conservative test. In the last example considered, we had $t = c = 5$ stages or trials. Then the reduced degrees of freedom for error (b) would be $72/4 = 18$ d.f. The stage (C) and anxiety $\times$ stage $(A \times C)$ mean squares each have 4 d.f. and the reduced degrees of freedom for these two mean squares would be 1 d.f. Then we would evaluate the F ratios for A and $A \times C$ in terms of the tabled value of F with 1 and 18 d.f. instead of the tabled value with 4 and 72 d.f. Similarly, the tests of significance of shock $\times$ stage $(B \times C)$ and of anxiety $\times$ shock $\times$ stage $(A \times B \times C)$, each with 8 d.f., would be evaluated in terms of the tabled value of F with 2 and 18 d.f.

If we find linear and quadratic components of a trial sum of squares, these sums of squares will each have 1 d.f. We suggest that for a conservative test of these comparisons, they be evaluated in terms of the tabled value of F with 1 and the reduced degrees of freedom for error (b). If we find the linear and quadratic components of an interaction sum of squares, these sums of squares may have more than 1 d.f. Again, for conservative

Table 17.20

EXPECTATION OF MEAN SQUARES FOR LINEAR AND QUADRATIC COMPONENTS

Source	Expectation of Mean Square
A	$\sigma^2[1 + (c-1)\rho] + c\sigma_{s(ab)}^2 + sbc\theta_a^2$
B	$\sigma^2[1 + (c-1)\rho] + c\sigma_{s(ab)}^2 + sac\theta_b^2$
$A \times B$	$\sigma^2[1 + (c-1)\rho] + c\sigma_{s(ab)}^2 + sc\theta_{ab}^2$
$S(AB)$: Error (a)	$\sigma^2[1 + (c-1)\rho] + c\sigma_{s(ab)}^2$
Linear	$\sigma^2(1-\rho) + \sigma_{s(ab)l}^2 + sab\theta_l^2$
$A \times L$	$\sigma^2(1-\rho) + \sigma_{s(ab)l}^2 + sb\theta_{al}^2$
$B \times L$	$\sigma^2(1-\rho) + \sigma_{s(ab)l}^2 + sa\theta_{bl}^2$
$A \times B \times L$	$\sigma^2(1-\rho) + \sigma_{s(ab)l}^2 + s\theta_{abl}^2$
$S(AB) \times L$: Error (l)	$\sigma^2(1-\rho) + \sigma_{s(ab)l}^2$
Quadratic	$\sigma^2(1-\rho) + \sigma_{s(ab)q}^2 + sab\theta_q^2$
$A \times Q$	$\sigma^2(1-\rho) + \sigma_{s(ab)q}^2 + sb\theta_{aq}^2$
$B \times Q$	$\sigma^2(1-\rho) + \sigma_{s(ab)q}^2 + sa\theta_{bq}^2$
$A \times B \times Q$	$\sigma^2(1-\rho) + \sigma_{s(ab)q}^2 + s\theta_{abq}^2$
$S(AB) \times Q$: Error (q)	$\sigma^2(1-\rho) + \sigma_{s(ab)q}^2$

tests of these comparisons, we suggest that they be evaluated in terms of the tabled value of F with 1 and the reduced degrees of freedom for error (b)

17.24 Trend Analysis with Binomial Variables

In some trend analysis experiments each trial may result in a response that can be classified as correct or incorrect. For example, in a discrimination experiment each subject may be given a series of trials in which he either makes a correct discrimination or fails to do so. In this instance, we have a variable that can take only the value of $X = 1$ or $X = 0$. The data obtained in an experiment of this kind would obviously violate the assumption underlying the analysis of variance.

The F test is, however, a robust test with respect to Type I errors under certain conditions, even when the assumptions underlying it are not true. Both Cochran (1950) and Seeger and Gabrielsson (1968) present evidence to indicate that for experiments of the kind described, the probability of a Type I error given by the tabled values of F corresponds quite well with the empirically determined probability, provided that at least $s = 10$ subjects are tested. With $s > 10$ subjects, one might have even more confidence that the probability of a Type I error is controlled at the significance level given by the F test.

QUESTIONS AND PROBLEMS

17.1 Learning scores for five subjects on each of four trials are given below:

	Trials			
Subjects	1	2	3	4
1	2	4	5	7
2	3	5	6	8
3	5	6	8	10
4	7	9	11	14
5	6	7	10	11

(a) Test the trial mean square for significance. (b) Determine whether the linear component of the trend of the trial means is significant. (c) Determine whether there is a significant curvature in the trend.

17.2 Fifteen subjects were randomly assigned to one of three treatments. The measures given below are recall scores for verbal material based on recall after one, two, and three days.

Groups	Subjects	Days 1	Days 2	Days 3
	1	28	25	22
	2	32	29	24
A_1	3	36	35	27
	4	45	42	40
	5	46	43	40
	1	27	24	20
	2	29	26	22
A_2	3	36	36	30
	4	42	43	39
	5	48	44	40
	1	40	38	33
	2	36	26	20
A_3	3	50	48	44
	4	45	43	35
	5	42	39	30

(*a*) Test the significance of the A, D, and $A \times D$ mean squares. (*b*) Find the linear and quadratic components of the day sum of squares and test them for significance. Note that the sum of the linear and quadratic components is equal to the day sum of squares. (*c*) Test the $A \times L$ and $A \times Q$ mean squares for significance. Note also that the sum of these two components is equal to the $A \times D$ sum of squares.

17.3 Let A_1 and A_2 correspond to two levels of anxiety. Within each level, subjects are randomly assigned to two treatments, B_1 and B_2. Each subject is then given four trials, C_1, C_2, C_3, and C_4. The outcome of the experiment is given below:

Groups	Subjects	C_1	C_2	C_3	C_4
	1	7	9	10	11
A_1B_1	2	8	9	11	12
	3	8	10	11	12
	1	5	6	7	8
A_1B_2	2	5	6	7	9
	3	3	4	9	10
	1	6	7	8	9
A_2B_1	2	7	8	10	11
	3	3	6	9	11
	1	2	3	4	5
A_2B_2	2	3	4	5	7
	3	2	4	5	7

(*a*) Analyze the results with the usual analysis of variance methods. (*b*) Find the linear and quadratic components of the C sum of squares and test each for significance. (*c*) Find the linear and quadratic components of the $A \times C$ and $B \times C$ sums of squares and test them for significance.

17.4 Twenty-one subjects are divided at random into three groups of seven subjects each. A complicated stylus maze has been constructed on a large board. On the face of the board are small brass disks arranged in columns and rows with a small space separating each disk. The back of the board is wired so that if particular disks are touched with the stylus, an electrical circuit operates. The subjects are instructed to take the stylus and to start at the upper left corner of the board and to move from disk to disk, one at a time, to the lower right corner. There is only one path that can be taken without operating the circuit. One group of subjects is told that ability to learn the maze is closely related to intelligence. Another group is told that the average college student makes only twenty errors on the fifth trial. The third group is told that they are simply to make as few errors as possible. Each group of subjects is given five trials. The errors are given below:

		Trials				
Groups	Subjects	1	2	3	4	5
	1	40	39	33	33	20
	2	40	33	31	23	22
"Intelligence"	3	38	34	30	28	26
Group	4	31	29	26	21	20
	5	38	37	36	32	26
	6	39	33	29	28	26
	7	38	32	28	25	21
	1	28	28	24	21	20
	2	39	25	23	23	17
"Average 20"	3	32	32	28	31	26
Group	4	34	27	26	25	23
	5	35	34	27	23	18
	6	35	27	23	21	21
	7	32	24	24	21	22
	1	40	40	30	30	29
	2	35	31	25	22	22
"Few Errors"	3	39	38	36	36	23
Group	4	36	24	21	23	21
	5	38	37	33	37	32
	6	39	38	35	34	34
	7	31	30	27	26	24

(*a*) Analyze the results with the usual analysis of variance methods. (*b*) Find the linear and quadratic components of the overall trend and test each for significance. (*c*) Find the linear and quadratic components of the interaction sum of squares and test each for significance.

17.5 We have treatment factors A and B with $a = 2$ levels and $b = 3$ levels. We use a factorial experiment with $s = 10$ subjects assigned to each treatment combination. Each subject is given $t = 4$ trials under the treatment combination to which he has been assigned. We initially analyze the total sum of squares into the trial, subject, and subject $\times$ trial sums of squares. (*a*) How many degrees of freedom will be associated with each of these sums of squares? (*b*) Into what components can the subject $\times$ trial sum of squares be analyzed and how many degrees of freedom will each of these components have? (*c*) Into what components can the subject sum of squares be analyzed and how many degrees of freedom will be associated with each of these components? (*d*) Give the usual analysis of variance summary table showing sources of variation and degrees of freedom for each source. (*e*) If we find the linear component of the $A \times$ trial sum of squares, how many degrees of freedom will this component have? (*f*) If we find the linear component of the $B \times$ trial sum of squares, how many degrees of freedom will this component have?

17.6 Twenty subjects were divided at random into two groups of $s = 10$ subjects each. Subjects in Group 1 were rewarded for each correct response

Groups	Subjects	T_1	T_2	T_3	T_4	T_5	T_6
Group 1	1	9	3	2	5	10	12
	2	10	7	3	11	11	11
	3	5	6	10	5	8	15
	4	4	11	9	13	5	11
	5	8	11	5	13	9	7
	6	6	3	4	10	10	12
	7	6	2	6	8	9	14
	8	4	4	6	10	14	6
	9	4	10	4	11	5	15
	10	1	6	4	5	11	13
Group 2	1	8	9	13	9	18	14
	2	5	10	9	7	10	12
	3	3	4	8	10	11	19
	4	5	8	13	11	15	18
	5	1	4	13	14	17	15
	6	4	8	6	12	9	20
	7	10	7	14	13	9	16
	8	10	9	9	16	9	11
	9	2	9	9	14	12	19
	10	9	10	8	12	16	11

made in a learning experiment and subjects in Group 2 were rewarded for each correct response and punished for each wrong response. Each subject was given six trials. (*a*) Complete the analysis of variance. (*b*) Is the linear component of the overall trial sum of squares significant? (*c*) Is the quadratic component of the overall trial sum of squares significant? (*d*) Do the linear components of the trial sums of squares for the two groups differ significantly?

17.7 We have $s = 10$ subjects and each subject is tested on three trials. The table below shows the measures for Subjects 1, 2, and 10 and the sums for each trial.

	Trials				
Subjects	1	2	3	D_1	D_2
1	2	4	7	5	1
2	2	6	10	8	0
.	.	.	.	.	.
.	.	.	.	.	.
.	.	.	.	.	.
10	11	10	8	−3	−1
Σ	35	40	55	20	10
L	−1	0	1		
Q	1	−2	1		

(*a*) What does the *sum, not the separate terms,* of $(5)^2/2 + (1)^2/6$ measure; that is, what sum of squared deviations is it? (*b*) If we calculate the sum of squares in (*a*) for each row of the table and add these sums of squares, what would this pooled sum of squares be called in the analysis of variance? How many degrees of freedom would this pooled sum of squares have? (*c*) What does $(20)^2/(2)(10)$ measure? (*d*) What does $(10)^2/(6)(10)$ measure? (*e*) We add the two sums of squares of (*c*) and (*d*) together and then subtract the result from the pooled sum of squares of (*b*). What would this remainder sum of squares be in the analysis of variance; that is, what is it called? (*f*) What does

$$\frac{(5)^2 + (8)^2 + \cdots + (-3)^2}{2} - \frac{(20)^2}{(2)(10)}$$

measure; that is, what is this sum of squares called? How many degrees of freedom does it have? (*g*) What does

$$\frac{(1)^2 + (0)^2 + \cdots + (-1)^2}{6} - \frac{(10)^2}{(6)(10)}$$

measure; that is, what is this sum of squares called? How many degrees of freedom does it have?

17.8 Ten subjects were randomly divided into two groups of $s = 5$ subjects each. Subjects in Group 1 were administered Treatment A_1 and those in Group 2 were administered Treatment A_2. The performance of the subjects in both groups was measured over $t = 3$ trials. The outcome of the experiment is given below:

		Trials		
Groups	Subjects	1	2	3
	1	1	3	6
	2	1	5	9
A_1	3	2	6	9
	4	6	8	10
	5	5	8	11
	6	3	3	7
	7	2	2	7
A_2	8	4	5	8
	9	5	6	8
	10	8	9	10

Find the various sums of squares involved in the analysis of variance. (*a*) What is the value of $MS_{S(A)}$? (*b*) What is the value of $MS_{S(A)T}$? (*c*) Find the linear and quadratic components of the trial sum of squares. (*d*) Find the linear and quadratic components of the $A \times T$ sum of squares. (*e*) Plot the trial means for A_1 and A_2 and explain the differences in the trends. (*f*) Complete the analysis of variance and interpret the various tests of significance.

17.9 In an experiment by Stevens, Stover, and Backus (1970), the subjects were thirty-six control boys and thirty-six boys with a primary symptom of hyperkinesis. Nine control and nine hyperkinetic boys were selected from each of four age groups, 8–11 years. The nine subjects in each age group were randomly divided into three treatment groups. Each subject was given three trials on the Whipple tapping board.

Let A correspond to the control and hyperkinetic factor and we have $a = 2$ levels of A. Let B correspond to the age factor and we have $b = 4$ levels of B. Let C correspond to the treatments and we have $c = 3$ levels of C. Let D correspond to the trials and we have $d = 3$ levels of D. We note that we have $s = 3$ subjects nested within the ABC combinations and that we have three repeated measures on each subject.

(*a*) Give the analysis of variance summary table showing sources of variation and degrees of freedom associated with each source. (*b*) Under the assumption that subjects are random and that A, B, C, and D represent fixed effects, give the expectations of the mean squares. (*c*) Would you expect $MS_{S(ABC)}$ to be larger or smaller than $MS_{S(ABC)D}$? Explain why.

CHAPTER 18

THE ANALYSIS OF COVARIANCE FOR A RANDOMIZED GROUP DESIGN

18.1 *Introduction*

In this chapter we consider the simplest form of the *analysis of covariance* as applied to a randomized group design. In the analysis of covariance we have two observations for each subject. One of these we designate as a *supplementary* measure X which is not itself of experimental interest.[1] The other we designate as Y. The Y measures are those obtained on the dependent variable of interest after the treatments have been applied. It is the significance of the differences between the Y means for the various treatments that is of interest.

The major objective of a covariance analysis is to obtain a reduced estimate of experimental error by taking into account the regression of the Y measures on the X measures. In general, if the Y measures are substantially correlated with the X measures, then the analysis of covariance will result in a smaller estimate of experimental error than would be obtained from the analysis of variance of the Y measures.

18.2 *Partitioning the Total Product Sum*

In the analysis of variance for a randomized group design with n observations for each group or treatment, we analyze

[1] A supplementary measure or observation is also referred to as a *concomitant* measure or observation.

the total sum of squares into the within treatment sum of squares and the treatment sum of squares.[2] We have found that we can express the deviation of a given value X_{kn} from the overall mean $\bar{X}..$ as

$$X_{kn} - \bar{X}.. = (X_{kn} - \bar{X}_{k}.) + (\bar{X}_{k}. - \bar{X}..)$$

Similarly, a deviation of Y_{kn} from the overall mean $\bar{Y}..$ can be expressed as

$$Y_{kn} - \bar{Y}.. = (Y_{kn} - \bar{Y}_{k}.) + (\bar{Y}_{k}. - \bar{Y}..)$$

Multiplying these two expressions and summing over the n observations for a single treatment group, we obtain the product sum

$$\begin{aligned}\sum_{1}^{n} (X_{kn} - \bar{X}..)(Y_{kn} - \bar{Y}..) &= \sum_{1}^{n} (X_{kn} - \bar{X}_{k}.)(Y_{kn} - \bar{Y}_{k}.) \\ &\quad + \sum_{1}^{n} (\bar{X}_{k}. - \bar{X}..)(Y_{kn} - \bar{Y}_{k}.) \\ &\quad + \sum_{1}^{n} (X_{kn} - \bar{X}_{k}.)(\bar{Y}_{k}. - \bar{Y}..) \\ &\quad + \sum_{1}^{n} (\bar{X}_{k}. - \bar{X}..)(\bar{Y}_{k}. - \bar{Y}..)\end{aligned}$$

Both $(\bar{X}_{k}. - \bar{X}..)$ and $(\bar{Y}_{k}. - \bar{Y}..)$ are constants and we know that the sum of the deviations of the n observations in the kth group from the mean of the group is equal to zero. Then

$$(\bar{X}_{k}. - \bar{X}..) \sum_{1}^{n} (Y_{kn} - \bar{Y}_{k}.) = 0$$

and

$$(\bar{Y}_{k}. - \bar{Y}..) \sum_{1}^{n} (X_{kn} - \bar{X}_{k}.) = 0$$

Therefore, we have

$$\begin{aligned}\sum_{1}^{n} (X_{kn} - \bar{X}..)(Y_{kn} - \bar{Y}..) &= \sum_{1}^{n} (X_{kn} - \bar{X}_{k}.)(Y_{kn} - \bar{Y}_{k}.) \\ &\quad + n(\bar{X}_{k}. - \bar{X}..)(\bar{Y}_{k}. - \bar{Y}..)\end{aligned}$$

[2] Just as the analysis of variance for a randomized group design does not require that we have equal n's for each of the treatments, so also this is not a necessary requirement for the analysis of covariance.

for a single group. Then, summing over the k groups, we have

$$\sum_{1}^{kn} (X_{kn} - \bar{X}_{..})(Y_{kn} - \bar{Y}_{..}) = \sum_{1}^{kn} (X_{kn} - \bar{X}_{k.})(Y_{kn} - \bar{Y}_{k.}) + n \sum_{1}^{k} (\bar{X}_{k.} - \bar{X}_{..})(\bar{Y}_{k.} - \bar{Y}_{..})$$

The term on the left in the above expression is called the *total product sum*. We have analyzed the total product sum into the two component parts shown on the right. The first term is the *product sum within treatments* and the second term is the *product sum between treatment groups*. We designate these product sums by Σxy_t, Σxy_w, and Σxy_b, respectively. The total product sum can be obtained by finding

$$\sum xy_t = \sum_{1}^{kn} X_{kn} Y_{kn} - \frac{(\sum X_{..})(\sum Y_{..})}{kn} \qquad (18.1)$$

The product sum between groups can be obtained by calculating

$$\sum xy_b = \sum_{1}^{k} \frac{(\sum X_{k.})(\sum Y_{k.})}{n} - \frac{(\sum X_{..})(\sum Y_{..})}{kn} \qquad (18.2)$$

For any single treatment group, the within treatment product sum will be given by

$$\sum_{1}^{n} x_{kn} y_{kn} = \sum_{1}^{n} X_{kn} Y_{kn} - \frac{(\sum X_{k.})(\sum Y_{k.})}{n} \qquad (18.3)$$

Table 18.1

MEASURES ON A SUPPLEMENTARY VARIABLE (X) AND A DEPENDENT VARIABLE (Y) FOR A RANDOMIZED GROUP DESIGN IN THE ABSENCE OF TREATMENT EFFECTS

	Treatment Groups					
	1		2		3	
	X	Y	X	Y	X	Y
	1	0	2	1	1	0
	6	7	3	2	4	3
	3	4	6	7	5	6
	4	3	4	3	3	2
	5	6	7	8	6	7
Σ	19	20	22	21	19	18

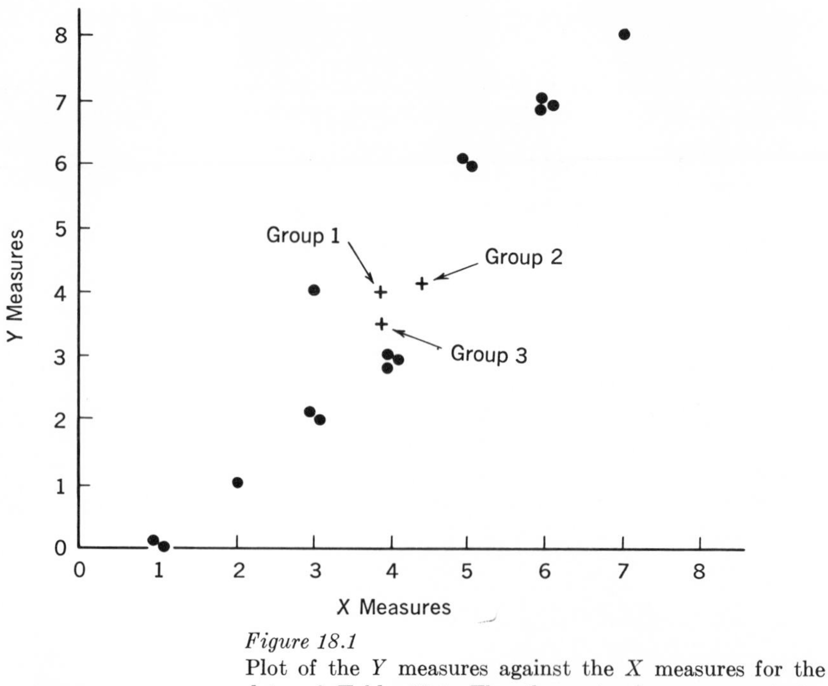

Figure 18.1
Plot of the Y measures against the X measures for the data of Table 18.1. The three + signs represent the means of the three groups on the X and Y variables.

and summing over all k groups, we obtain the product sum within treatments or

$$\sum xy_w = \sum_1^k \left[\sum_1^n X_{kn}Y_{kn} - \frac{(\sum X_{k\cdot})(\sum Y_{k\cdot})}{n} \right]$$

or

$$\sum xy_w = \sum_1^{kn} X_{kn}Y_{kn} - \sum_1^k \frac{(\sum X_{k\cdot})(\sum Y_{k\cdot})}{n} \tag{18.4}$$

Because the sum of the product sums for treatments and within treatments must be equal to the total product sum, we can also obtain the within treatment product sum by subtraction. Thus,

$$\sum xy_w = \sum xy_t - \sum xy_b \tag{18.5}$$

18.3 Relationship between X and Y in the Absence of Treatment Effects

In Table 18.1 we give the X and Y measures for a randomized group design in which we have $n = 5$ subjects assigned to each of $k = 3$ treatments. Let us assume that the X measures are on the same variable as the Y measures, the only difference being that the X measure is obtained prior to the application of the treatments and the Y measure after the application of the treatments. In Table 18.1, however, the Y measures were derived in such a way that there are no significant differences between the treatment (Y) means.

Figure 18.1 shows the plot of the Y measures against the X measures. With fairly reliable measurements and in the *absence* of treatment effects, we should expect to obtain a plot of the X and Y values that is similar to that of Figure 18.1.[3] With a randomized group design, the X values on the baseline of the figure will be divided at random into three sets of five each. Because these measures are obtained prior to the application of the treatments, we should not expect to find any significant differences between the treatment groups on the X measures. In the absence of treatment effects we should not expect to find any significant differences between the groups on the Y measures either.

18.4 Relationship between X and Y When Treatment Effects Are Present

We now consider the case where we do have treatment effects that are additive. Let us increase the Y values in Table 18.1

Table 18.2

MEASURES ON A SUPPLEMENTARY VARIABLE (X) AND A DEPENDENT VARIABLE (Y) FOR A RANDOMIZED GROUP DESIGN WITH TREATMENT EFFECTS PRESENT

	Treatment Groups					
	1		2		3	
	X	Y	X	Y	X	Y
	1	5	2	1	1	10
	6	12	3	2	4	13
	3	9	6	7	5	16
	4	8	4	3	3	12
	5	11	7	8	6	17
Σ	19	45	22	21	19	68

[3] The important feature of Figure 18.1 is that the trend of the points can be represented by a single regression line.

for each subject in Group 1 by five. We leave the Y values for each subject in Group 2 unchanged. For the subjects in Group 3 we increase the Y values by ten. Making these changes in the Y measures, we obtain the results shown in Table 18.2. The X values in Table 18.2 are unchanged and are the same as those given in Table 18.1.

In Figure 18.2 we show the plot of the Y measures against the X measures for the data of Table 18.2. In this figure it is clear that for each treatment group, the points cluster about relatively parallel lines at different heights or with different Y-intercepts. This is the sort of graph we should expect to obtain under the following conditions:

(1) We have treatment effects that are additive.

(2) In the absence of treatment effects the relationship between X and Y is positive and linear.[4]

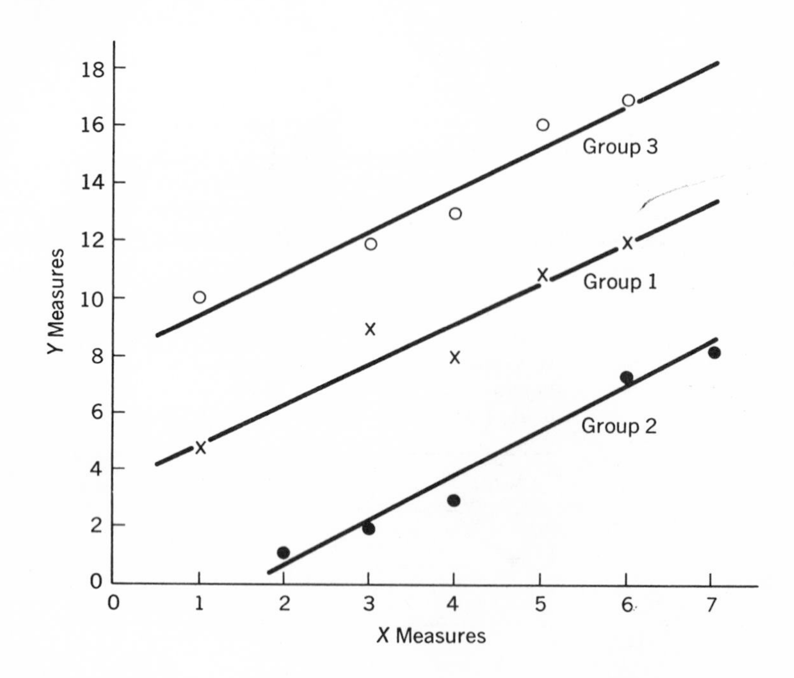

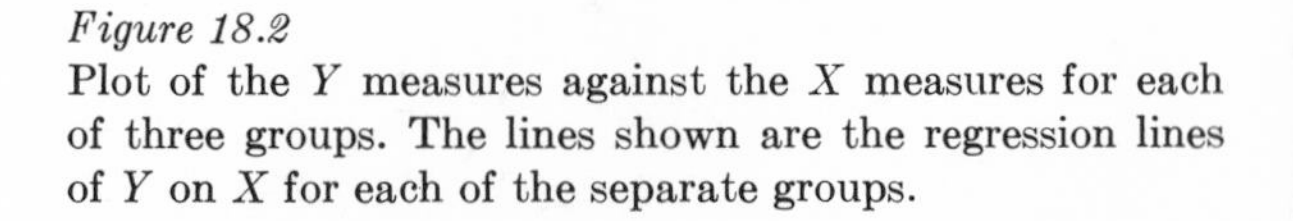
Figure 18.2
Plot of the Y measures against the X measures for each of three groups. The lines shown are the regression lines of Y on X for each of the separate groups.

[4] Analysis of covariance techniques can be applied to the case of nonlinear relationships. However, the methods of analysis given in this chapter are not appropriate if the relationship between X and Y is not linear.

(3) We have randomly assigned subjects to the treatment groups.
(4) The X measures are uninfluenced by the treatments.

18.5 *Sums of Squares and Product Sums for a Randomized Group Design*

To apply the analysis of covariance to the data of Table 18.2 we need to partition the total sum of squares of both the X and Y variable and also the total product sum. Let the total sum of squares for the X measures be Σx_t^2. Then

$$\sum x_t^2 = (1)^2 + (6)^2 + \cdots + (6)^2 - \frac{(60)^2}{15} = 48.00$$

We also let Σx_b^2 be the between treatment sum of squares and

$$\sum x_b^2 = \frac{(19)^2}{5} + \frac{(22)^2}{5} + \frac{(19)^2}{5} - \frac{(60)^2}{15} = 1.20$$

We let $\sum_1^n x_k^2$ be the sum of squares within the kth treatment group and, for the $k = 3$ groups, we have

$$\sum_1^n x_1^2 = (1)^2 + (6)^2 + \cdots + (5)^2 - \frac{(19)^2}{5} = 14.80$$

$$\sum_1^n x_2^2 = (2)^2 + (3)^2 + \cdots + (7)^2 - \frac{(22)^2}{5} = 17.20$$

$$\sum_1^n x_3^2 = (1)^2 + (4)^2 + \cdots + (6)^2 - \frac{(19)^2}{5} = 14.80$$

Then the within treatment sum of squares (Σx_w^2) will be

$$\sum x_w^2 = 14.80 + 17.20 + 14.80 = 46.80$$

Similarly, for the Y measures, we obtain

$$\sum y_t^2 = (5)^2 + (12)^2 + \cdots + (17)^2 - \frac{(134)^2}{15} = 322.93$$

and

$$\sum y_b^2 = \frac{(45)^2}{5} + \frac{(21)^2}{5} + \frac{(68)^2}{5} - \frac{(134)^2}{15} = 220.93$$

For the sum of squares within the $k = 3$ groups, we have

$$\sum_1^n y_1^2 = (5)^2 + (12)^2 + \cdots + (11)^2 - \frac{(45)^2}{5} = 30.00$$

$$\sum_1^n y_2^2 = (1)^2 + (2)^2 + \cdots + (8)^2 - \frac{(21)^2}{5} = 38.80$$

$$\sum_1^n y_3^2 = (10)^2 + (13)^2 + \cdots + (17)^2 - \frac{(68)^2}{5} = 33.20$$

Then the within treatment sum of squares (Σy_w^2) will be

$$\sum y_w^2 = 30.00 + 38.80 + 33.20 = 102.00$$

We now find the total product sum. Thus,

$$\sum xy_t = (1)(5) + (6)(12) + \cdots + (6)(17) - \frac{(60)(134)}{15} = 53.00$$

For the product sum between groups, we have

$$\sum xy_b = \frac{(19)(45)}{5} + \frac{(22)(21)}{5} + \frac{(19)(68)}{5} - \frac{(60)(134)}{15} = -14.20$$

The product sums for each of the $k = 3$ groups are

$$\sum_1^n xy_1 = (1)(5) + (6)(12) + \cdots + (5)(11) - \frac{(19)(45)}{5} = 20.00$$

$$\sum_1^n xy_2 = (2)(1) + (3)(2) + \cdots + (7)(8) - \frac{(22)(21)}{5} = 25.60$$

$$\sum_1^n xy_3 = (1)(10) + (4)(13) + \cdots + (6)(17) - \frac{(19)(68)}{5} = 21.60$$

and the product sum within groups will be equal to

$$\sum xy_w = 20.00 + 25.60 + 21.60 = 67.20$$

18.6 Variation within Each Group about the Regression Line for the Group

Consider only one of the k treatment groups. A "best fitting" straight line can be drawn to represent the regression of the Y variable on the X variable for this group. The line will give the "best fit" in the sense that the sum of squared deviations of the Y values from this line will be less than from any other straight line.[5] The slope of the line of best fit will be given by the regression coefficient of Y on X which we designate by b_k. Then

$$b_k = \frac{\sum_1^n (X_{kn} - \bar{X}_{k\cdot})(Y_{kn} - \bar{Y}_{k\cdot})}{\sum_1^n (X_{kn} - \bar{X}_{k\cdot})^2} \tag{18.6}$$

If we let $x_k = X_{kn} - \bar{X}_{k\cdot}$ and $y_k = Y_{kn} - \bar{Y}_{k\cdot}$, then (18.6) can be written

$$b_k = \frac{\sum_1^n xy_k}{\sum_1^n x_k^2} \tag{18.7}$$

where $\sum_1^n xy_k$ is the product sum for the kth group and $\sum_1^n x_k^2$ is the sum of squares within the kth group on the X variable.

The equation for the regression line will be

$$\tilde{Y}_{kn} = \bar{Y}_{k\cdot} + b_k(X_{kn} - \bar{X}_{k\cdot}) \tag{18.8}$$

We let $\tilde{y}_k = \tilde{Y}_{kn} - \bar{Y}_{k\cdot}$ and subtracting $\bar{Y}_{k\cdot}$ from both sides of (18.8), we have

$$\tilde{y}_k = b_k x_k \tag{18.9}$$

The sum of squared deviations of the actual $y_k = Y_{kn} - \bar{Y}_{k\cdot}$ values about the regression line with slope b_k will be equal to

$$\sum_1^n (y_k - b_k x_k)^2 = \sum_1^n y_k^2 - 2b_k \sum_1^n xy_k + b_k^2 \sum_1^n x_k^2$$

Simplifying this expression, we obtain

$$\sum_1^n (y_k - b_k x_k)^2 = \sum_1^n y_k^2 - \frac{\left(\sum_1^n xy_k\right)^2}{\sum_1^n x_k^2} \tag{18.10}$$

The sum of squares defined by (18.10) has $n - 2$ d.f., the first term on the right having $n - 1$ d.f. and the second term 1 d.f.

[5] The proof is given in Chapter 19.

We can obviously apply (18.10) to each of the k groups. Then, summing these sums of squares over the k groups, we obtain

$$\sum_{1}^{k}\sum_{1}^{n}(y_k - b_k x_k)^2 = \sum_{1}^{k}\sum_{1}^{n} y_k^2 - \sum_{1}^{k}\frac{\left(\sum_{1}^{n} xy_k\right)^2}{\sum_{1}^{n} x_k^2} \tag{18.11}$$

We designate this sum of squares as S_1. Because the first term on the right is the sum of squares within treatments, on the Y variable, we have

$$S_1 = \sum y_w^2 - \sum_{1}^{k}\frac{\left(\sum_{1}^{n} xy_k\right)^2}{\sum_{1}^{n} x_k^2} \tag{18.12}$$

S_1 is a sum of k sums of squares, each with $n - 2$ d.f. Therefore, S_1 will have $k(n - 2)$ d.f.

Taking the appropriate values from Table 18.3, we have

$$S_1 = 102.00 - \left[\frac{(20.00)^2}{14.8} + \frac{(25.60)^2}{17.2} + \frac{(21.60)^2}{14.8}\right]$$

$$= 5.35$$

with $3(5 - 2) = 9$ d.f.

The regression coefficients for each of the three treatment groups, in the example under consideration, can be calculated from the data of Table 18.3. Thus, we have

$$b_1 = \frac{20.00}{14.80} = 1.35$$

$$b_2 = \frac{25.60}{17.20} = 1.49$$

$$b_3 = \frac{21.60}{14.80} = 1.46$$

The regression line for each group, with slope as given by the regression coefficient for the group, is shown in Figure 18.2. S_1 measures the variation of the n observations within each group about the regression line for the group. Because the group regression coefficients are not exactly equal to each other, the regression lines in the figure are not exactly parallel.

Table 18.3

SUMS OF SQUARES AND PRODUCT SUMS FOR THE DATA OF TABLE 18.2

	$\sum x^2$	$\sum y^2$	$\sum xy$
Group 1	14.80	30.00	20.00
Group 2	17.20	38.80	25.60
Group 3	14.80	33.20	21.60
Within	46.80	102.00	67.20
Between	1.20	220.93	−14.20
Total	48.00	322.93	53.00

18.7 Variation within Groups about a Regression Line with Slope b_w

Let us assume that the separate regression coefficients for the k groups are all estimates of the same common population regression coefficient. Then the best estimate of this common regression coefficient, which we designate as b_w, will be

$$b_w = \frac{\sum_1^n xy_1 + \sum_1^n xy_2 + \cdots + \sum_1^n xy_k}{\sum_1^n x_1^2 + \sum_1^n x_2^2 + \cdots + \sum_1^n x_k^2} \tag{18.13}$$

The numerator of the above expression is the product sum within treatment groups and the denominator is the sum of squares within treatments on the X variable. Thus, we have

$$b_w = \frac{\sum xy_w}{\sum x_w^2} \tag{18.14}$$

and the equation for the regression line with slope b_w will be

$$\tilde{Y}_{kn} = \bar{Y}_{k\cdot} + b_w(X_{kn} - \bar{X}_{k\cdot}) \tag{18.15}$$

or, if we subtract $\bar{Y}_{k\cdot}$ from both sides of (18.15), we have

$$\tilde{y}_k = b_w x_k \tag{18.16}$$

The sum of squared deviations of the actual y_k values about the regression line with common slope equal to b_w will be

$$\sum_1^k \sum_1^n (y_k - b_w x_k)^2 = \sum_1^k \sum_1^n y_k^2 - 2b_w \sum_1^k \sum_1^n xy_k + b_w^2 \sum_1^k \sum_1^n x_k^2$$

We designate this sum of squares as S_2. Simplifying the above expression, we have

$$S_2 = \sum y_w{}^2 - \frac{(\sum xy_w)^2}{\sum x_w{}^2} \tag{18.17}$$

The first term on the right has $k(n - 1)$ d.f. and the last term has 1 d.f. S_2, therefore, will have $k(n - 1) - 1$ d.f. Taking the appropriate values from Table 18.3, we have

$$S_2 = 102.00 - \frac{(67.20)^2}{46.80} = .51$$

with $3(5 - 1) - 1 = 11$ d.f.

From the data of Table 18.3, we have

$$b_w = \frac{67.20}{46.80} = 1.44$$

The three regression lines, each with slope equal to b_w, are shown in Figure 18.3. S_2 is a measure of the variation of the observations within each treatment group about the regression lines with common slope equal to b_w.

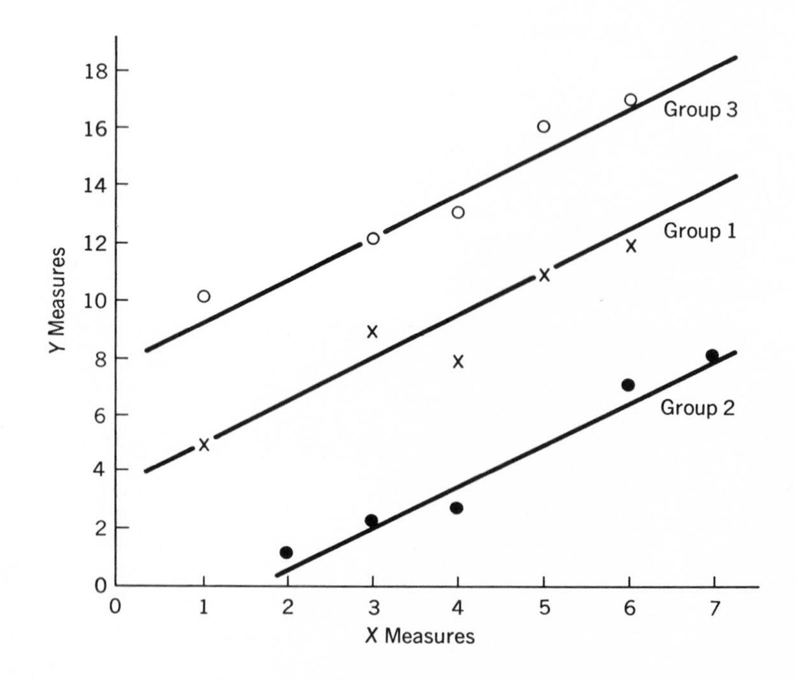

Figure 18.3
Plot of the Y measures against the X measures for each of three groups. The three regression lines have a common slope equal to b_w.

18.8 Test of Homogeneity of the Regression Coefficients

Now S_2 can never be smaller than S_1, because S_1 is based on the squared deviations within each group from a regression line with slope b_k fitted separately for each group and the values of b_k were found in such a way as to minimize the sum of squared deviations within each of the groups. S_2, on the other hand, is based on the squared deviations from a regression line with slope b_w for each group. Thus, if b_k is not equal to b_w, the sum of squared deviations for a given group with b_k as the slope of the regression line will be smaller than the sum of squared deviations obtained by using b_w as the slope of the regression line. If $b_1 = b_2 = b_3 = b_w$, then S_2 will be exactly equal to S_1. If the b_k values do show considerable variation, then S_2 will be considerably larger than S_1.

To determine whether the regression coefficients differ significantly, we find

$$S_3 = S_2 - S_1 \tag{18.18}$$

Because S_2 has $k(n - 1) - 1$ d.f. and because S_1 has $k(n - 2)$ d.f., we will have

$$k(n - 1) - 1 - k(n - 2) = k - 1$$

degrees of freedom for S_3. For the present problem, we have

$$S_3 = 5.51 - 5.35 = 0.16$$

with 2 d.f.

Then, for the test of significance of the differences among the group regression coefficients, we have

$$F = \frac{S_3/(k - 1)}{S_1/k(n - 2)} \tag{18.19}$$

or, for the present problem,

$$F = \frac{0.16/2}{5.35/9} = 0.13$$

with 2 and 9 d.f. and this is, obviously, a nonsignificant value of F.

If the F defined by (18.19) is significant, we would conclude that the separate regression lines are not parallel within the limits of random variation; that is, they have significantly different slopes. This may occur because the treatment effects are not additive. In this instance, one of the transformations, described previously, applied to the Y measures may result in a scale on which the treatment effects are additive. It is important to emphasize that the application of the analysis of covariance does assume that the regression lines for the various treatment groups all can be assumed to have a common slope equal to b_w. Because, in the example under consideration, we have not obtained a significant F, we can proceed to use the analysis of covariance to test the differences between the treatment means for significance.

18.9 Test of Significance of the Adjusted Treatment Mean Square

The regression coefficient based on the total product sum and total sum of squares on the X variable will be given by

$$b_t = \frac{\sum_{1}^{kn} (X_{kn} - \bar{X}..)(Y_{kn} - \bar{Y}..)}{\sum_{1}^{kn} (X_{kn} - \bar{X}..)^2} \tag{18.20}$$

or

$$b_t = \frac{\sum xy_t}{\sum x_t^2} \tag{18.21}$$

and the equation for the regression line will be

$$\tilde{Y}_{kn} = \bar{Y}.. + b_t(X_{kn} - \bar{X}..) \tag{18.22}$$

If we let $\tilde{y} = \tilde{Y}_{kn} - \bar{Y}..$ and $x = X_{kn} - \bar{X}..$, then the equation for the regression line will be

$$\tilde{y} = b_t x \tag{18.23}$$

The sum of the squared deviations of the actual y values about the regression line with slope equal to b_t will be

$$\sum_{1}^{kn} (y - b_t x)^2 = \sum_{1}^{kn} y^2 - 2b_t \sum_{1}^{kn} xy + b_t^2 \sum_{1}^{kn} x^2$$

and we designate this sum of squares as S_4. Simplifying the expression for S_4, we have

$$S_4 = \sum y_t^2 - \frac{(\sum xy_t)^2}{\sum x_t^2} \tag{18.24}$$

and S_4 will have $kn - 2$ d.f.

For the present example,

$$S_4 = 322.93 - \frac{(53.00)^2}{48.00} = 264.41$$

with 13 d.f.

Taking the difference between S_4 and S_2, we have an "adjusted" treatment sum of squares or

$$S_5 = S_4 - S_2 \tag{18.25}$$

The first term on the right has $kn - 2$ d.f. and the second term has $k(n - 1) - 1$ d.f. Then we have

$$kn - 2 - [k(n - 1) - 1] = k - 1$$

degrees of freedom for S_5. For the present problem, we have

$$S_5 = 264.41 - 5.51 = 258.90$$

We note that (18.25) can also be written as

$$S_5 = \left[\sum y_t^2 - \frac{(\sum xy_t)^2}{\sum x_t^2}\right] - \left[\sum y_w^2 - \frac{(\sum xy_w)^2}{\sum x_w^2}\right]$$

or

$$S_5 = \sum y_b^2 - \left[\frac{(\sum xy_t)^2}{\sum x_t^2} - \frac{(\sum xy_w)^2}{\sum x_w^2}\right] \tag{18.26}$$

In general, with a randomized group design, we would not expect to have very large differences in the X means of the various treatment groups. If the X means are identical, then $\Sigma x_b^2 = 0$ and we would have $\Sigma x_t^2 = \Sigma x_w^2$. Also, if the X means are identical, then $\Sigma xy_b = 0$ and we would have $\Sigma xy_t = \Sigma xy_w$. Under this condition, the adjusted sum of squares defined by S_5 will be exactly equal to the unadjusted treatment sum of squares.

In the present example, we find that the adjusted treatment sum of squares is 258.90, whereas the unadjusted treatment sum of squares is 220.93. The reason that the adjusted sum of squares is larger than the unadjusted is because Σxy_b is negative and Σxy_w is positive.[6] The adjusted treatment sum of squares will always be larger than the unadjusted, if Σxy_b is negative and Σxy_w is positive. If Σxy_b is positive and Σxy_w is positive, then the adjusted treatment sum of squares can be smaller than the unadjusted treatment sum of squares. This is why we do not want the X means to be influenced by the treatments in the same way in which the Y means are influenced. If the X means are influenced by the treatments so as to result in the means being positively correlated, then the adjusted treatment sum of squares will, in essence, partial out some of the treatment effects in the sense that the adjusted treatment sum of squares will be smaller than the unadjusted treatment sum of squares.

[6] Because $\Sigma xy_t = \Sigma xy_w + \Sigma xy_b$, if Σxy_b is negative and Σxy_w is positive, then we must have $\Sigma xy_t < \Sigma xy_w$. Because it is also true that $\Sigma x_t^2 \geq \Sigma x_w^2$, then we must have $(\Sigma xy_t)^2/\Sigma x_t^2 < (\Sigma xy_w)^2/\Sigma x_w^2$ and the term within brackets in (18.26) will be negative. Because the brackets are preceded by a negative sign, when the brackets are removed the term becomes positive and therefore we must have $S_5 > \Sigma y_b^2$.

With a randomized group design and with the X measures obtained prior to the application of the treatments, it is obvious that the X means cannot be influenced by the treatments. In this instance, any covariation, positive or negative, between the X and Y means can only be attributed to the random variation in the X means. In some cases, by chance, the treatment groups with low X means may have high Y means and the treatment groups with high X means may have low Y means, so that Σxy_b is negative, as in the present example. In other cases, by chance, groups with high X means may also have high Y means and groups with low X means may also have low Y means, so that Σxy_b is positive. In the randomized group design, the adjusted treatment sum of squares, as defined by (19.26), "corrects" or "adjusts" the treatment sum of squares on the Y variable for the random covariation of the X and Y means. The adjusted treatment sum of squares, in other words, represents the variation in the treatment or Y means that would presumably be obtained if all of the treatment groups had the same mean on the X variable.

Table 18.4 summarizes the analysis of covariance. The estimate of experimental error is the mean square for S_2. For the test of significance of the adjusted treatment mean square, we have

$$F = \frac{S_5/(k-1)}{S_2/[k(n-1)-1]} \tag{18.27}$$

or, for the data of Table 18.4,

$$F = \frac{258.90/2}{5.51/11} = 258.38$$

with 2 and 11 d.f. This is a highly significant value and we conclude that the adjusted treatment mean square is significant. In the next section we consider tests of significance for comparisons made on the *adjusted* treatment means.

Table 18.4
SUMMARY OF THE COVARIANCE ANALYSIS OF THE DATA OF TABLE 18.2

Source of Variation	Sum of Squares	d.f.	Mean Square	F
S_5: Treatments	258.90	2	129.450	258.38
S_2: Error	5.51	11	0.501	
S_4: Total	264.41	13		

18.10 Adjusted Treatment Means and Tests of Comparisons

We let an adjusted treatment mean on the Y variable be

$$\bar{Y}'_{k\cdot} = \bar{Y}_{k\cdot} - b_w(\bar{X}_{k\cdot} - \bar{X}_{\cdot\cdot}) \tag{18.28}$$

Then the standard error of the difference between any two such adjusted means will be

$$s_{\bar{Y}'_{1\cdot}-\bar{Y}'_{2\cdot}} = \sqrt{s_2^2\left[\frac{1}{n_1} + \frac{1}{n_2} + \frac{(\bar{X}_{1\cdot} - \bar{X}_{2\cdot})^2}{\sum x_w^2}\right]} \tag{18.29}$$

where s_2^2 is the error mean square of the analysis of covariance. We note that if $n_1 = n_2 = n$ and if $\bar{X}_{1\cdot} = \bar{X}_{2\cdot}$, then

$$s_{\bar{Y}'_{1\cdot}-\bar{Y}'_{2\cdot}} = \sqrt{\frac{2s_2^2}{n}}$$

As a test of significance of the difference between two adjusted means, we have

$$t = \frac{\bar{Y}'_{1\cdot} - \bar{Y}'_{2\cdot}}{s_{\bar{Y}'_{1\cdot}-\bar{Y}'_{2\cdot}}} \tag{18.30}$$

with degrees of freedom equal to $k(n - 1) - 1$.

Because (18.29) takes a different value for each comparison of two adjusted means, Finney (1946) has suggested that

$$s_e^2 = s_2^2\left(1 + \frac{\sum x_b^2}{(k - 1)\sum x_w^2}\right) \tag{18.31}$$

be used as an effective residual error mean square for all comparisons on the adjusted treatment means. In this case, we have

$$t = \frac{\bar{Y}'_{1\cdot} - \bar{Y}'_{2\cdot}}{\sqrt{s_e^2\left(\frac{1}{n_1} + \frac{1}{n_2}\right)}} \tag{18.32}$$

as a test of significance of the difference between any two adjusted means. The t defined by (18.32) also has $k(n - 1) - 1$ d.f.

More generally, for any comparison on the adjusted treatment means, we have

$$s_{d_i} = \sqrt{s_e^2 \sum_{1}^{k} \frac{a_{\cdot i}^2}{n_k}} \tag{18.33}$$

as the standard error of the comparison. If all treatment means are based on the same number of observations, so that $n_1 = n_2 = \cdots = n_k = n$, then

$$s_{d_i} = \sqrt{\frac{s_e^2}{n} \sum_1^k a_{\cdot i}^2} \qquad (18.34)$$

where $\sum_1^k a_{\cdot i}^2$ is the sum of the squared coefficients for the comparison.

In the analysis of covariance problem we have considered, we have, substituting in (18.31),

$$s_e^2 = 0.501 \left[1 + \frac{1.20}{(3-1)(46.80)} \right]$$

$$= 0.508$$

Then, substituting in (18.28), we obtain the values of the three adjusted treatment means:

$$\bar{Y}'_{1\cdot} = 9.0 - 1.44(3.8 - 4.0) = 9.29$$

$$\bar{Y}'_{2\cdot} = 4.2 - 1.44(4.4 - 4.0) = 3.62$$

$$\bar{Y}'_{3\cdot} = 13.6 - 1.44(3.8 - 4.0) = 13.89$$

Taking the two adjusted means, $\bar{Y}'_{1\cdot}$ and $\bar{Y}'_{3\cdot}$, with the smallest difference we have

$$t = \frac{13.89 - 9.29}{\sqrt{\frac{2(0.508)}{5}}}$$

$$= 10.20$$

with $k(n-1) - 1 = 11$ d.f. and this is a highly significant value.

18.11 Analysis of Variance of Difference Measures

The arithmetic of the analysis of covariance is somewhat more involved than that of the analysis of variance in that we must deal with the product sums as well as with the sums of squares. In some cases, the experimenter may avoid a covariance analysis by dealing with the difference between the X and Y values for each subject. We define a difference measure as

$$D_{kn} = Y_{kn} - X_{kn} \qquad (18.35)$$

and the analysis of variance may be applied directly to the D measures. In

Table 18.5

DIFFERENCE SCORES FOR THE DATA OF TABLE 18.2

	Treatment Groups		
	1	2	3
	4	−1	9
	6	−1	9
	6	1	11
	4	−1	9
	6	1	11
$\sum$	26	−1	49

Table 18.5 we show the D measures for the data of Table 18.2 which we have treated by the analysis of covariance.

For the total sum of squares, we have

$$\text{Total} = (4)^2 + (6)^2 + \cdots + (11)^2 - \frac{(74)^2}{15} = 264.93$$

and the treatment sum of squares will be equal to

$$\text{Treatment} = \frac{(26)^2}{5} + \frac{(-1)^2}{5} + \frac{(49)^2}{5} - \frac{(74)^2}{15} = 250.53$$

The within treatment sum of squares can be obtained by subtraction. Thus,

$$\text{Within} = 264.93 - 250.53 = 14.40$$

Table 18.6 summarizes the analysis of variance of the D measures. We have $F = 104.38$ with 2 and 12 d.f. and this is a highly significant value.

It should be clear that, in taking $D_{kn} = Y_{kn} - X_{kn}$, we have in essence assumed b_w to be equal to 1.00. If b_w does not differ too greatly from 1.00,

Table 18.6

SUMMARY OF THE ANALYSIS OF VARIANCE OF THE DIFFERENCE SCORES OF TABLE 18.5

Source of Variation	Sum of Squares	d.f.	Mean Square	F
Treatments	250.53	2	125.26	104.38
Error	14.40	12	1.20	
Total	264.93	14		

then the analysis of covariance and the analysis of variance of the D measures will give very similar results.[7] On the other hand, if b_w differs considerably from 1.00, then we can expect the estimate of experimental error based on the analysis of variance of the D measures to be considerably larger than the estimate of experimental error obtained if we used the correct slope b_w. In the present example, we have $b_w = 1.44$ and the error mean square for the analysis of covariance is equal to 0.501. The error mean square for the analysis of variance of the difference measures is 1.20, an estimate that is more than twice as large as the error estimate of the analysis of covariance.

Before undertaking the analysis of variance of the D measures, it is wise to plot the Y measures against the X measures, keeping the observations for different treatments identified in some manner. In this way it is possible to determine from the graph whether the various regression lines are parallel and also, if they are parallel, if b_w can be assumed to be approximately equal to 1.00.

18.12 A Randomized Block Design as an Alternative to the Analysis of Covariance

Assume that the X measures of Table 18.2 were available for each subject prior to assigning the subjects to the treatments. Under this condition, we may consider a randomized block design as an alternative to the analysis of covariance. Using the X measures to form five blocks of three subjects each, we obtain Table 18.7 which also shows the random assignment of each of the subjects in each block to the three treatments. In Table 18.8 we show the Y measures for each treatment group rearranged so that the columns correspond to treatments.

Table 18.7

ARRANGEMENT OF SUBJECTS IN BLOCKS ON THE BASIS OF THE X MEASURES OF TABLE 18.2 AND RANDOMIZATION OF TREATMENTS WITHIN BLOCKS

Blocks Based on X			Randomization of Treatments		
1	1	2	*A*	*C*	*B*
3	3	3	*B*	*A*	*C*
4	4	4	*B*	*C*	*A*
5	5	6	*C*	*A*	*B*
6	6	7	*A*	*C*	*B*

[7] For a further discussion, see Cox (1957).

Table 18.8
MEASURES ON THE DEPENDENT VARIABLE (Y) FOR SUBJECTS IN A RANDOMIZED BLOCK DESIGN BASED ON THE RANDOMIZED BLOCKS OF TABLE 18.7 AND THE CORRESPONDING Y MEASURES OF TABLE 18.2

	A	B	C	Σ
Block 1	5	1	10	16
Block 2	9	2	12	23
Block 3	8	3	13	24
Block 4	11	7	16	34
Block 5	12	8	17	37
Σ	45	21	68	134

Applying the analysis of variance to the data of Table 18.8, we have the following sums of squares:

$$\text{Total} = (5)^2 + (9)^2 + \cdots + (17)^2 - \frac{(134)^2}{15} = 322.93$$

$$\text{Block} = \frac{(16)^2}{3} + \frac{(23)^2}{3} + \cdots + \frac{(37)^2}{3} - \frac{(134)^2}{15} = 98.26$$

$$\text{Treatment} = \frac{(45)^2}{5} + \frac{(21)^2}{5} + \frac{(68)^2}{5} - \frac{(134)^2}{15} = 220.93$$

and

$$\text{Block} \times \text{treatment} = 322.93 - 98.26 - 220.93 = 3.74$$

The summary of the analysis of variance is given in Table 18.9. We note

Table 18.9
SUMMARY OF THE ANALYSIS OF VARIANCE FOR THE RANDOMIZED BLOCK DESIGN OF TABLE 18.8

Source of Variation	Sum of Squares	d.f.	Mean Square	F
Treatments	220.93	2	110.46	235.02
Blocks	98.26	4	24.56	
$B \times T$	3.74	8	0.47	
Total	322.93	14		

that the estimate of experimental error of the randomized block design (0.47) does not differ too greatly from the estimate obtained with the analysis of covariance (0.50) for the same data.[8]

The randomized block design is a useful alternative to the analysis of covariance and it is easy to see that the arithmetic for the randomized block design is considerably simpler than that for the analysis of covariance. The major difference between the two methods of analysis is in terms of how we use the X measures. In the randomized block design they are used to group subjects into blocks. Once the blocks are formed, we make no further use of the actual numerical values of the X measures in our analysis. With the analysis of covariance, on the other hand, we do consider the actual numerical values of the X measures in our analysis.

To use a randomized block design, we must have available in advance the X measures for all subjects to be used in the experiment. This is necessary so that we can arrange the subjects into blocks. We then randomize the treatments within each block. In some cases we may not be able to obtain the X measures prior to assigning the treatments to the subjects. For example, we may have an experimental procedure such that each subject appears in the laboratory for only one session. At this session the X measure is obtained, the treatment is applied, and the Y measure is obtained after the application of the treatment. Under this condition we cannot use the X measures to arrange the subjects into blocks, because we do not know what these measures will be for all of the subjects involved in the experiment. We shall know the X measure for each subject at the time he is tested, but we do not know what these measures will be for those subjects who have not yet appeared in the laboratory. Under these circumstances, if we wish to use the X measures in an attempt to reduce our estimate of experimental error, we must use either the analysis of covariance or the analysis of variance of the difference measures in analyzing the data.

18.13 Analysis of Covariance and Other Experimental Designs

The discussion of the analysis of covariance has been limited to a randomized group design. The applications of the analysis of covariance, however, are not restricted to this design. If we use measures of one variable to group subjects into blocks and if in addition we have another supplementary measure X for each subject, we can use the analysis of covariance with a randomized block design. Similarly, the analysis of

[8] For a further discussion of the relationship between a randomized block design and a covariance analysis, see Cox (1957) and Feldt (1958).

covariance can be used with a Latin square design. Examples and methods of analysis for these designs can be found in Snedecor (1956) and Federer (1955).

QUESTIONS AND PROBLEMS

18.1 A randomized group design was used in an experiment with $n = 5$ subjects assigned to each of three treatments. A supplementary measure X was available for each subject. The supplementary measures and the measures on the dependent variable Y are given below:

Treatment 1		Treatment 2		Treatment 3	
X	Y	X	Y	X	Y
3	6	2	11	2	20
9	6	7	14	9	21
16	8	13	18	14	25
19	13	19	18	20	21
24	12	23	20	23	29

(*a*) Make a plot of the observations retaining the identity of the treatment groups. What are some of the things you can determine from the plot?

(*b*) Express each value of X and Y as deviations from the means of the treatment groups to which they belong. Make a plot of these deviations and compare the plot with that of (*a*). How would you account for the difference between the two plots?

(*c*) By means of the analysis of variance, determine if the means for the groups differ significantly on the X measures.

(*d*) By means of the analysis of variance, determine if the means for the groups differ significantly on the Y measures.

(*e*) Determine whether the separate regression coefficients for the groups differ significantly.

(*f*) Analyze the data using the analysis of covariance. Compare the estimates of experimental error of the analysis of covariance with that obtained in (*d*).

(*g*) Note that the observations given above are arranged in such a way that across rows the X measures are fairly homogeneous. Assume that each row corresponds to a block. Analyze the Y measures assuming the design is a randomized block design. Compare the results of this analysis with that of (*f*).

18.2 In an experiment by Mowrer (1934), previously unrotated pigeons were tested for clockwise postrotational nystagmus. The rate of rotation was one revolution in 1½ seconds. An average initial score for each pigeon, based on two tests, is indicated by the symbol X. The twenty-four pigeons were then divided into four groups of six each. Each group was then subjected to ten daily periods of rotation under one of the experimental conditions indicated below. The rotation speed was the same as during the initial test and the rotation periods lasted 30 seconds, with a 30-second rest interval between each period. Groups 1, 2, and 3 were practiced in a clockwise direction only. For Group 4 the environment was rotated in a counterclockwise direction. At the end of twenty-four days of practice, each group was tested again under the same conditions as on the initial test. These records are called Y.

Group 1 Rotation of body only—vision excluded		Group 2 Rotation of body only—vision permitted		Group 3 Rotation of body and environment		Group 4 Rotation of environment only	
Initial X	Final Y	Initial X	Final Y	Initial X	Final Y	Initial X	Final Y
23.8	7.9	28.5	25.1	27.5	20.1	22.9	19.9
23.8	7.1	18.5	20.7	28.1	17.7	25.2	28.2
22.6	7.7	20.3	20.3	35.7	16.8	20.8	18.1
22.8	11.2	26.6	18.9	13.5	13.5	27.7	30.5
22.0	6.4	21.2	25.4	25.9	21.0	19.1	19.3
19.6	10.0	24.0	30.0	27.9	29.3	32.2	35.1

(*a*) Use the analysis of variance to determine whether the means on the X variable for the groups differ significantly. (*b*) Use the analysis of variance to determine whether the means of the groups on the Y variable differ significantly. (*c*) Analyze the results of the experiment using the analysis of covariance.

18.3 We have a randomized group design with $n = 5$ subjects assigned at random to each of $k = 3$ treatments. We have available a supplementary measure X for each subject. After the treatments have been applied, we obtain for each subject a measure on a dependent variable Y. The basic data are given below:

	$\sum x^2$	$\sum y^2$	$\sum xy$
Group 1	20.0	30.0	25.0
Group 2	25.0	35.0	20.0
Group 3	20.0	30.0	25.0
Pooled within	65.0	95.0	70.0
Total	70.0	195.0	60.0

(*a*) Find the value of F for a test of significance of the differences among the k regression coefficients. (*b*) Find the value of F for the test of significance of the adjusted treatment mean square. (*c*) Find the values of the regression coefficients for each group. (*d*) Is the adjusted treatment sum of squares in the analysis of covariance larger than or smaller than the unadjusted treatment sum of squares? Why is this so? (*e*) Suppose that for each subject we obtain $D = Y - X$ and then analyze these difference measures by means of the analysis of variance. Do you think the error mean square for this analysis would be comparable to the error mean square of the analysis of covariance? Why? (*f*) If for each value of Y we subtract the treatment mean of the group to which it belongs and then plot these deviations against the paired values of X, what would the resulting plot look like? Why?

CHAPTER 19

THE METHOD OF LEAST SQUARES

19.1 *Introduction*

In the analysis of variance designs we have discussed, we assumed that a given observation is a sum of a number of additive components plus a component that represents a random error. The additive components are unknown constants or parameters, whereas the random error is a variable. For example, in a randomized group design, we assumed that

$$X = \mu + t_k + e_{kn}$$

where μ is a constant for all kn observations, t_k is a constant for those n observations obtained for the kth treatment, and e_{kn} is a random error associated with each of the kn observations.

The method of least squares is a technique for using the obtained data of an experiment to provide estimates of the unknown parameters or constants in such a way as to minimize the sum of the squared errors. In the sections that follow, we state, without proof, the elementary rules of the differential calculus that we need in order to apply the method of least squares to problems in the analysis of variance. The review is not complete but it is sufficient for our purpose. The student who has no knowledge of the differential calculus can verify the rules by consulting any elementary calculus text.

19.2 *The Derivative of a Power Function*

Let $y = x^n$. Then y is a function of the power of x. The derivative of this function, written dy/dx, is

$$\frac{dy}{dx} = nx^{n-1} \qquad (19.1)$$

and (19.1) states the general rule for the derivative of $y = x^n$. We multiply x by its original exponent and decrease the original exponent by 1.

19.3 The Derivative of a Power Function Involving an Added Constant

Let $y = a + x^n$, where a is an added constant. Then the derivative of this function with respect to x is

$$\frac{dy}{dx} = nx^{n-1} \tag{19.2}$$

and (19.2) defines the general rule that an added constant does not appear in the derivative.

19.4 The Derivative of a Power Function When a Is a Coefficient

If $y = ax^n$, where a is a constant and a coefficient of x, then the derivative of this function with respect to x is

$$\frac{dy}{dx} = anx^{n-1} \tag{19.3}$$

and (19.3) defines the general rule that if a is a coefficient of the term with respect to which we are differentiating, it appears as a coefficient of that term in the derivative.

19.5 The Method of Least Squares

Suppose that $e = X - a$ and that we have a sample of n values of X. Then we let

$$Q = \sum e^2 = \sum (X - a)^2 \tag{19.4}$$

and we want to find the value of a that will minimize Q. To do so, using the method of least squares, we differentiate Q with respect to a and set the derivative equal to zero. If we expand the right side of (19.4) and sum, we have

$$Q = \sum X^2 - 2a \sum X + na^2$$

and, using the rules we have stated, we have[1]

$$\frac{dQ}{da} = -2 \sum X + 2na$$

Setting the above derivative equal to zero, we have

$$-2 \sum X + 2na = 0$$

or

$$2na = 2 \sum X$$

Then, dividing both sides of the above expression by $2n$, we obtain

$$a = \bar{X}$$

and the mean, $\bar{X}$, is the value of a for which Q will be a minimum. In other words, we have just proved that

$$\sum (X - \bar{X})^2 < \sum (X - a)^2$$

for any value of $a \neq \bar{X}$.

19.6 Another Method for Finding the Derivative of $Q = \Sigma(X - a)^2$

We show another method of finding the derivative of $Q = \Sigma(X - a)^2$ that is useful. Again we let

$$e = X - a \qquad (19.5)$$

and

$$Q = \sum e^2 = \sum (X - a)^2 \qquad (19.6)$$

Differentiating (19.5) with respect to a, we have

$$\frac{de}{da} = -1$$

and if we differentiate (19.6) with respect to e, we obtain

$$\frac{dQ}{de} = 2 \sum e$$

[1] For any given set of observations ΣX and ΣX^2 are constants but a may conceivably take any one of an infinite number of values. The least squares solution for a, however, requires that we find the *single* value of a that will minimize Q for a set of n observations. If we were to let a take a different value for each possible value of X, then we would have the trivial solution that Q is minimized when $a = X$ for each value of X.

Then we have the general rule that

$$\frac{dQ}{da} = \frac{de}{da} \times \frac{dQ}{de} \tag{19.7}$$

and substituting in (19.7) we obtain

$$\begin{aligned}\frac{dQ}{da} &= -2 \sum e \\ &= -2 \sum (X - a) \\ &= -2 \sum X + 2na\end{aligned}$$

as before.

19.7 *Least Squares Estimates of t_k and μ for a Randomized Group Design*

If we have a randomized group design, with k treatments and n observations for each treatment, then we have assumed that

$$X_{kn} = \mu + t_k + e_{kn} \tag{19.8}$$

where μ is common to all kn observations, t_k is a treatment effect that is common to the n observations receiving a given treatment, and e_{kn} is a random error associated with each of the kn observations. Then

$$e_{kn} = X_{kn} - t_k - \mu \tag{19.9}$$

We let

$$Q = \sum e_{kn}^2 = \sum (X_{kn} - t_k - \mu)^2 \tag{19.10}$$

and we want to find the estimates of μ and of the k values of t_k that will minimize Q. In our discussion of the randomized group design we took $\bar{X}_{..}$ as an estimate of μ and $\bar{X}_{k.} - \bar{X}_{..}$ as an estimate of t_k. We now show that these estimates are in fact least squares estimates.

Differentiating (19.9) with respect to μ and (19.10) with respect to e, we have -1 and $2\Sigma e$ as the respective derivatives. Then, multiplying the two derivatives, we obtain

$$\frac{dQ}{d\mu} = -2 \sum e$$

or

$$\frac{dQ}{d\mu} = -2(\sum X_{..} - n \sum t_k - kn\mu) \tag{19.11}$$

In the analysis of variance model for the randomized group design we have shown that, without any loss in generality, we can let $\Sigma t_k = 0$. Then, with $\Sigma t_k = 0$, if we set the derivative equal to zero and solve for μ, we have

$$2kn\mu = 2 \sum X_{..}$$

or

$$\mu = \bar{X}_{..}$$

and $\bar{X}_{..}$ is the least squares estimate of μ.

If we differentiate (19.9) with respect to t_k and (19.10) with respect to e and multiply the two derivatives, we have

$$\frac{dQ}{dt_k} = -2 \sum e$$

But, in this case, we have only n observation equations in which a given t_k appears, so that[2]

$$\frac{dQ}{dt_k} = -2(\sum X_{k\cdot} - nt_k - n\mu) \qquad (19.12)$$

Setting this derivative equal to zero and solving for t_k, we have

$$2nt_k = 2 \sum X_{k\cdot} - 2n\mu$$

or

$$t_k = \bar{X}_{k\cdot} - \mu$$

Substituting in the above expression with the least squares estimate of μ, we obtain

$$t_k = \bar{X}_{k\cdot} - \bar{X}_{..}$$

and $\bar{X}_{k\cdot} - \bar{X}_{..}$ is the least squares estimate of t_k.

If we now substitute the least squares estimates of μ and t_k in (19.10), we have

$$\sum_{1}^{kn} e_{kn}^2 = \sum_{1}^{k} \sum_{1}^{n} [X_{kn} - (\bar{X}_{k\cdot} - \bar{X}_{..}) - \bar{X}_{..}]^2$$

$$= \sum_{1}^{k} \sum_{1}^{n} (X_{kn} - \bar{X}_{k\cdot})^2 \qquad (19.13)$$

and (19.13) is the sum of squares within treatments.

[2] If an observation equation for e_{kn}^2 contains the t_k with respect to which we are differentiating, then its derivative is

$$-2(X_{kn} - t_k - \mu)$$

If the observation equation does not contain t_k, then its derivative is zero.

19.8 Least Squares Estimates for a Two-Factor Experiment with a Randomized Group Design

For a two-factor experiment with a randomized group design, we have

$$X_{abn} = \mu + t_a + t_b + t_{ab} + e_{abn} \tag{19.14}$$

and

$$e_{abn} = X_{abn} - t_a - t_b - t_{ab} - \mu \tag{19.15}$$

with

$$Q = \sum e_{abn}^2 = \sum (X_{abn} - t_a - t_b - t_{ab} - \mu)^2 \tag{19.16}$$

Differentiating (19.15) with respect to any one of the unknowns, t_a, t_b, t_{ab}, or μ, we obtain -1 as the derivative and the derivative of (19.16) with respect to e is $2\Sigma e$. Multiplying the two derivatives, we obtain

$$-2\sum e = -2\sum (X_{abn} - t_a - t_b - t_{ab} - \mu) \tag{19.17}$$

where the summation will be over those observation equations that contain the term with respect to which we have differentiated. For example, all abn observation equations will contain μ and we have

$$\frac{dQ}{d\mu} = -2(\sum X_{\cdots} - bn\sum_{1}^{a} t_a - an\sum_{1}^{b} t_b - n\sum_{1}^{ab} t_{ab} - abn\mu) \tag{19.18}$$

But, as we have stated previously, we can, without any loss in generality, impose the side conditions that

$$\sum_{1}^{a} t_a = \sum_{1}^{b} t_b = \sum_{1}^{a} t_{ab} = \sum_{1}^{b} t_{ab} = \sum_{1}^{ab} t_{ab} = 0 \tag{19.19}$$

and with these side conditions (19.18) becomes

$$\frac{dQ}{d\mu} = -2\sum X_{\cdots} + 2abn\mu \tag{19.20}$$

Setting this derivative equal to zero and solving for μ, we obtain

$$\mu = \bar{X}_{\cdots}$$

and $\bar{X}_{\cdots}$ is the least squares estimate of μ.

To find the least squares estimate of a given t_a we note that we have bn observation equations containing t_a and therefore (19.17) becomes

$$\frac{dQ}{dt_a} = -2\left(\sum X_{a\cdot\cdot} - bnt_a - n\sum_{1}^{b} t_b - n\sum_{1}^{b} t_{ab} - bn\mu\right) \tag{19.21}$$

and because of the side conditions of (19.19), we have

$$\frac{dQ}{dt_a} = -2\sum X_{a\cdot\cdot} + 2bnt_a + 2bn\mu \tag{19.22}$$

Setting this derivative equal to zero and substituting the least squares estimate $\bar{X}_{...}$ for μ, we have

$$t_a = \bar{X}_{a..} - \bar{X}_{...}$$

and $\bar{X}_{a..} - \bar{X}_{...}$ is the least squares estimate of t_a. Similarly, we would find that the least squares estimate of t_b is

$$t_b = \bar{X}_{.b.} - \bar{X}_{...}$$

To find the least squares estimate of t_{ab}, we have

$$\frac{dQ}{dt_{ab}} = -2(\sum X_{ab.} - nt_a - nt_b - nt_{ab} - n\mu) \tag{19.23}$$

Substituting in (19.23) with the least squares estimates of t_a, t_b, and μ, and setting the derivative equal to zero, we obtain

$$-2[\sum X_{ab.} - n(\bar{X}_{a..} - \bar{X}_{...}) - n(\bar{X}_{.b.} - \bar{X}_{...}) - nt_{ab} - n\bar{X}_{...}] = 0$$

or

$$-2(\sum X_{ab.} - n\bar{X}_{a..} - n\bar{X}_{.b.} - nt_{ab} + n\bar{X}_{...}) = 0$$

Solving for t_{ab}, we have

$$t_{ab} = \bar{X}_{ab.} - \bar{X}_{a..} - \bar{X}_{.b.} + \bar{X}_{...} \tag{19.24}$$

and the right side of (19.24) is the least squares estimate of t_{ab}.

If the least squares estimates of μ, t_a, t_b, and t_{ab} are substituted in (19.16), then we have

$$\sum e_{abn}^2 = \sum [X_{abn} - (\bar{X}_{a..} - \bar{X}_{...}) - (\bar{X}_{.b.} - \bar{X}_{...}) - (\bar{X}_{ab.} - \bar{X}_{a..} - \bar{X}_{.b.} + \bar{X}_{...}) - \bar{X}_{...}]^2$$

or

$$\sum e_{abn}^2 = \sum_{1}^{abn} (X_{abn} - \bar{X}_{ab.})^2 \tag{19.25}$$

and (19.25) is the within treatment sum of squares.

19.9 Least Squares Estimates for a Randomized Block Design

For a randomized block design with b blocks and t treatments, we have as a model

$$X_{bt} = \mu + t_b + t_t + e_{bt} \tag{19.26}$$

where t_b is a block effect, t_t is a treatment effect, and e_{bt} is a random error.

Then

$$e_{bt} = X_{bt} - t_b - t_t - \mu \tag{19.27}$$

and

$$Q = \sum e_{bt}^2 = \sum (X_{bt} - t_b - t_t - \mu)^2 \tag{19.28}$$

Following the same procedures we have used in the previous section, we would find that if Q is to be minimized, then the least squares estimates of t_b, t_t, and μ, with the side conditions that $\Sigma t_b = \Sigma t_t = 0$, are as follows:

$$t_b = \bar{X}_{b.} - \bar{X}_{..}$$

$$t_t = \bar{X}_{.t} - \bar{X}_{..}$$

$$\mu = \bar{X}_{..}$$

Substituting in (19.28) with the least squares estimate of t_b, t_t, and μ, we have

$$\sum e_{bt}^2 = \sum (X_{bt} - \bar{X}_{b.} - \bar{X}_{.t} + \bar{X}_{..})^2 \tag{19.29}$$

and (19.29) is the error sum of squares for the randomized block design.

19.10 Least Squares Estimates for a Latin Square Design

For a Latin square design with r rows, c columns, and t treatments, with $r = c = t$, we have as a model

$$X_{rct} = \mu + t_r + t_c + t_t + e_{rct} \tag{19.30}$$

where t_r is a row effect, t_c is a column effect, t_t is a treatment effect, and e_{rct} is a random error. Then

$$e_{rct} = X_{rct} - t_r - t_c - t_t - \mu \tag{19.31}$$

and

$$Q = \sum e_{rct}^2 = \sum (X_{rct} - t_r - t_c - t_t - \mu)^2 \tag{19.32}$$

With side conditions that

$$\sum_1^r t_r = \sum_1^t t_t = \sum_1^c t_c = 0$$

we have the following least squares estimates:

$$t_r = \bar{X}_{r..} - \bar{X}_{...}$$

$$t_c = \bar{X}_{.c.} - \bar{X}_{...}$$

$$t_t = \bar{X}_{..t} - \bar{X}_{...}$$

$$\mu = \bar{X}_{...}$$

If these least squares estimates are substituted in (19.32), then we obtain

$$\sum e_{rct}^2 = \sum (X_{rct} - \bar{X}_{r\cdot\cdot} - \bar{X}_{\cdot c \cdot} - \bar{X}_{\cdot\cdot t} + 2\bar{X}_{\cdots})^2 \tag{19.33}$$

as the error sum of squares for a Latin square design.

19.11 Least Squares Estimate of b_k

In the analysis of covariance for a randomized group design, we pointed out that

$$\sum_{1}^{n} (y_k - b_k x_k)^2 \tag{19.34}$$

would be at a minimum when b_k is equal to $\Sigma xy_k/\Sigma x_k^2$. We now prove that this is the case.

We let

$$e = (y_k - b_k x_k) \tag{19.35}$$

and

$$Q = \sum e^2 = \sum (y_k - b_k x_k)^2 \tag{19.36}$$

Then, differentiating (19.35) with respect to b_k and (19.36) with respect to e, we have

$$\frac{de}{db_k} = -x_k \qquad \text{and} \qquad \frac{dQ}{de} = 2\sum e$$

and multiplying these two derivatives, we obtain

$$\frac{dQ}{db_k} = (-x_k)2\sum e \tag{19.37}$$

or

$$\frac{dQ}{db_k} = -2(\sum xy_k - b_k \sum x_k^2) \tag{19.38}$$

Setting this derivative equal to zero, we have

$$-2\sum xy_k + 2b_k \sum x_k^2 = 0 \tag{19.39}$$

and solving for b_k we obtain

$$b_k = \frac{\sum xy_k}{\sum x_k^2} \tag{19.40}$$

and the right side of (19.40) is the least squares estimate of b_k. Thus, $\Sigma(y_k - b_k x_k)^2$ will be at a minimum if and only if $b_k = \Sigma xy_k/\Sigma x_k^2$.

REFERENCES

Anderson, R. L., and T. A. Bancroft. *Statistical theory in research.* New York: McGraw-Hill, 1952.

Archer, E. J. Some Greco-Latin analysis of variance designs for learning studies. *Psychological Bulletin*, 1952, **49,** 521–537.

Bartlett, M. S. Square-root transformation in analysis of variance. *Journal of the Royal Statistical Society Supplement*, 1936, **3,** 68–78.

Bartlett, M. S. The use of transformations. *Biometrics*, 1947, **3,** 39–52.

Bliss, C. I. The analysis of field experimental data expressed in percentages. *Plant Protection*, 1937, Leningrad, No. 12, 67–77.

Bliss, C. I., and C. L. Rose. The assay of parathyroid extract from the serum calcium of dogs. *American Journal of Hygiene*, 1940, **31,** 79–98.

Boneau, C. A. The effects of violations of assumptions underlying the *t* test. *Psychological Bulletin*, 1960, **57,** 49–64.

Box, G. E. P. Non-normality and tests on variances. *Biometrika*, 1953, **40,** 318–335.

Child, I. L. Children's preference for goals easy or difficult to obtain. *Psychological Monographs*, 1946, No. 280.

Clark, M., and D. A. Worcester. A comparison of the results obtained from the teaching of shorthand by the word unit method and the sentence method. *Journal of Educational Psychology*, 1932, **23,** 122–131.

Cochran, W. G. The comparison of percentages in matched samples. *Biometrika*, 1950, **37,** 256–266.

Cochran, W. G. Testing a linear relation among variances. *Biometrics*, 1951, **7,** 17–32.

Cochran, W. G. Some methods for strengthening the common χ^2 tests. *Biometrics*, 1954, **10,** 417–451.

Cochran, W. G., and Gertrude M. Cox. *Experimental designs.* (2d ed.) New York: Wiley, 1957.

Cox, D. R. The use of a concomitant variable in selecting an experimental design. *Biometrika*, 1957, **44,** 150–158.

Cox, D. R. *Planning of experiments.* New York: Wiley, 1958.

Cramér, H. *Mathematical methods of statistics.* Princeton: Princeton University Press, 1946.

Crespi, L. P. Quantitative variation of incentive and performance in the white rat. *American Journal of Psychology*, 1942, **55,** 467–517.

Crump, S. L. The estimation of variance components in analysis of variance. *Biometrics*, 1946, **2**, 7–11.

Crutchfield, R. S. Efficient factorial design. *Journal of Psychology*, 1938, **5**, 339–346.

Curtiss, J. H. On transformations used in the analysis of variance. *Annals of Mathematical Statistics*, 1945, **14**, 107–122.

De Lury, D. B. The analysis of Latin squares when some observations are missing. *Journal of the American Statistical Association*, 1946, **41**, 370–389.

Donaldson, T. S. Robustness of the F-test to errors of both kinds and the correlation between the numerator and denominator of the F-ratio. *Journal of the American Statistical Association*, 1968, **63**, 660–676.

Duncan, D. B. Multiple range and multiple F tests. *Biometrics*, 1955, **11**, 1–42.

Duncan, D. B. Multiple range tests for correlated and heteroscedastic means. *Biometrics*, 1957, **13**, 164–176.

Dunnett, C. W. A multiple comparison procedure for comparing several treatments with a control. *Journal of the American Statistical Association*, 1955, **50**, 1096–1121.

Edwards, A. L. *Expected values of discrete random variables and elementary statistics.* New York: Wiley, 1964.

Edwards, A. L. *Edwards Personality Inventory.* Chicago: Science Research Associates, 1967a.

Edwards, A. L. *Statistical methods.* (2d ed.) New York: Holt, Rinehart and Winston, 1967b.

Edwards, A. L. *Probability and statistics.* New York: Holt, Rinehart and Winston, 1971.

Edwards, A. L., and P. Horst. The calculation of sums of squares for interactions in the analysis of variance. *Psychometrika*, 1950, **15**, 17–24.

Federer, W. T. *Experimental design.* New York: Macmillan, 1955.

Feldt, L. S. A comparison of the precision of three experimental designs employing a concomitant variable. *Psychometrika*, 1958, **23**, 335–354.

Festinger, L., and D. Katz. *Research methods in the behavioral sciences.* New York: Dryden, 1953.

Finney, D. J. Standard errors of yields adjusted for regression on an independent measurement. *Biometrics Bulletin*, 1946, **2**, 53–55.

Fisher, R. A. Discussion on Dr. Wishart's paper. *Journal of the Royal Statistical Society Supplement*, 1934, **1**, 51–53.

Fisher, R. A. *Statistical methods for research workers.* (6th ed.) Edinburgh: Oliver & Boyd, 1936.

Fisher, R. A. *The design of experiments.* (3d ed.) Edinburgh: Oliver & Boyd, 1942.

Fisher, R. A., and F. Yates. *Statistical tables for biological, agricultural and medical research.* (3d ed.) Edinburgh: Oliver & Boyd, 1948.

Freeman, M. F., and J. W. Tukey. Transformations related to the angular and the square root. *Annals of Mathematical Statistics*, 1950, **21**, 607–611.

French, Elizabeth G., and F. H. Thomas. The relation of achievement motivation to problem-solving effectiveness. *Journal of Abnormal and Social Psychology*, 1958, **56**, 45–48.

Gaito, J. Unequal intervals and unequal n in trend analyses. *Psychological Bulletin*, 1965, **63**, 125–127.

Geisser, S., and W. W. Greenhouse. An extension of Box's results on the use of the *F* distribution in multivariate analysis. *Annals of Mathematical Statistics*, 1958, **29**, 885–891.

Glanville, A. D., G. L. Kreezer, and D. M. Dallenbach. The effect of type size on accuracy of apprehension and speed of localizing words. *American Journal of Psychology*, 1946, **59**, 220–235.

Goodman, L. A., and W. H. Kruskal. Measures of association for cross classifications. *Journal of the American Statistical Association*, 1954, **49**, 723–764.

Goodman, L. A., and W. H. Kruskal. Measures of association for cross classifications. II: Further discussion and references. *Journal of the American Statistical Association*, 1959, **54**, 123–163.

Graham, F. K., and B. S. Kendall. Performance of brain-damaged cases on a memory for designs test. *Journal of Abnormal and Social Psychology*, 1946, **41**, 303–314.

Grandage, A. Orthogonal coefficients for unequal intervals. *Biometrics*, 1958, **14**, 287–289.

Grant, D. A. The Latin square principle in the design and analysis of psychological experiments. *Psychological Bulletin*, 1948, **45**, 427–442.

Grant, D. A. Analysis-of-variance tests in the analysis and comparison of curves. *Psychological Bulletin*, 1956, **53**, 141–154.

Grant, D. A., and A. S. Patel. Effect of an electric shock stimulus upon the conceptual behavior of "anxious" and "non-anxious" subjects. *Journal of General Psychology*, 1957, **57**, 247–256.

Guilford, J. P. *Psychometric methods*. (2d ed.) New York: McGraw-Hill, 1954.

Haggard, E. A. Experimental studies in affective processes: II. On the quantification and evaluation of "measured" changes in skin resistance. *Journal of Experimental Psychology*, 1945, **35**, 46–56.

Hartman, G. Application of individual taste differences towards phenylthiocarbamide in genetic investigations. *Annals of Eugenics*. Cambridge, 1939, **9**, 123–135.

Hellman, M. A study of some etiological factors of malocclusion. *Dental Cosmos*, 1914, **56**, 1017–1032.

Hsu, Tse-Chi, and Feldt, L. S. The effect of limitations on the number of criterion scores on the significance level of the *F*-test. *American Educational Research Journal*, 1969, **6**, 515–527.

Hyman, R., and E. Z. Vogt. Water witching: Magical ritual in contemporary United States. *Psychology Today*, 1967, **1**, 35–42.

Kempthorne, O. *The design and analysis of experiments*. New York: Wiley, 1952.

Kendall, M. G., and B. B. Smith. Second paper on random sampling numbers. *Journal of the Royal Statistical Society Supplement*, 1939, **6**, 51–61.

Kendler, H. H. Drive interaction: I. Learning as a function of the simultaneous presence of the hunger and thirst drives. *Journal of Experimental Psychology*, 1945, **35**, 96–109.

Kramer, C. Y. Extension of multiple range tests to group means with unequal numbers of replications. *Biometrics*, 1956, **12**, 307–310.

Kuenne, M. R. Experimental investigation of the relation of language to transposition behavior in young children. *Journal of Experimental Psychology*, 1946, **36**, 471–490.

Lewis, H. B. An experimental study of the role of the ego in work. I. The role of the ego in cooperative work. *Journal of Experimental Psychology*, 1944, **34**, 113–126.

Lewis, H. B., and M. Franklin. An experimental study of the role of the ego in work. II. The significance of task-orientation in work. *Journal of Experimental Psychology*, 1944, **34**, 195–215.

Lindzey, G. (Ed.) *Handbook of social psychology*. (Vol. 1) Cambridge: Addison-Wesley, 1954.

Maier, N. R. F. Reasoning in humans. III. The mechanisms of equivalent stimuli and of reasoning. *Journal of Experimental Psychology*, 1945, **35**, 349–360.

McNemar, Q. Note on the sampling error of the difference between correlated proportions or percentages. *Psychometrika*, 1947, **12**, 153–157.

Megargee, E. L. (Ed.) *Research in clinical assessment*. New York: Harper and Row, 1966.

Merrington, Maxine, and Catherine M. Thompson. Table of percentage points of the inverted beta (F) distribution. *Biometrika*, 1943, **33**, 73–88.

Merritt, C. B., and R. G. Fowler. The pecuniary honesty of the public at large. *Journal of Abnormal and Social Psychology*, 1948, **43**, 90–93.

Moore, K. The effect of controlled temperature changes on the behavior of the white rat. *Journal of Experimental Psychology*, 1944, **34**, 70–79.

Morgan, C. T. The statistical treatment of hoarding data. *Journal of Comparative Psychology*, 1945, **38**, 247–256.

Morgan, C. T. *Introduction to psychology*. New York: McGraw-Hill, 1956.

Morgan, J. J. B. Value of wrong responses in inductive reasoning. *Journal of Experimental Psychology*, 1945, **35**, 141–146.

Mosteller, F., and R. R. Bush. Selected quantitative techniques. In G. Lindzey (Ed.) *Handbook of social psychology*. Cambridge: Addison-Wesley, 1954, 289–334.

Mosteller, F., R. E. K. Rourke, and G. B. Thomas, Jr. *Probability and statistics*. Reading, Mass.: Addison-Wesley, 1961.

Mowrer, O. H. The modification of vestibular nystagmus by means of repeated elicitation. *Comparative Psychological Monograph*, 1934, No. 5.

Mueller, C. G. Numerical transformations in the analysis of experimental data. *Psychological Bulletin*, 1949, **46**, 198–223.

Pearce, S. C. Experimenting with organisms as blocks. *Biometrika*, 1957, **44**, 141–149.

Pfungst, O. *Clever Hans*. New York: Holt, 1911.

Rosenthal, R. *Experimenter effects in behavioral research*. New York: Appleton-Century-Crofts, 1966.

Rosenthal, R. Covert communication in the psychological experiment. *Psychological Bulletin*, 1967, **67**, 356–367.

Rosenzweig, S. An experimental study of "repression" with special reference to need-persistive and ego-defensive reactions to frustration. *Journal of Experimental Psychology*, 1943, **32**, 64–74.

Ryan, T. A. Multiple comparisons in psychological research. *Psychological Bulletin*, 1959, **56**, 26–47.

Satterthwaite, F. E. An approximate distribution of estimates of variance components. *Biometrics*, 1946, **2**, 110–114.

Scheffé, H. A method for judging all contrasts in the analysis of variance. *Biometrika*, 1953, **40**, 87–104.

Schroeder, Elinor M. *On measurement of motor skills*. New York: King's Crown Press, 1945.

Schultz, E. F., Jr. Rules of thumb for determining expectations of mean squares. *Biometrics*, 1955, **11**, 123–135.

Seeger, P., and Gabrielsson, A. Applicability of the Cochran Q test and the F test for statistical analysis of dichotomous data for dependent samples. *Psychological Bulletin*, 1968, **69**, 269–277.

Selltiz, Clarie, Marie Jahoda, M. Deutsch, and S. Cook. *Research methods in social relations*. New York: Holt, Rinehart and Winston, 1964.

Shontz, F. C. *Research methods in personality*. New York: Appleton-Century-Crofts, 1965.

Sidowski, J. B. (Ed.) *Experimental methods and instrumentation in psychology*. New York: McGraw-Hill, 1966.

Sleight, R. B. The effect of instrument dial shape on legibility. *Journal of Applied Psychology*, 1948, **32**, 170–188.

Snedecor, G. W. *Statistical methods*. (5th ed.) Ames: Iowa State College Press, 1956.

Solomon, R. L., and Lessac, M. S. A control group design for experimental studies of developmental processes. *Psychological Bulletin*, 1968, **70**, 145–150.

Stelmachers, Z. T., and McHugh, R. B. Contribution of stereotyped and individualized information to predictive accuracy. *Journal of Consulting Psychology*, 1964, **28**, 234–242.

Stevens, D. A., Stover, C. E., and Backus, J. T. The hyperkinetic child: Effect of incentives on speed of rapid tapping. *Journal of Consulting and Clinical Psychology*, 1970, **34**, 56–59.

Stollak, G. E., B. J. Guerney, Jr., and M. Rothberg (Eds.) *Psychotherapy research*. Chicago: Rand McNally, 1966.

Underwood, B. J. *Psychological research*. New York: Appleton-Century-Crofts, 1957.

Villars, D. S. *Statistical design and analysis of experiments for development research*. Dubuque, Iowa: Brown, 1951.

Wagenaar, W. A. Note on the construction of digram-balanced Latin squares. *Psychological Bulletin*, 1969, **72**, 384–386.

Williams, E. J. Experimental designs balanced for the estimation of residual effects of treatments. *Australian Journal of Scientific Research*, 1949, **2**, 149–168.

Winer, B. J. *Statistical principles in experimental design*. New York: McGraw-Hill, 1962.

Wishart, J. Statistics in agricultural research. *Journal of the Royal Statistical Society Supplement*, 1934, **1**, 26–51.

Woodworth, R. S. *Experimental psychology*. New York: Holt, 1938.

ANSWERS TO SELECTED PROBLEMS

Chapter 2

2.1 (*a*) $P = 0.04$
(*b*) $P = 0.02$

2.2 (*a*) $P = 1/70 = 0.01$
(*b*) $P = 16/70 = 0.23$
(*c*) $P = 36/70 = 0.51$
(*d*) 6 Right, 0 Wrong: $P = 1/924 = 0.0011$
5 Right, 1 Wrong: $P = 36/924 = 0.0390$
4 Right, 2 Wrong: $P = 225/924 = 0.2435$
3 Right, 3 Wrong: $P = 400/924 = 0.4329$
2 Right, 4 Wrong: $P = 225/924 = 0.2435$
1 Right, 5 Wrong: $P = 36/924 = 0.0390$
0 Right, 6 Wrong: $P = 1/924 = 0.0011$

2.3 (*a*) $P = \frac{1}{4} = 0.25$
(*b*) $P = 12/256 = 0.05$
(*c*) $P = 27/64 = 0.42$
(*d*) 2
(*e*) 20

2.4 105

2.7 (*a*) $P = 0.03$

2.9 (*a*) $P = \frac{1}{2}$
(*b*) 70
(*c*) $P = 1/70$
(*d*) $P = 1/6720$
(*e*) $P = 120/6720$

2.10 (*a*) $P = 1/70$
(*b*) $P = 16/70$
(*c*) $P = 36/70$
(*d*) $P = 16/70$
(*e*) $P = 1/70$

2.12 120

2.13 120

Chapter 3

3.1 $Z = 1.92$

3.2 $Z = 1.79$

3.3 $Z = 2.01$

3.4 $Z = 1.74$

3.5 $Z = 1.75$

3.6 $Z = 1.68$

3.7 $Z = 1.56$

3.8 $Z = 1.88$

3.9 $Z = 0.92$

3.10 $Z = 1.54$

3.11 $E(T) = 5 \quad E(T - \mu_T)^2 = 3.33$

3.12 $E(T) = 5 \quad E(T - \mu_T)^2 = 2.27$

3.13 $\mu_T = 1.33 \quad \sigma_T{}^2 = 0.89$

3.14 $E(T) = 12 \quad E(T - \mu_T)^2 = 9$

3.15 $Z = 1.8$

3.18 $n = 25$

3.19 $Z = 2.9$

3.20 $Z = 1.8$

3.21 $Z = 2.4$

3.23 $E(X) = 1 \quad E(X - \mu)^2 = 2$

Chapter 4

4.1 $\chi^2 = 13.16$

4.2 $\chi^2 = 1.14$

4.3 $\chi^2 = 44.72$

4.4 $\chi^2 = 54.58$

4.5 $\chi^2 = 23.82$

4.6 $\chi^2 = 46.65$

4.7 $\chi^2 = 9.27$

4.8 (*a*) $\chi^2 = 20.0$
(*b*) $\chi^2 = 16.0$

4.9 $\chi^2 = 26.00$

4.10 $Z^2 = \chi^2 = 3.6$

4.11 $\chi^2 = 17.5$

Chapter 5

5.1 $20.11 \leq \mu \leq 24.69$

5.2 $t = 2.80$

5.3 $t = 4.38$

5.4 $t = 3.01$

5.5 $t = 2.46$

5.6 $t = 1.11$

5.7 $t = 1.74$

5.8 (*a*) Approximately twenty-seven subjects in each group
(*b*) Approximately thirteen subjects in each group

5.12 The probability of rejecting H_0 is approximately 0.979.

5.13 (*a*) $t = -3.0$
(*b*) $-10.05 \leq \mu_1 - \mu_2 \leq -1.95$

5.14 (*a*) $t = -2.0$
(*b*) $-4.306 \leq \mu_1 - \mu_2 \leq 0.306$

5.15 $58.02 \geq \mu \geq 49.98$

5.16 (*a*) $t = 5.0$
(*b*) $14.05 \geq \mu_1 - \mu_2 \geq 5.95$

Chapter 6

6.1 (*a*) $F = 3.16$
(*b*) $t = 1.60$

6.2 (*a*) $F = 4.14$
(*b*) $t = 2.30$

6.3 (*a*) $F = 3.32$
(*b*) $t = 2.29$

6.4 (*a*) $F = 3.27$
(*b*) $t = 2.72$

6.5 (*a*) $F = 8.19$
(*b*) $t = 1.59$

6.6 (*a*) $F = 2.31$
(*b*) $t = 7.70$

6.7 (*a*) $F = 1.52$
(*b*) $t = 7.03$

6.8 (*a*) $t = 2.093$
(*b*) $t = 2.025$

Chapter 7

7.1 $MS_T = 31.67 \quad MS_W = 82.67 \quad F < 1.00$

7.2 $F = 9.06$

7.3 $MS_T = 4.95 \quad MS_W = 8.60 \quad F < 1.00$

7.4 $MS_T = 53.79 \quad MS_W = 93.59 \quad F < 1.00$

7.5 $F = 6.52$

7.6 $F = 11.09$

7.7 (*a*)

Operator	*A*	*B*	*C*	*D*
$\bar{X}$	7.48	3.11	6.85	6.27
s	3.11	1.24	3.02	2.33

(*b*)

Operator	*A*	*B*	*C*	*D*
$\bar{X}$	0.901	0.585	0.865	0.839
s	0.160	0.181	0.169	0.151

(*c*) $F = 13.8$

7.8 (*a*)

Treatment	1	2	3
$\bar{X}$	15.30	2.40	4.70
s^2	26.23	4.27	9.12

(*b*) $F = 35.84$

(*c*)

Treatment	1	2	3
$\bar{X}$	3.93	1.57	2.14
s^2	0.39	0.47	0.68

(*d*) $F = 29.7$

7.10 $F = 1.50$

7.11 (*a*) $t = -2.0$
(*b*) $F = 4.0$

Chapter 8

8.1 *A B C D E F G H*

8.2 For the standard error of the difference, we have

$$s_{\bar{X}_k - \bar{X}_0} = \sqrt{(2)(36)/10} = 2.68$$

With $k = 5$ and d.f. $= 54$, for a one-sided test with joint confidence coefficient of $P = 0.95$, t is approximately 2.29. For a difference to be significant, we must have

$$\bar{X}_{k\cdot} - \bar{X}_{0\cdot} \geq (2.68)(2.29)$$

With this test, the means for Treatments *D* and *E* are significantly greater than the mean of the control group and the means for Treatments *A*, *B*, and *C* are not. For a one-sided test with joint confidence coefficient of $P = 0.99$, t is approximately 2.96. Then, for a difference to be significant, we must have

$$\bar{X}_{k\cdot} - \bar{X}_{0\cdot} \geq (2.68)(2.96)$$

and with this test only the mean for Treatment D is significantly greater than the mean for the control group.

8.5 (*a*) $F = 4.67$
(*b*) $F = 9.00$

8.7 (*a*) 2.53
(*b*) 1.79
(*c*) 4.38

Chapter 9

9.1

Type:	$F = 337.47$
Background:	$F = 6.17$
Time:	$F = 988.59$
Type × background:	$F < 1.00$
Type × time:	$F = 155.29$
Background × time:	$F < 1.00$
Type × background × time:	$F = 1.63$

9.2

A:	$F = 3.77$
B:	$F < 1.00$
$A \times B$:	$F = 1.60$

9.3

A:	$F < 1.00$
B:	$F = 7.39$
C:	$F < 1.00$
$A \times B$:	$F < 1.00$
$A \times C$:	$F < 1.00$
$B \times C$:	$F = 2.41$
$A \times B \times C$:	$F < 1.00$

9.6

A:	$F < 1.00$
B:	$F < 1.00$
C:	$F < 1.00$
$A \times B$:	$F = 4.05$
$A \times C$:	$F < 1.00$
$B \times C$:	$F < 1.00$
$A \times B \times C$:	$F < 1.00$

9.7 (*a*) $F = 4.67$
(*b*)

A:	$F = 9.00$
B:	$F = 4.00$
$A \times B$:	$F = 1.00$

9.8 A: $F = 6.75$
B: $F < 1.00$
$A \times B$: $F < 1.00$

9.9 A: $F = 3.81$
B: $F = 34.28$
$A \times B$: $F < 1.00$

9.10 A: $F = 10.29$
B: $F = 26.70$
C: $F = 8.13$
$A \times B$: $F = 2.04$
$A \times C$: $F = 3.85$
$B \times C$: $F = 2.04$
$A \times B \times C$: $F < 1.00$

9.11 (*b*) A: $F = 13.74$
B: $F = 47.07$
C: $F = 82.24$
$A \times B$: $F < 1.00$
$A \times C$: $F < 1.00$
$B \times C$: $F = 2.49$
$A \times B \times C$: $F < 1.00$

Chapter 10

10.2 A: $F = 1.34$
B: $F = 3.49$
$A \times B$: $F < 1.00$

10.3 A: $F = 35.37$
B: $F = 290.21$
$A \times B$: $F = 3.03$

10.4 (*a*) Treatment = 28.33
(*b*) $A = 1.66$
$B = 20.83$
$A \times B = 5.84$
(*d*) $MS_{D_1} = 1.66$
$MS_{D_2} = 5.21$
$MS_{D_3} = 15.62$
$MS_{D_4} = 0.21$
$MS_{D_5} = 5.63$

10.5 (*a*) Treatment = 45.16
(*b*) Drive = 11.36
Intensity = 28.16
Drive × intensity = 5.64

(d) $MS_{D_1} = 10.76$
$MS_{D_2} = 0.60$
$MS_{D_3} = 22.76$
$MS_{D_4} = 5.40$
$MS_{D_5} = 4.44$
$MS_{D_6} = 1.20$
$MS_{D_7} = 0.00$
$MS_{D_8} = 0.00$

Chapter 12

12.1 A: $F = 79.33$
B: $F = 20.72$
C: $F = 66.84$
$A \times B$: $F < 1.00$
$A \times C$: $F = 11.48$
$B \times C$: $F = 3.16$
$A \times B \times C$: $F = 1.20$

12.4 Sex of subjects: $F = 15.80$
Instructions: $F < 1.00$
Barrier: $F = 7.22$
Sex of experimenter: $F = 6.06$

Chapter 13

13.1 $F = 22.73$

13.2 $F = 16.26$

13.3 (a) $F = 5.10$
(b) $t = 2.26$

13.4 $F = 2.20$

13.6 $F = 2.75$

13.7 (a) $B \times T = 41.33$
$A = 105.00$
$C = 7.50$
$A \times C = 5.00$
(b) $B \times A = 28.33$
$B \times C = 5.33$
$B \times A \times C = 7.67$

13.8 (a) $MS_{D_1} = 60.00$ $MS_{D_2} = 45.00$
(b) $D_1 \times C = 0.00$ $D_2 \times C = 5.00$
(c) $B \times D_1 = 6.83$ $B \times D_2 = 21.50$
(d) $B \times D_1 \times C = 4.17$ $B \times D_2 \times C = 3.50$

13.10 (a) A: $F = 235.64$
C: $F = 58.91$
$A \times C$: $F < 1.00$
(b) A: $F = 152.47$
C: $F = 49.85$
$A \times C$: $F = 2.67$

Chapter 14

14.1 (a) $MS_W = MS_{ST} = 2.5$
(b) $MS_{ST} > MS_W = 2.5$

14.2 (a) $F = 12.0$
(b) $t^2 = 12.0$

14.4 $F = MS_T/MS_{ST} = 4.50$

14.8 (a) $F = MS_A/MS_{S(A)} = 43.35$
(b) $F = MS_B/MS_{S(A)B} = 17.41$
(c) $F = MS_{AB}/MS_{S(A)B} = 1.59$

14.9 $F = MS_A/MS_{ST} = 235.64$
$F = MS_B/MS_{ST} = 58.91$
$F = MS_{AB}/MS_{ST} < 1.00$

14.11 (a) $r_{12} = r_{13} = r_{23} = 0$
(b) $s_1^2 = s_2^2 = s_3^2 = MS_W = 2/7$
(c) $MS_{ST} = 2/7$

14.12 (a) $MS_{ST} = 2/7$
(b) $\overline{MS_W} = 140/21$
(c) $\overline{r_{ij}s_is_j} = 134/21$

Chapter 15

15.1 $F = 2.00$

15.4 $F = 13.50$

15.5 $F = 27.16$

Chapter 16

16.1 (a) $F = 8.87$
(b) $F = 11.92$

16.2 Square 1: $F = 18.11$
Square 2: $F = 2.77$
Square 3: $F = 26.28$
Square 4: $F = 2.46$

16.3 Orders: $F < 1.00$
Screen size: $F = 9.22$
Periods: $F = 1.43$

16.9 (*a*) $MS_{S(G)} = 26.37$
(*b*) $MS_{res} = 24.11$
(*c*) $MS_{S(G)P} = 4.90$

16.10 (*a*) $MS_{R(S)} = 2.49$
(*b*) $MS_{ST} = 0.11$
(*c*) $MS_{SC} = 0.20$
(*d*) $MS_{res} = 0.78$

16.11 (*a*) $MS_{S(G)} = 15.63$
(*b*) $MS_{res} = 1.78$
(*c*) $MS_{S(G)P} = 6.20$

Chapter 17

17.1 (*a*) Trials: $F = 98.4$
(*b*) Linear: $F = 294.5$
(*c*) Quadratic: $F < 1.0$

17.2 (*a*) Treatments (A): $F < 1.0$
Days (D): $F = 81.2$
$A \times D$: $F = 2.1$
(*b*) Linear: $F = 157.9$
Quadratic: $F = 4.5$
(*c*) $A \times L$: $F = 4.1$
$A \times Q$: $F < 1.0$

17.3 (*a*) A: $F = 33.45$
B: $F = 88.59$
$A \times B$: $F < 1.00$
C: $F = 54.37$
$A \times C$: $F < 1.00$
$B \times C$: $F < 1.00$
$A \times B \times C$: $F = 1.04$
(*b*) Linear: $F = 162.80$
Quadratic: $F < 1.00$
(*c*) $A \times L$: $F < 1.00$
$A \times Q$: $F < 1.00$
$B \times L$: $F < 1.00$
$B \times Q$: $F < 1.00$

17.4 (*a*) Instructions: $F = 4.67$
Trials: $F = 68.83$
$I \times T$: $F = 1.23$
(*b*) Linear: $F = 270.22$
Quadratic: $F = 2.68$

(*c*) $I \times L$: $F = 3.01$
$I \times Q$: $F = 1.08$

17.6 (*a*) Groups: $F = 43.35$
Trials: $F = 17.41$
$G \times T$: $F = 1.59$
(*b*) Linear: $F = 84.05$
(*c*) Quadratic: $F < 1.00$
(*d*) $G \times L$: $F = 3.88$

17.8 (*a*) $MS_{S(A)} = 12.55$
(*b*) $MS_{S(A)T} = 0.68$
(*c*) Linear $= 115.20$
Quadratic $= 2.40$
(*d*) $A \times L = 7.20$
$A \times Q = 2.40$

Chapter 18

18.1 (*c*) $F < 1.00$
(*d*) $F = 19.69$
(*e*) $F < 1.00$
(*f*) $F = 62.75$
(*g*) $F = 61.92$

18.2 (*a*) $F < 1.00$
(*b*) $F = 13.72$
(*c*) $F = 15.34$

18.3 (*a*) $F < 1.00$
(*b*) $F = 34.82$
(*c*) $b_1 = 1.25$
$b_2 = 0.80$
$b_3 = 1.25$

APPENDIX

Table I. Table of Random Numbers*

Row	Column Number							
	00000 01234	00000 56789	11111 01234	11111 56789	22222 01234	22222 56789	33333 01234	33333 56789
				1st Thousand				
00	23157	54859	01837	25993	76249	70886	95230	36744
01	05545	55043	10537	43508	90611	83744	10962	21343
02	14871	60350	32404	36223	50051	00322	11543	80834
03	38976	74951	94051	75853	78805	90194	32428	71695
04	97312	61718	99755	30870	94251	25841	54882	10513
05	11742	69381	44339	30872	32797	33118	22647	06850
06	43361	28859	11016	45623	93009	00499	43640	74036
07	93806	20478	38268	04491	55751	18932	58475	52571
08	49540	13181	08429	84187	69538	29661	77738	09527
09	36768	72633	37948	21569	41959	68670	45274	83880
10	07092	52392	24627	12067	06558	45344	67338	45320
11	43310	01081	44863	80307	52555	16148	89742	94647
12	61570	06360	06173	63775	63148	95123	35017	46993
13	31352	83799	10779	18941	31579	76448	62584	86919
14	57048	86526	27795	93692	90529	56546	35065	32254
15	09243	44200	68721	07137	30729	75756	09298	27650
16	97957	35018	40894	88329	52230	82521	22532	61587
17	93732	59570	43781	98885	56671	66826	95996	44569
18	72621	11225	00922	68264	35666	59434	71687	58167
19	61020	74418	45371	20794	95917	37866	99536	19378
20	97839	85474	33055	91718	45473	54144	22034	23000
21	89160	97192	22232	90637	35055	45489	88438	16361
22	25966	88220	62871	79265	02823	52862	84919	54883
23	81443	31719	05049	54806	74690	07567	65017	16543
24	11322	54931	42362	34386	08624	97687	46245	23245

* Table I is reproduced from M. G. Kendall and B. B. Smith. Randomness and random sampling numbers. *Journal of the Royal Statistical Society*, **101** (1938), 147–166, by permission of the Royal Statistics Society.

Table I. Table of Random Numbers*—Continued

Row	Column Number							
	00000 01234	00000 56789	11111 01234	11111 56789	22222 01234	22222 56789	33333 01234	33333 56789
				2nd Thousand				
00	64755	83885	84122	25920	17696	15655	95045	95947
01	10302	52289	77436	34430	38112	49067	07348	23328
02	71017	98495	51308	50374	66591	02887	53765	69149
03	60012	55605	88410	34879	79655	90169	78800	03666
04	37330	94656	49161	42802	48274	54755	44553	65090
05	47869	87001	31591	12273	60626	12822	34691	61212
06	38040	42737	64167	89578	39323	49324	88434	38706
07	73508	30908	83054	80078	86669	30295	56460	45336
08	32623	46474	84061	04324	20628	37319	32356	43969
09	97591	99549	36630	35106	62069	92975	95320	57734
10	74012	31955	59790	96982	66224	24015	96749	07589
11	56754	26457	13351	05014	90966	33674	69096	33488
12	49800	49908	54831	21998	08528	26372	92923	65026
13	43584	89647	24878	56670	00221	50193	99591	62377
14	16653	79664	60325	71301	35742	83636	73058	87229
15	48502	69055	65322	58748	31446	80237	31252	96367
16	96765	54692	36316	86230	48296	38352	23816	64094
17	38923	61550	80357	81784	23444	12463	33992	28128
18	77958	81694	25225	05587	51073	01070	60218	61961
19	17928	28065	25586	08771	02641	85064	65796	48170
20	94036	85978	02318	04499	41054	10531	87431	21596
21	47460	60479	56230	48417	14372	85167	27558	00368
22	47856	56088	51992	82439	40644	17170	13463	18288
23	57616	34653	92298	62018	10375	76515	62986	90756
24	08300	92704	66752	66610	57188	79107	54222	22013

* Table I is reproduced from M. G. Kendall and B. B. Smith. Randomness and random sampling numbers. *Journal of the Royal Statistical Society*, **101** (1938), 147–166, by permission of the Royal Statistical Society.

Table I. Table of Random Numbers*—Continued

Row	Column Number							
	00000 01234	00000 56789	11111 01234	11111 56789	22222 01234	22222 56789	33333 01234	33333 56789
	3rd Thousand							
00	89221	02362	65787	74733	51272	30213	92441	39651
01	04005	99818	63918	29032	94012	42363	01261	10650
02	98546	38066	50856	75045	40645	22841	53254	44125
03	41719	84401	59226	01314	54581	40398	49988	65579
04	28733	72489	00785	25843	24613	49797	85567	84471
05	65213	83927	77762	03086	80742	24395	68476	83792
06	65553	12678	90906	90466	43670	26217	69900	31205
07	05668	69080	73029	85746	58332	78231	45986	92998
08	39302	99718	49757	79519	27387	76373	47262	91612
09	64592	32254	45879	29431	38320	05981	18067	87137
10	07513	48792	47314	83660	68907	05336	82579	91582
11	86593	68501	56638	99800	82839	35148	56541	07232
12	83735	22599	97977	81248	36838	99560	32410	67614
13	08595	21826	54655	08204	87990	17033	56258	05384
14	41273	27149	44293	69458	16828	63962	15864	35431
15	00473	75908	56238	12242	72631	76314	47252	06347
16	86131	53789	81383	07868	89132	96182	07009	86432
17	33849	78359	08402	03586	03176	88663	08018	22546
18	61870	41657	07468	08612	98083	97349	20775	45091
19	43898	65923	25078	86129	78491	97653	91500	80786
20	29939	39123	04548	45985	60952	06641	28726	46473
21	38505	85555	14388	55077	18657	94887	67831	70819
22	31824	38431	67125	25511	72044	11562	53279	82268
23	91430	03767	13561	15597	06750	92552	02391	38753
24	38635	68976	25498	97526	96458	03805	04116	63514

* Table I is reproduced from M. G. Kendall and B. B. Smith. Randomness and random sampling numbers. *Journal of the Royal Statistical Society*, **101** (1938), 147–166, by permission of the Royal Statistical Society.

Table I. Table of Random Numbers*—Continued

Row	Column Number 00000 01234	00000 56789	11111 01234	11111 56789	22222 01234	22222 56789	33333 01234	33333 56789
				4th Thousand				
00	02490	54122	27944	39364	94239	72074	11679	54082
01	11967	36469	60627	83701	09253	30208	01385	37482
02	48256	83465	49699	24079	05403	35154	39613	03136
03	27246	73080	21481	23536	04881	89977	49484	93071
04	32532	77265	72430	70722	86529	18457	92657	10011
05	66757	98955	92375	93431	43204	55825	45443	69265
06	11266	34545	76505	97746	34668	26999	26742	97516
07	17872	39142	45561	80146	93137	48924	64257	59284
08	62561	30365	03408	14754	51798	08133	61010	97730
09	62796	30779	35497	70501	30105	08133	00997	91970
10	75510	21771	04339	33660	42757	62223	87565	48468
11	87439	01691	63517	26590	44437	07217	98706	39032
12	97742	02621	10748	78803	38337	65226	92149	59051
13	98811	06001	21571	02875	21828	83912	85188	61624
14	51264	01852	64607	92553	29004	26695	78583	62998
15	40239	93376	10419	68610	49120	02941	80035	99317
16	26936	59186	51667	27645	46329	44681	94190	66647
17	88502	11716	98299	40974	42394	62200	69094	81646
18	63499	38093	25593	61995	79867	80569	01023	38374
19	36379	81206	03317	78710	73828	31083	60509	44091
20	93801	22322	47479	57017	59334	30647	43061	26660
21	29856	87120	56311	50053	25365	81265	22414	02431
22	97720	87931	88265	13050	71017	15177	06957	92919
23	85237	09105	74601	46377	59938	15647	34177	92753
24	75746	75268	31727	95773	72364	87324	36879	06802

* Table I is reproduced from M. G. Kendall and B. B. Smith. Randomness and random sampling numbers. *Journal of the Royal Statistical Society*, **101** (1938), 147–166, by permission of the Royal Statistical Society.

Table I. Table of Random Numbers*—Concluded

Row	Column Number 00000 01234	00000 56789	11111 01234	11111 56789	22222 01234	22222 56789	33333 01234	33333 56789
				5th Thousand				
00	29935	06971	63175	52579	10478	89379	61428	21363
01	15114	07126	51890	77787	75510	13103	42942	48111
02	03870	43225	10589	87629	22039	94124	38127	65022
03	79390	39188	40756	45269	65959	20640	14284	22960
04	30035	06915	79196	54428	64819	52314	48721	81594
05	29039	99861	28759	79802	68531	39198	38137	24373
06	78196	08108	24107	49777	09599	43569	84820	94956
07	15847	85493	91442	91351	80130	73752	21539	10986
08	36614	62248	49194	97209	92587	92053	41021	80064
09	40549	54884	91465	43862	35541	44466	88894	74180
10	40878	08997	14286	09982	90308	78007	51587	16658
11	10229	49282	41173	31468	59455	18756	08908	06660
12	15918	76787	30624	25928	44124	25088	31137	71614
13	13403	18796	49909	94404	64979	41462	18155	98335
14	66523	94596	74908	90271	10009	98648	17640	68909
15	91665	36469	68343	17870	25975	04662	21272	50620
16	67415	87515	08207	73729	73201	57593	96917	69699
17	76527	96996	23724	33448	63392	32394	60887	90617
18	19815	47789	74348	17147	10954	34355	81194	54407
19	25592	53587	76384	72575	84347	68918	05739	57222
20	55902	45539	63646	31609	95999	82887	40666	66692
21	02470	58376	79794	22482	42423	96162	47491	17264
22	18630	53263	13319	97619	35859	12350	14632	87659
23	89673	38230	16063	92007	59503	38402	76450	33333
24	62986	67364	06595	17427	84623	14565	82860	57300

* Table I is reproduced from M. G. Kendall and B. B. Smith. Randomness and random sampling numbers. *Journal of the Royal Statistical Society*, **101** (1938), 147–166, by permission of the Royal Statistical Society.

Table II. Table of Squares, Square Roots, and Reciprocals of Numbers from 1 to 1000*

N	N^2	$\sqrt{N}$	$1/N$	N	N^2	$\sqrt{N}$	$1/N$
1	1	1.0000	1.000000	41	1681	6.4031	.024390
2	4	1.4142	.500000	42	1764	6.4807	.023810
3	9	1.7321	.333333	43	1849	6.5574	.023256
4	16	2.0000	.250000	44	1936	6.6332	.022727
5	25	2.2361	.200000	45	2025	6.7082	.022222
6	36	2.4495	.166667	46	2116	6.7823	.021739
7	49	2.6458	.142857	47	2209	6.8557	.021277
8	64	2.8284	.125000	48	2304	6.9282	.020833
9	81	3.0000	.111111	49	2401	7.0000	.020408
10	100	3.1623	.100000	50	2500	7.0711	.020000
11	121	3.3166	.090909	51	2601	7.1414	.019608
12	144	3.4641	.083333	52	2704	7.2111	.019231
13	169	3.6056	.076923	53	2809	7.2801	.018868
14	196	3.7417	.071429	54	2916	7.3485	.018519
15	225	3.8730	.066667	55	3025	7.4162	.018182
16	256	4.0000	.062500	56	3136	7.4833	.017857
17	289	4.1231	.058824	57	3249	7.5498	.017544
18	324	4.2426	.055556	58	3364	7.6158	.017241
19	361	4.3589	.052632	59	3481	7.6811	.016949
20	400	4.4721	.050000	60	3600	7.7460	.016667
21	441	4.5826	.047619	61	3721	7.8102	.016393
22	484	4.6904	.045455	62	3844	7.8740	.016129
23	529	4.7958	.043478	63	3969	7.9373	.015873
24	576	4.8990	.041667	64	4096	8.0000	.015625
25	625	5.0000	.040000	65	4225	8.0623	.015385
26	676	5.0990	.038462	66	4356	8.1240	.015152
27	729	5.1962	.037037	67	4489	8.1854	.014925
28	784	5.2915	.035714	68	4624	8.2462	.014706
29	841	5.3852	.034483	69	4761	8.3066	.014493
30	900	5.4772	.033333	70	4900	8.3666	.014286
31	961	5,5678	.032258	71	5041	8.4261	.014085
32	1024	5.6569	.031250	72	5184	8.4853	.013889
33	1089	5.7446	.030303	73	5329	8.5440	.013699
34	1156	5.8310	.029412	74	5476	8.6023	.013514
35	1225	5.9161	.028571	75	5625	8.6603	.013333
36	1296	6.0000	.027778	76	5776	8.7178	.013158
37	1369	6.0828	.027027	77	5929	8.7750	.012987
38	1444	6.1644	.026316	78	6084	8.8318	.012821
39	1521	6.2450	.025641	79	6241	8.8882	.012658
40	1600	6.3246	.025000	80	6400	8.9443	.012500

* Portions of Table II have been reproduced from J. W. Dunlap and A. K. Kurtz. *Handbook of Statistical Nomographs, Tables, and Formulas*, World Book Company, New York (1932), by permission of the authors and publishers.

Table II. Table of Squares, Square Roots, and Reciprocals of Numbers from 1 to 1000*—Continued

N	N^2	$\sqrt{N}$	$1/N$	N	N^2	$\sqrt{N}$	$1/N$
81	6561	9.0000	.012346	121	14641	11.0000	.00826446
82	6724	9.0554	.012195	122	14884	11.0454	.00819672
83	6889	9.1104	.012048	123	15129	11.0905	.00813008
84	7056	9.1652	.011905	124	15376	11.1355	.00806452
85	7225	9.2195	.011765	125	15625	11.1803	.00800000
86	7396	9.2736	.011628	126	15876	11.2250	.00793651
87	7569	9.3274	.011494	127	16129	11.2694	.00787402
88	7744	9.3808	.011364	128	16384	11.3137	.00781250
89	7921	9.4340	.011236	129	16641	11.3578	.00775194
90	8100	9.4868	.011111	130	16900	11.4018	.00769231
91	8281	9.5394	.010989	131	17161	11.4455	.00763359
92	8464	9.5917	.010870	132	17424	11.4891	.00757576
93	8649	9.6437	.010753	133	17689	11.5326	.00751880
94	8836	9.6954	.010638	134	17956	11.5758	.00746269
95	9025	9.7468	.010526	135	18225	11.6190	.00740741
96	9216	9.7980	.010417	136	18496	11.6619	.00735294
97	9409	9.8489	.010309	137	18769	11.7047	.00729927
98	9604	9.8995	.010204	138	19044	11.7473	.00724638
99	9801	9.9499	.010101	139	19321	11.7898	.00719424
100	10000	10.0000	.010000	140	19600	11.8322	.00714286
101	10201	10.0499	.00990099	141	19881	11.8743	.00709220
102	10404	10.0995	.00980392	142	20164	11.9164	.00704225
103	10609	10.1489	.00970874	143	20449	11.9583	.00699301
104	10816	10.1980	.00961538	144	20736	12.0000	.00694444
105	11025	10.2470	.00952381	145	21025	12.0416	.00689655
106	11236	10.2956	.00943396	146	21316	12.0830	.00684932
107	11449	10.3441	.00934579	147	21609	12.1244	.00680272
108	11664	10.3923	.00925926	148	21904	12.1655	.00675676
109	11881	10.4403	.00917431	149	22201	12.2066	.00671141
110	12100	10.4881	.00909091	150	22500	12.2474	.00666667
111	12321	10.5357	.00900901	151	22801	12.2882	.00662252
112	12544	10.5830	.00892857	152	23104	12.3288	.00657895
113	12769	10.6301	.00884956	153	23409	12.3693	.00653595
114	12996	10.6771	.00877193	154	23716	12.4097	.00649351
115	13225	10.7238	.00869565	155	24025	12.4499	.00645161
116	13456	10.7703	.00862069	156	24336	12.4900	.00641026
117	13689	10.8167	.00854701	157	24649	12.5300	.00636943
118	13924	10.8628	.00847458	158	24964	12.5698	.00632911
119	14161	10.9087	.00840336	159	25281	12.6095	.00628931
120	14400	10.9545	.00833333	160	25600	12.6491	.00625000

* Portions of Table II have been reproduced from J. W. Dunlap and A. K. Kurtz. *Handbook of Statistical Nomographs, Tables, and Formulas*, World Book Company, New York (1932), by permission of the authors and publishers.

Table II. Table of Squares, Square Roots, and Reciprocals of Numbers from 1 to 1000*—Continued

N	N^2	$\sqrt{N}$	$1/N$	N	N^2	$\sqrt{N}$	$1/N$
161	25921	12.6886	.00621118	201	40401	14.1774	.00497512
162	26244	12.7279	.00617284	202	40804	14.2127	.00495050
163	26569	12.7671	.00613497	203	41209	14.2478	.00492611
164	26896	12.8062	.00609756	204	41616	14.2829	.00490196
165	27225	12.8452	.00606061	205	42025	14.3178	.00487805
166	27556	12.8841	.00602410	206	42436	14.3527	.00485437
167	27889	12.9228	.00598802	207	42849	14.3875	.00483092
168	28224	12.9615	.00595238	208	43264	14.4222	.00480769
169	28561	13.0000	.00591716	209	43681	14.4568	.00478469
170	28900	13.0384	.00588235	210	44100	14.4914	.00476190
171	29241	13.0767	.00584795	211	44521	14.5258	.00473934
172	29584	13.1149	.00581395	212	44944	14.5602	.00471698
173	29929	13.1529	.00578035	213	45369	14.5945	.00469484
174	30276	13.1909	.00574713	214	45796	14.6287	.00467290
175	30625	13.2288	.00571429	215	46225	14.6629	.00465116
176	30976	13.2665	.00568182	216	46656	14.6969	.00462963
177	31329	13.3041	.00564972	217	47089	14.7309	.00460829
178	31684	13.3417	.00561798	218	47524	14.7648	.00458716
179	32041	13.3791	.00558659	219	47961	14.7986	.00456621
180	32400	13.4164	.00555556	220	48400	14.8324	.00454545
181	32761	13.4536	.00552486	221	48841	14.8661	.00452489
182	33124	13.4907	.00549451	222	49284	14.8997	.00450450
183	33489	13.5277	.00546448	223	49729	14.9332	.00448430
184	33856	13.5647	.00543478	224	50176	14.9666	.00446429
185	34225	13.6015	.00540541	225	50625	15.0000	.00444444
186	34596	13.6382	.00537634	226	51076	15.0333	.00442478
187	34969	13.6748	.00534759	227	51529	15.0665	.00440529
188	35344	13.7113	.00531915	228	51984	15.0997	.00438596
189	35721	13.7477	.00529101	229	52441	15.1327	.00436681
190	36100	13.7840	.00526316	230	52900	15.1658	.00434783
191	36481	13.8203	.00523560	231	53361	15.1987	.00432900
192	36864	13.8564	.00520833	232	53824	15.2315	.00431034
193	37249	13.8924	.00518135	233	54289	15.2643	.00429185
194	37636	13.9284	.00515464	234	54756	15.2971	.00427350
195	38025	13.9642	.00512821	235	55225	15.3297	.00425532
196	38416	14.0000	.00510204	236	55696	15.3623	.00423729
197	38809	14.0357	.00507614	237	56169	15.3948	.00421941
198	39204	14.0712	.00505051	238	56644	15.4272	.00420168
199	39601	14.1067	.00502513	239	57121	15.4596	.00418410
200	40000	14.1421	.00500000	240	57600	15.4919	.00416667

* Portions of Table II have been reproduced from J. W. Dunlap and A. K. Kurtz. *Handbook of Statistical Nomographs, Tables, and Formulas*, World Book Company, New York (1932), by permission of the authors and publishers.

Table II. Table of Squares, Square Roots, and Reciprocals of Numbers from 1 to 1000*—Continued

N	N^2	$\sqrt{N}$	$1/N$	N	N^2	$\sqrt{N}$	$1/N$
241	58081	15.5242	.00414938	281	78961	16.7631	.00355872
242	58564	15.5563	.00413223	282	79524	16.7929	.00354610
243	59049	15.5885	.00411523	283	80089	16.8226	.00353357
244	59536	15.6205	.00409836	284	80656	16.8523	.00352113
245	60025	15.6525	.00408163	285	81225	16.8819	.00350877
246	60516	15.6844	.00406504	286	81796	16.9115	.00349650
247	61009	15.7162	.00404858	287	82369	16.9411	.00348432
248	61504	15.7480	.00403226	288	82944	16.9706	.00347222
249	62001	15.7797	.00401606	289	83521	17.0000	.00346021
250	62500	15.8114	.00400000	290	84100	17 0294	.00344828
251	63001	15.8430	.00398406	291	84681	17.0587	.00343643
252	63504	15.8745	.00396825	292	85264	17.0880	.00342466
253	64009	15.9060	.00395257	293	85849	17.1172	.00341297
254	64516	15.9374	.00393701	294	86436	17.1464	.00340136
255	65025	15.9687	.00392157	295	87025	17.1756	.00338983
256	65536	16.0000	.00390625	296	87616	17.2047	.00337838
257	66049	16.0312	.00389105	297	88209	17.2337	.00336700
258	66564	16.0624	.00387597	298	88804	17.2627	.00335570
259	67081	16.0935	.00386100	299	89401	17.2916	.00334448
260	67600	16.1245	.00384615	300	90000	17.3205	.00333333
261	68121	16.1555	.00383142	301	90601	17.3494	.00332226
262	68644	16.1864	.00381679	302	91204	17.3781	.00331126
263	69169	16.2173	.00380228	303	91809	17.4069	.00330033
264	69696	16.2481	.00378788	304	92416	17.4356	.00328947
265	70225	16.2788	.00377358	305	93025	17.4642	.00327869
266	70756	16.3095	.00375940	306	93636	17.4929	.00326797
267	71289	16.3401	.00374532	307	94249	17.5214	.00325733
268	71824	16.3707	.00373134	308	94864	17.5499	.00324675
269	72361	16.4012	.00371747	309	95481	17.5784	.00323625
270	72900	16.4317	.00370370	310	96100	17.6068	.00322581
271	73441	16.4621	.00369004	311	96721	17.6352	.00321543
272	73984	16.4924	.00367647	312	97344	17.6635	.00320513
273	74529	16.5227	.00366300	313	97969	17.6918	.00319489
274	75076	16.5529	.00364964	314	98596	17.7200	.00318471
275	75625	16.5831	.00363636	315	99225	17.7482	.00317460
276	76176	16.6132	.00362319	316	99856	17.7764	.00316456
277	76729	16.6433	.00361011	317	100489	17.8045	.00315457
278	77284	16.6733	.00359712	318	101124	17.8326	.00314465
279	77841	16.7033	.00358423	319	101761	17.8606	.00313480
280	78400	16.7332	.00357143	320	102400	17.8885	.00312500

* Portions of Table II have been reproduced from J. W. Dunlap and A. K. Kurtz. *Handbook of Statistical Nomographs, Tables, and Formulas*, World Book Company, New York (1932), by permission of the authors and publishers.

Table II. Table of Squares, Square Roots, and Reciprocals of Numbers from 1 to 1000*—Continued

N	N^2	$\sqrt{N}$	$1/N$	N	N^2	$\sqrt{N}$	$1/N$
321	103041	17.9165	.00311526	361	130321	19.0000	.00277008
322	103684	17.9444	.00310559	362	131044	19.0263	.00276243
323	104329	17.9722	.00309598	363	131769	19.0526	.00275482
324	104976	18.0000	.00308642	364	132496	19.0788	.00274725
325	105625	18.0278	.00307692	365	133225	19.1050	.00273973
326	106276	18.0555	.00306748	366	133956	19.1311	.00273224
327	106929	18.0831	.00305810	367	134689	19.1572	.00272480
328	107584	18.1108	.00304878	368	135424	19.1833	.00271739
329	108241	18.1384	.00303951	369	136161	19.2094	.00271003
330	108900	18.1659	.00303030	370	136900	19.2354	.00270270
331	109561	18.1934	.00302115	371	137641	19.2614	.00269542
332	110224	18.2209	.00301205	372	138384	19.2873	.00268817
333	110889	18.2483	.00300300	373	139129	19.3132	.00268097
334	111556	18.2757	.00299401	374	139876	19.3391	.00267380
335	112225	18.3030	.00298507	375	140625	19.3649	.00266667
336	112896	18.3303	.00297619	376	141376	19.3907	.00265957
337	113569	18.3576	.00296736	377	142129	19.4165	.00265252
338	114244	18.3848	.00295858	378	142884	19.4422	.00264550
339	114921	18.4120	.00294985	379	143641	19.4679	.00263852
340	115600	18.4391	.00294118	380	144400	19.4936	.00263158
341	116281	18.4662	.00293255	381	145161	19.5192	.00262467
342	116964	18.4932	.00292398	382	145924	19.5448	.00261780
343	117649	18.5203	.00291545	383	146689	19.5704	.00261097
344	118336	18.5472	.00290698	384	147456	19.5959	.00260417
345	119025	18.5742	.00289855	385	148225	19.6214	.00259740
346	119716	18.6011	.00289017	386	148996	19.6469	.00259067
347	120409	18.6279	.00288184	387	149769	19.6723	.00258398
348	121104	18.6548	.00287356	388	150544	19.6977	.00257732
349	121801	18.6815	.00286533	389	151321	19.7231	.00257069
350	122500	18.7083	.00285714	390	152100	19.7484	.00256410
351	123201	18.7350	.00284900	391	152881	19.7737	.00255754
352	123904	18.7617	.00284091	392	153664	19.7990	.00255102
353	124609	18.7883	.00283286	393	154449	19.8242	.00254453
354	125316	18.8149	.00282486	394	155236	19.8494	.00253807
355	126025	18.8414	.00281690	395	156025	19.8746	.00253165
356	126736	18.8680	.00280899	396	156816	19.8997	.00252525
357	127449	18.8944	.00280112	397	157609	19.9249	.00251889
358	128164	18.9209	.00279330	398	158404	19.9499	.00251256
359	128881	18.9473	.00278552	399	159201	19.9750	.00250627
360	129600	18.9737	.00277778	400	160000	20.0000	.00250000

* Portions of Table II have been reproduced from J. W. Dunlap and A. K. Kurtz. *Handbook of Statistical Nomographs, Tables, and Formulas*, World Book Company, New York (1932), by permission of the authors and publishers.

Table II. Table of Squares, Square Roots, and Reciprocals of Numbers from 1 to 1000*—Continued

N	N^2	$\sqrt{N}$	$1/N$	N	N^2	$\sqrt{N}$	$1/N$
401	160801	20.0250	.00249377	441	194481	21.0000	.00226757
402	161604	20.0499	.00248756	442	195364	21.0238	.00226244
403	162409	20.0749	.00248139	443	196249	21.0476	.00225734
404	163216	20.0998	.00247525	444	197136	21.0713	.00225225
405	164025	20.1246	.00246914	445	198025	21.0950	.00224719
406	164836	20.1494	.00246305	446	198916	21.1187	.00224215
407	165649	20.1742	.00245700	447	199809	21.1424	.00223714
408	166464	20.1990	.00245098	448	200704	21.1660	.00223214
409	167281	20.2237	.00244499	449	201601	21.1896	.00222717
410	168100	20.2485	.00243902	450	202500	21.2132	.00222222
411	168921	20.2731	.00243309	451	203401	21.2368	.00221729
412	169744	20.2978	.00242718	452	204304	21.2603	.00221239
413	170569	20.3224	.00242131	453	205209	21.2838	.00220751
414	171396	20.3470	.00241546	454	206116	21.3073	.00220264
415	172225	20.3715	.00240964	455	207025	21.3307	.00219780
416	173056	20.3961	.00240385	456	207936	21.3542	.00219298
417	173889	20.4206	.00239808	457	208849	21.3776	.00218818
418	174724	20.4450	.00239234	458	209764	21.4009	.00218341
419	175561	20.4695	.00238663	459	210681	21.4243	.00217865
420	176400	20.4939	.00238095	460	211600	21.4476	.00217391
421	177241	20.5183	.00237530	461	212521	21.4709	.00216920
422	178084	20.5426	.00236967	462	213444	21.4942	.00216450
423	178929	20.5670	.00236407	463	214369	21.5174	.00215983
424	179776	20.5913	.00235849	464	215296	21.5407	.00215517
425	180625	20.6155	.00235294	465	216225	21.5639	.00215054
426	181476	20.6398	.00234742	466	217156	21.5870	.00214592
427	182329	20.6640	.00234192	467	218089	21.6102	.00214133
428	183184	20.6882	.00233645	468	219024	21.6333	.00213675
429	184041	20.7123	.00233100	469	219961	21.6564	.00213220
430	184900	20.7364	.00232558	470	220900	21.6795	.00212766
431	185761	20.7605	.00232019	471	221841	21.7025	.00212314
432	186624	20.7846	.00231481	472	222784	21.7256	.00211864
433	187489	20.8087	.00230947	473	223729	21.7486	.00211416
434	188356	20.8327	.00230415	474	224676	21.7715	.00210970
435	189225	20.8567	.00229885	475	225625	21.7945	.00210526
436	190096	20.8806	.00229358	476	226576	21.8174	.00210084
437	190969	20.9045	.00228833	477	227529	21.8403	.00209644
438	191844	20.9284	.00228311	478	228484	21.8632	.00209205
439	192721	20.9523	.00227790	479	229441	21.8861	.00208768
440	193600	20.9762	.00227273	480	230400	21.9089	.00208333

* Portions of Table II have been reproduced from J. W. Dunlap and A. K. Kurtz. *Handbook of Statistical Nomographs, Tables, and Formulas*, World Book Company, New York (1932), by permission of the authors and publishers.

Table II. Table of Squares, Square Roots, and Reciprocals of Numbers from 1 to 1000*—Continued

N	N^2	$\sqrt{N}$	$1/N$	N	N^2	$\sqrt{N}$	$1/N$
481	231361	21.9317	.00207900	521	271441	22.8254	.00191939
482	232324	21.9545	.00207469	522	272484	22.8473	.00191571
483	233289	21.9773	.00207039	523	273529	22.8692	.00191205
484	234256	22.0000	.00206612	524	274576	22.8910	.00190840
485	235225	22.0227	.00206186	525	275625	22.9129	.00190476
486	236196	22.0454	.00205761	526	276676	22.9347	.00190114
487	237169	22.0681	.00205339	527	277729	22.9565	.00189753
488	238144	22.0907	.00204918	528	278784	22.9783	.00189394
489	239121	22.1133	.00204499	529	279841	23.0000	.00189036
490	240100	22.1359	.00204082	530	280900	23.0217	.00188679
491	241081	22.1585	.00203666	531	281961	23.0434	.00188324
492	242064	22.1811	.00203252	532	283024	23.0651	.00187970
493	243049	22.2036	.00202840	533	284089	23.0868	.00187617
494	244036	22.2261	.00202429	534	285156	23.1084	.00187266
495	245025	22.2486	.00202020	535	286225	23.1301	.00186916
496	246016	22.2711	.00201613	536	287296	23.1517	.00186567
497	247009	22.2935	.00201207	537	288369	23.1733	.00186220
498	248004	22.3159	.00200803	538	289444	23.1948	.00185874
499	249001	22.3383	.00200401	539	290521	23.2164	.00185529
500	250000	22.3607	.00200000	540	291600	23.2379	.00185185
501	251001	22.3830	.00199601	541	292681	23.2594	.00184843
502	252004	22.4054	.00199203	542	293764	23.2809	.00184502
503	253009	22.4277	.00198807	543	294849	23.3024	.00184162
504	254016	22.4499	.00198413	544	295936	23.3238	.00183824
505	255025	22.4722	.00198020	545	297025	23.3452	.00183486
506	256036	22.4944	.00197628	546	298116	23.3666	.00183150
507	257049	22.5167	.00197239	547	299209	23.3880	.00182815
508	258064	22.5389	.00196850	548	300304	23.4094	.00182482
509	259081	22.5610	.00196464	549	301401	23.4307	.00182149
510	260100	22.5832	.00196078	550	302500	23.4521	.00181818
511	261121	22.6053	.00195695	551	303601	23.4734	.00181488
512	262144	22.6274	.00195312	552	304704	23.4947	.00181159
513	263169	22.6495	.00194932	553	305809	23.5160	.00180832
514	264196	22.6716	.00194553	554	306916	23.5372	.00180505
515	265225	22.6936	.00194175	555	308025	23.5584	.00180180
516	266256	22.7156	.00193798	556	309136	23.5797	.00179856
517	267289	22.7376	.00193424	557	310249	23.6008	.00179533
518	268324	22.7596	.00193050	558	311364	23.6220	.00179211
519	269361	22.7816	.00192678	559	312481	23.6432	.00178891
520	270400	22.8035	.00192308	560	313600	23.6643	.00178571

* Portions of Table II have been reproduced from J. W. Dunlap and A. K. Kurtz. *Handbook of Statistical Nomographs, Tables, and Formulas*, World Book Company, New York (1932), by permission of the authors and publishers.

Table II. Table of Squares, Square Roots, and Reciprocals of Numbers from 1 to 1000*—Continued

N	N^2	$\sqrt{N}$	$1/N$	N	N^2	$\sqrt{N}$	$1/N$
561	314721	23.6854	.00178253	601	361201	24.5153	.00166389
562	315844	23.7065	.00177936	602	362404	24.5357	.00166113
563	316969	23.7276	.00177620	603	363609	24.5561	.00165837
564	318096	23.7487	.00177305	604	364816	24.5764	.00165563
565	319225	23.7697	.00176991	605	366025	24.5967	.00165289
566	320356	23.7908	.00176678	606	367236	24.6171	.00165017
567	321489	23.8118	.00176367	607	368449	24.6374	.00164745
568	322624	23.8328	.00176056	608	369664	24.6577	.00164474
569	323761	23.8537	.00175747	609	370881	24.6779	.00164204
570	324900	23.8747	.00175439	610	372100	24.6982	.00163934
571	326041	23.8956	.00175131	611	373321	24.7184	.00163666
572	327184	23.9165	.00174825	612	374544	24.7386	.00163399
573	328329	23.9374	.00174520	613	375769	24.7588	.00163132
574	329476	23.9583	.00174216	614	376996	24.7790	.00162866
575	330625	23.9792	.00173913	615	378225	24.7992	.00162602
576	331776	24.0000	.00173611	616	379456	24.8193	.00162338
577	332929	24.0208	.00173310	617	380689	24.8395	.00162075
578	334084	24.0416	.00173010	618	381924	24.8596	.00161812
579	335241	24.0624	.00172712	619	383161	24.8797	.00161551
580	336400	24.0832	.00172414	620	384400	24.8998	.00161290
581	337561	24.1039	.00172117	621	385641	24.9199	.00161031
582	338724	24.1247	.00171821	622	386884	24.9399	.00160772
583	339889	24.1454	.00171527	623	388129	24.9600	.00160514
584	341056	24.1661	.00171233	624	389376	24.9800	.00160256
585	342225	24.1868	.00170940	625	390625	25.0000	.00160000
586	343396	24.2074	.00170648	626	391876	25.0200	.00159744
587	344569	24.2281	.00170358	627	393129	25.0400	.00159490
588	345744	24.2487	.00170068	628	394384	25.0599	.00159236
589	346921	24.2693	.00169779	629	395641	25.0799	.00158983
590	348100	24.2899	.00169492	630	396900	25.0998	.00158730
591	349281	24.3105	.00169205	631	398161	25.1197	.00158479
592	350464	24.3311	.00168919	632	399424	25.1396	.00158228
593	351649	24.3516	.00168634	633	400689	25.1595	.00157978
594	352836	24.3721	.00168350	634	401956	25.1794	.00157729
595	354025	24.3926	.00168067	635	403225	25.1992	.00157480
596	355216	24.4131	.00167785	636	404496	25.2190	.00157233
597	356409	24.4336	.00167504	637	405769	25.2389	.00156986
598	357604	24.4540	.00167224	638	407044	25.2587	.00156740
599	358801	24.4745	.00166945	639	408321	25.2784	.00156495
600	360000	24.4949	.00166667	640	409600	25.2982	.00156250

* Portions of Table II have been reproduced from J. W. Dunlap and A. K. Kurtz. *Handbook of Statistical Nomographs, Tables, and Formulas*, World Book Company, New York (1932), by permission of the authors and publishers.

Table II. Table of Squares, Square Roots, and Reciprocals of Numbers from 1 to 1000*—Continued

N	N^2	$\sqrt{N}$	$1/N$	N	N^2	$\sqrt{N}$	$1/N$
641	410881	25.3180	.00156006	681	463761	26.0960	.00146843
642	412164	25.3377	.00155763	682	465124	26.1151	.00146628
643	413449	25.3574	.00155521	683	466489	26.1343	.00146413
644	414736	25.3772	.00155280	684	467856	26.1534	.00146199
645	416025	25.3969	.00155039	685	469225	26.1725	.00145985
646	417316	25.4165	.00154799	686	470596	26.1916	.00145773
647	418609	25.4362	.00154560	687	471969	26.2107	.00145560
648	419904	25.4558	.00154321	688	473344	26.2298	.00145349
649	421201	25.4755	.00154083	689	474721	26.2488	.00145138
650	422500	25.4951	.00153846	690	476100	26.2679	.00144928
651	423801	25.5147	.00153610	691	477481	26.2869	.00144718
652	425104	25.5343	.00153374	692	478864	26.3059	.00144509
653	426409	25.5539	.00153139	693	480249	26.3249	.00144300
654	427716	25.5734	.00152905	694	481636	26.3439	.00144092
655	429025	25.5930	.00152672	695	483025	26.3629	.00143885
656	430336	25.6125	.00152439	696	484416	26.3818	.00143678
657	431649	25.6320	.00152207	697	485809	26.4008	.00143472
658	432964	25.6515	.00151976	698	487204	26.4197	.00143266
659	434281	25.6710	.00151745	699	488601	26.4386	.00143062
660	435600	25.6905	.00151515	700	490000	26.4575	.00142857
661	436921	25.7099	.00151286	701	491401	26.4764	.00142653
662	438244	25.7294	.00151057	702	492804	26.4953	.00142450
663	439569	25.7488	.00150830	703	494209	26.5141	.00142248
664	440896	25.7682	.00150602	704	495616	26.5330	.00142045
665	442225	25.7876	.00150376	705	497025	26.5518	.00141844
666	443556	25.8070	.00150150	706	498436	26.5707	.00141643
667	444889	25.8263	.00149925	707	499849	26.5895	.00141443
668	446224	25.8457	.00149701	708	501264	26.6083	.00141243
669	447561	25.8650	.00149477	709	502681	26.6271	.00141044
670	448900	25.8844	.00149254	710	504100	26.6458	.00140845
671	450241	25.9037	.00149031	711	505521	26.6646	.00140647
672	451584	25.9230	.00148810	712	506944	26.6833	.00140449
673	452929	25.9422	.00148588	713	508369	26.7021	.00140252
674	454276	25.9615	.00148368	714	509796	26.7208	.00140056
675	455625	25.9808	.00148148	715	511225	26.7395	.00139860
676	456976	26.0000	.00147929	716	512656	26.7582	.00139665
677	458329	26.0192	.00147710	717	514089	26.7769	.00139470
678	459684	26.0384	.00147493	718	515524	26.7955	.00139276
679	461041	26.0576	.00147275	719	516961	26.8142	.00139082
680	462400	26.0768	.00147059	720	518400	26.8328	.00138889

* Portions of Table II have been reproduced from J. W. Dunlap and A. K. Kurtz. *Handbook of Statistical Nomographs, Tables, and Formulas*, World Book Company, New York (1932), by permission of the authors and publishers.

Table II. Table of Squares, Square Roots, and Reciprocals of Numbers from 1 to 1000*—Continued

N	N^2	$\sqrt{N}$	$1/N$	N	N^2	$\sqrt{N}$	$1/N$
721	519841	26.8514	.00138696	761	579121	27.5862	.00131406
722	521284	26.8701	.00138504	762	580644	27.6043	.00131234
723	522729	26.8887	.00138313	763	582169	27.6225	.00131062
724	524176	26.9072	.00138122	764	583696	27.6405	.00130890
725	525625	26.9258	.00137931	765	585225	27.6586	.00130719
726	527076	26.9444	.00137741	766	586756	27.6767	.00130548
727	528529	26.9629	.00137552	767	588289	27.6948	.00130378
728	529984	26.9815	.00137363	768	589824	27.7128	.00130208
729	531441	27.0000	.00137174	769	591361	27.7308	.00130039
730	532900	27.0185	.00136986	770	592900	27.7489	.00129870
731	534361	27.0370	.00136799	771	594441	27.7669	.00129702
732	535824	27.0555	.00136612	772	595984	27.7849	.00129534
733	537289	27.0740	.00136426	773	597529	27.8029	.00129366
734	538756	27.0924	.00136240	774	599076	27.8209	.00129199
735	540225	27.1109	.00136054	775	600625	27.8388	.00129032
736	541696	27.1293	.00135870	776	602176	27.8568	.00128866
737	543169	27.1477	.00135685	777	603729	27.8747	.00128700
738	544644	27.1662	.00135501	778	605284	27.8927	.00128535
739	546121	27.1846	.00135318	779	606841	27.9106	.00128370
740	547600	27.2029	.00135135	780	608400	27.9285	.00128205
741	549081	27.2213	.00134953	781	609961	27.9464	.00128041
742	550564	27.2397	.00134771	782	611524	27.9643	.00127877
743	552049	27.2580	.00134590	783	613089	27.9821	.00127714
744	553536	27.2764	.00134409	784	614656	28.0000	.00127551
745	555025	27.2947	.00134228	785	616225	28.0179	.00127389
746	556516	27.3130	.00134048	786	617796	28.0357	.00127226
747	558009	27.3313	.00133869	787	619369	28.0535	.00127065
748	559504	27.3496	.00133690	788	620944	28.0713	.00126904
749	561001	27.3679	.00133511	789	622521	28.0891	.00126743
750	562500	27.3861	.00133333	790	624100	28.1069	.00126582
751	564001	27.4044	.00133156	791	625681	28.1247	.00126422
752	565504	27.4226	.00132979	792	627264	28.1425	.00126263
753	567009	27.4408	.00132802	793	628849	28.1603	.00126103
754	568516	27.4591	.00132626	794	630436	28.1780	.00125945
755	570025	27.4773	.00132450	795	632025	28.1957	.00125786
756	571536	27.4955	.00132275	796	633616	28.2135	.00125628
757	573049	27.5136	.00132100	797	635209	28.2312	.00125471
758	574564	27.5318	.00131926	798	636804	28.2489	.00125313
759	576081	27.5500	.00131752	799	638401	28.2666	.00125156
760	577600	27.5681	.00131579	800	640000	28.2843	.00125000

* Portions of Table II have been reproduced from J. W. Dunlap and A. K. Kurtz. *Handbook of Statistical Nomographs, Tables, and Formulas*, World Book Company, New York (1932), by permission of the authors and publishers.

Table II. Table of Squares, Square Roots, and Reciprocals of Numbers from 1 to 1000*—Continued

N	N^2	$\sqrt{N}$	$1/N$	N	N^2	$\sqrt{N}$	$1/N$
801	641601	28.3019	.00124844	841	707281	29.0000	.00118906
802	643204	28.3196	.00124688	842	708964	29.0172	.00118765
803	644809	28.3373	.00124533	843	710649	29.0345	.00118624
804	646416	28.3549	.00124378	844	712336	29.0517	.00118483
805	648025	28.3725	.00124224	845	714025	29.0689	.00118343
806	649636	28.3901	.00124069	846	715716	29.0861	.00118203
807	651249	28.4077	.00123916	847	717409	29.1033	.00118064
808	652864	28.4253	.00123762	848	719104	29.1204	.00117925
809	654481	28.4429	.00123609	849	720801	29.1376	.00117786
810	656100	28.4605	.00123457	850	722500	29.1548	.00117647
811	657721	28.4781	.00123305	851	724201	29.1719	.00117509
812	659344	28.4956	.00123153	852	725904	29.1890	.00117371
813	660969	28.5132	.00123001	853	727609	29.2062	.00117233
814	662596	28.5307	.00122850	854	729316	29.2233	.00117096
815	664225	28.5482	.00122699	855	731025	29.2404	.00116959
816	665856	28.5657	.00122549	856	732736	29.2575	.00116822
817	667489	28.5832	.00122399	857	734449	29.2746	.00116686
818	669124	28.6007	.00122249	858	736164	29.2916	.00116550
819	670761	28.6182	.00122100	859	737881	29.3087	.00116414
820	672400	28.6356	.00121951	860	739600	29.3258	.00116279
821	674041	28.6531	.00121803	861	741321	29.3428	.00116144
822	675684	28.6705	.00121655	862	743044	29.3598	.00116009
823	677329	28.6880	.00121507	863	744769	29.3769	.00115875
824	678976	28.7054	.00121359	864	746496	29.3939	.00115741
825	680625	28.7228	.00121212	865	748225	29.4109	.00115607
826	682276	28.7402	.00121065	866	749956	29.4279	.00115473
827	683929	28.7576	.00120919	867	751689	29.4449	.00115340
828	685584	28.7750	.00120773	868	753424	29.4618	.00115207
829	687241	28.7924	.00120627	869	755161	29.4788	.00115075
830	688900	28.8097	.00120482	870	756900	29.4958	.00114943
831	690561	28.8271	.00120337	871	758641	29.5127	.00114811
832	692224	28.8444	.00120192	872	760384	29.5296	.00114679
833	693889	28.8617	.00120048	873	762129	29.5466	.00114548
834	695556	28.8791	.00119904	874	763876	29.5635	.00114416
835	697225	28.8964	.00119760	875	765625	29.5804	.00114286
836	698896	28.9137	.00119617	876	767376	29.5973	.00114155
837	700569	28.9310	.00119474	877	769129	29.6142	.00114025
838	702244	28.9482	.00119332	878	770884	29.6311	.00113895
839	703921	28.9655	.00119190	879	772641	29.6479	.00113766
840	705600	28.9828	.00119048	880	774400	29.6648	.00113636

* Portions of Table II have been reproduced from J. W. Dunlap and A. K. Kurtz. *Handbook of Statistical Nomographs, Tables, and Formulas*, World Book Company, New York (1932), by permission of the authors and publishers.

Table II. Table of Squares, Square Roots, and Reciprocals of Numbers from 1 to 1000*—Continued

N	N^2	$\sqrt{N}$	$1/N$	N	N^2	$\sqrt{N}$	$1/N$
881	776161	29.6816	.00113507	921	848241	30.3480	.00108578
882	777924	29.6985	.00113379	922	850084	30.3645	.00108460
883	779689	29.7153	.00113250	923	851929	30.3809	.00108342
884	781456	29.7321	.00113122	924	853776	30.3974	.00108225
885	783225	29.7489	.00112994	925	855625	30.4138	.00108108
886	784996	29.7658	.00112867	926	857476	30.4302	.00107991
887	786769	29.7825	.00112740	927	859329	30.4467	.00107875
888	788544	29.7993	.00112613	928	861184	30.4631	.00107759
889	790321	29.8161	.00112486	929	863041	30.4795	.00107643
890	792100	29.8329	.00112360	930	864900	30.4959	.00107527
891	793881	29.8496	.00112233	931	866761	30.5123	.00107411
892	795664	29.8664	.00112108	932	868624	30.5287	.00107296
893	797449	29.8831	.00111982	933	870489	30.5450	.00107181
894	799236	29.8998	.00111857	934	872356	30.5614	.00107066
895	801025	29.9166	.00111732	935	874225	30.5778	.00106952
896	802816	29.9333	.00111607	936	876096	30.5941	.00106838
897	804609	29.9500	.00111483	937	877969	30.6105	.00106724
898	806404	29.9666	.00111359	938	879844	30.6268	.00106610
899	808201	29.9833	.00111235	939	881721	30.6431	.00106496
900	810000	30.0000	.00111111	940	883600	30.6594	.00106383
901	811801	30.0167	.00110988	941	885481	30.6757	.00106270
902	813604	30.0333	.00110865	942	887364	30.6920	.00106157
903	815409	30.0500	.00110742	943	889249	30.7083	.00106045
904	817216	30.0666	.00110619	944	891136	30.7246	.00105932
905	819025	30.0832	.00110497	945	893025	30.7409	.00105820
906	820836	30.0998	.00110375	946	894916	30.7571	.00105708
907	822649	30.1164	.00110254	947	896809	30.7734	.00105597
908	824464	30.1330	.00110132	948	898704	30.7896	.00105485
909	826281	30.1496	.00110011	949	900601	30.8058	.00105374
910	828100	30.1662	.00109890	950	902500	30.8221	.00105263
911	829921	30.1828	.00109769	951	904401	30.8383	.00105152
912	831744	30.1993	.00109649	952	906304	30.8545	.00105042
913	833569	30.2159	.00109529	953	908209	30.8707	.00104932
914	835396	30.2324	.00109409	954	910116	30.8869	.00104822
915	837225	30.2490	.00109290	955	912025	30.9031	.00104712
916	839056	30.2655	.00109170	956	913936	30.9192	.00104603
917	840889	30.2820	.00109051	957	915849	30.9354	.00104493
918	842724	30.2985	.00108932	958	917764	30.9516	.00104384
919	844561	30.3150	.00108814	959	919681	30.9677	.00104275
920	846400	30.3315	.00108696	960	921600	30.9839	.00104167

* Portions of Table II have been reproduced from J. W. Dunlap and A. K. Kurtz. *Handbook of Statistical Nomographs, Tables, and Formulas*, World Book Company, New York (1932), by permission of the authors and publishers.

Table II. Table of Squares, Square Roots, and Reciprocals of Numbers from 1 to 1000*—Concluded

N	N^2	$\sqrt{N}$	$1/N$	N	N^2	$\sqrt{N}$	$1/N$
961	923521	31.0000	.00104058	981	962361	31.3209	.00101937
962	925444	31.0161	.00103950	982	964324	31.3369	.00101833
963	927369	31.0322	.00103842	983	966289	31.3528	.00101729
964	929296	31.0483	.00103734	984	968256	31.3688	.00101626
965	931225	31.0644	.00103627	985	970225	31.3847	.00101523
966	933156	31.0805	.00103520	986	972196	31.4006	.00101420
967	935089	31.0966	.00103413	987	974169	31.4166	.00101317
968	937024	31.1127	.00103306	988	976144	31.4325	.00101215
969	938961	31.1288	.00103199	989	978121	31.4484	.00101112
970	940900	31.1448	.00103093	990	980100	31.4643	.00101010
971	942841	31.1609	.00102987	991	982081	31.4802	.00100908
972	944784	31.1769	.00102881	992	984064	31.4960	.00100806
973	946729	31.1929	.00102775	993	986049	31.5119	.00100705
974	948676	31.2090	.00102669	994	988036	31.5278	.00100604
975	950625	31.2250	.00102564	995	990025	31.5436	.00100503
976	952576	31.2410	.00102459	996	992016	31.5595	.00100402
977	954529	31.2570	.00102354	997	994009	31.5753	.00100301
978	956484	31.2730	.00102249	998	996004	31.5911	.00100200
979	958441	31.2890	.00102145	999	998001	31.6070	.00100100
980	960400	31.3050	.00102041	1000	1000000	31.6228	.00100000

* Portions of Table II have been reproduced from J. W. Dunlap and A. K. Kurtz: *Handbook of Statistical Nomographs, Tables, and Formulas*, World Book Company, New York (1932), by permission of the authors and publishers.

Table III. Areas and Ordinates of the Normal Curve in Terms of $Z = (X - \mu)/\sigma$

(1) Z	(2) A AREA FROM MEAN TO Z	(3) B AREA IN LARGER PORTION	(4) C AREA IN SMALLER PORTION	(5) y ORDINATE AT Z
0.00	.0000	.5000	.5000	.3989
0.01	.0040	.5040	.4960	.3989
0.02	.0080	.5080	.4920	.3989
0.03	.0120	.5120	.4880	.3988
0.04	.0160	.5160	.4840	.3986
0.05	.0199	.5199	.4801	.3984
0.06	.0239	.5239	.4761	.3982
0.07	.0279	.5279	.4721	.3980
0.08	.0319	.5319	.4681	.3977
0.09	.0359	.5359	.4641	.3973
0.10	.0398	.5398	.4602	.3970
0.11	.0438	.5438	.4562	.3965
0.12	.0478	.5478	.4522	.3961
0.13	.0517	.5517	.4483	.3956
0.14	.0557	.5557	.4443	.3951
0.15	.0596	.5596	.4404	.3945
0.16	.0636	.5636	.4364	.3939
0.17	.0675	.5675	.4325	.3932
0.18	.0714	.5714	.4286	.3925
0.19	.0753	.5753	.4247	.3918
0.20	.0793	.5793	.4207	.3910
0.21	.0832	.5832	.4168	.3902
0.22	.0871	.5871	.4129	.3894
0.23	.0910	.5910	.4090	.3885
0.24	.0948	.5948	.4052	.3876
0.25	.0987	.5987	.4013	.3867
0.26	.1026	.6026	.3974	.3857
0.27	.1064	.6064	.3936	.3847
0.28	.1103	.6103	.3897	.3836
0.29	.1141	.6141	.3859	.3825
0.30	.1179	.6179	.3821	.3814
0.31	.1217	.6217	.3783	.3802
0.32	.1255	.6255	.3745	.3790
0.33	.1293	.6293	.3707	.3778
0.34	.1331	.6331	.3669	.3765

Table III. Areas and Ordinates of the Normal Curve in Terms of $Z = (X - \mu)/\sigma$—Continued

(1) Z	(2) A Area from Mean to Z	(3) B Area in Larger Portion	(4) C Area in Smaller Portion	(5) y Ordinate at Z
0.35	.1368	.6368	.3632	.3752
0.36	.1406	.6406	.3594	.3739
0.37	.1443	.6443	.3557	.3725
0.38	.1480	.6480	.3520	.3712
0.39	.1517	.6517	.3483	.3697
0.40	.1554	.6554	.3446	.3683
0.41	.1591	.6591	.3409	.3668
0.42	.1628	.6628	.3372	.3653
0.43	.1664	.6664	.3336	.3637
0.44	.1700	.6700	.3300	.3621
0.45	.1736	.6736	.3264	.3605
0.46	.1772	.6772	.3228	.3589
0.47	.1808	.6808	.3192	.3572
0.48	.1844	.6844	.3156	.3555
0.49	.1879	.6879	.3121	.3538
0.50	.1915	.6915	.3085	.3521
0.51	.1950	.6950	.3050	.3503
0.52	.1985	.6985	.3015	.3485
0.53	.2019	.7019	.2981	.3467
0.54	.2054	.7054	.2946	.3448
0.55	.2088	.7088	.2912	.3429
0.56	.2123	.7123	.2877	.3410
0.57	.2157	.7157	.2843	.3391
0.58	.2190	.7190	.2810	.3372
0.59	.2224	.7224	.2776	.3352
0.60	.2257	.7257	.2743	.3332
0.61	.2291	.7291	.2709	.3312
0.62	.2324	.7324	.2676	.3292
0.63	.2357	.7357	.2643	.3271
0.64	.2389	.7389	.2611	.3251
0.65	.2422	.7422	.2578	.3230
0.66	.2454	.7454	.2546	.3209
0.67	.2486	.7486	.2514	.3187
0.68	.2517	.7517	.2483	.3166
0.69	.2549	.7549	.2451	.3144

Table III. Areas and Ordinates of the Normal Curve in Terms of $Z = (X - \mu)/\sigma$—Continued

(1) Z	(2) A Area from Mean to Z	(3) B Area in Larger Portion	(4) C Area in Smaller Portion	(5) y Ordinate at Z
0.70	.2580	.7580	.2420	.3123
0.71	.2611	.7611	.2389	.3101
0.72	.2642	.7642	.2358	.3079
0.73	.2673	.7673	.2327	.3056
0.74	.2704	.7704	.2296	.3034
0.75	.2734	.7734	.2266	.3011
0.76	.2764	.7764	.2236	.2989
0.77	.2794	.7794	.2206	.2966
0.78	.2823	.7823	.2177	.2943
0.79	.2852	.7852	.2148	.2920
0.80	.2881	.7881	.2119	.2897
0.81	.2910	.7910	.2090	.2874
0.82	.2939	.7939	.2061	.2850
0.83	.2967	.7967	.2033	.2827
0.84	.2995	.7995	.2005	.2803
0.85	.3023	.8023	.1977	.2780
0.86	.3051	.8051	.1949	.2756
0.87	.3078	.8078	.1922	.2732
0.88	.3106	.8106	.1894	.2709
0.89	.3133	.8133	.1867	.2685
0.90	.3159	.8159	.1841	.2661
0.91	.3186	.8186	.1814	.2637
0.92	.3212	.8212	.1788	.2613
0.93	.3238	.8238	.1762	.2589
0.94	.3264	.8264	.1736	.2565
0.95	.3289	.8289	.1711	.2541
0.96	.3315	.8315	.1685	.2516
0.97	.3340	.8340	.1660	.2492
0.98	.3365	.8365	.1635	.2468
0.99	.3389	.8389	.1611	.2444
1.00	.3413	.8413	.1587	.2420
1.01	.3438	.8438	.1562	.2396
1.02	.3461	.8461	.1539	.2371
1.03	.3485	.8485	.1515	.2347
1.04	.3508	.8508	.1492	.2323

Table III. Areas and Ordinates of the Normal Curve in Terms of $Z = (X - \mu)/\sigma$—Continued

(1) Z	(2) A Area from Mean to Z	(3) B Area in Larger Portion	(4) C Area in Smaller Portion	(5) y Ordinate at Z
1.05	.3531	.8531	.1469	.2299
1.06	.3554	.8554	.1446	.2275
1.07	.3577	.8577	.1423	.2251
1.08	.3599	.8599	.1401	.2227
1.09	.3621	.8621	.1379	.2203
1.10	.3643	.8643	.1357	.2179
1.11	.3665	.8665	.1335	.2155
1.12	.3686	.8686	.1314	.2131
1.13	.3708	.8708	.1292	.2107
1.14	.3729	.8729	.1271	.2083
1.15	.3749	.8749	.1251	.2059
1.16	.3770	.8770	.1230	.2036
1.17	.3790	.8790	.1210	.2012
1.18	.3810	.8810	.1190	.1989
1.19	.3830	.8830	.1170	.1965
1.20	.3849	.8849	.1151	.1942
1.21	.3869	.8869	.1131	.1919
1.22	.3888	.8888	.1112	.1895
1.23	.3907	.8907	.1093	.1872
1.24	.3925	.8925	.1075	.1849
1.25	.3944	.8944	.1056	.1826
1.26	.3962	.8962	.1038	.1804
1.27	.3980	.8980	.1020	.1781
1.28	.3997	.8997	.1003	.1758
1.29	.4015	.9015	.0985	.1736
1.30	.4032	.9032	.0968	.1714
1.31	.4049	.9049	.0951	.1691
1.32	.4066	.9066	.0934	.1669
1.33	.4082	.9082	.0918	.1647
1.34	.4099	.9099	.0901	.1626
1.35	.4115	.9115	.0885	.1604
1.36	.4131	.9131	.0869	.1582
1.37	.4147	.9147	.0853	.1561
1.38	.4162	.9162	.0838	.1539
1.39	.4177	.9177	.0823	.1518

Table III. Areas and Ordinates of the Normal Curve in Terms of $Z = (X - \mu)/\sigma$—Continued

(1) Z	(2) A Area from Mean to Z	(3) B Area in Larger Portion	(4) C Area in Smaller Portion	(5) y Ordinate at Z
1.40	.4192	.9192	.0808	.1497
1.41	.4207	.9207	.0793	.1476
1.42	.4222	.9222	.0778	.1456
1.43	.4236	.9236	.0764	.1435
1.44	.4251	.9251	.0749	.1415
1.45	.4265	.9265	.0735	.1394
1.46	.4279	.9279	.0721	.1374
1.47	.4292	.9292	.0708	.1354
1.48	.4306	.9306	.0694	.1334
1.49	.4319	.9319	.0681	.1315
1.50	.4332	.9332	.0668	.1295
1.51	.4345	.9345	.0655	.1276
1.52	.4357	.9357	.0643	.1257
1.53	.4370	.9370	.0630	.1238
1.54	.4382	.9382	.0618	.1219
1.55	.4394	.9394	.0606	.1200
1.56	.4406	.9406	.0594	.1182
1.57	.4418	.9418	.0582	.1163
1.58	.4429	.9429	.0571	.1145
1.59	.4441	.9441	.0559	.1127
1.60	.4452	.9452	.0548	.1109
1.61	.4463	.9463	.0537	.1092
1.62	.4474	.9474	.0526	.1074
1.63	.4484	.9484	.0516	.1057
1.64	.4495	.9495	.0505	.1040
1.65	.4505	.9505	.0495	.1023
1.66	.4515	.9515	.0485	.1006
1.67	.4525	.9525	.0475	.0989
1.68	.4535	.9535	.0465	.0973
1.69	.4545	.9545	.0455	.0957
1.70	.4554	.9554	.0446	.0940
1.71	.4564	.9564	.0436	.0925
1.72	.4573	.9573	.0427	.0909
1.73	.4582	.9582	.0418	.0893
1.74	.4591	.9591	.0409	.0878

Table III. Areas and Ordinates of the Normal Curve in Terms of $Z = (X - \mu)/\sigma$—Continued

(1) Z	(2) A Area from Mean to Z	(3) B Area in Larger Portion	(4) C Area in Smaller Portion	(5) y Ordinate at Z
1.75	.4599	.9599	.0401	.0863
1.76	.4608	.9608	.0392	.0848
1.77	.4616	.9616	.0384	.0833
1.78	.4625	.9625	.0375	.0818
1.79	.4633	.9633	.0367	.0804
1.80	.4641	.9641	.0359	.0790
1.81	.4649	.9649	.0351	.0775
1.82	.4656	.9656	.0344	.0761
1.83	.4664	.9664	.0336	.0748
1.84	.4671	.9671	.0329	.0734
1.85	.4678	.9678	.0322	.0721
1.86	.4686	.9686	.0314	.0707
1.87	.4693	.9693	.0307	.0694
1.88	.4699	.9699	.0301	.0681
1.89	.4706	.9706	.0294	.0669
1.90	.4713	.9713	.0287	.0656
1.91	.4719	.9719	.0281	.0644
1.92	.4726	.9726	.0274	.0632
1.93	.4732	.9732	.0268	.0620
1.94	.4738	.9738	.0262	.0608
1.95	.4744	.9744	.0256	.0596
1.96	.4750	.9750	.0250	.0584
1.97	.4756	.9756	.0244	.0573
1.98	.4761	.9761	.0239	.0562
1.99	.4767	.9767	.0233	.0551
2.00	.4772	.9772	.0228	.0540
2.01	.4778	.9778	.0222	.0529
2.02	.4783	.9783	.0217	.0519
2.03	.4788	.9788	.0212	.0508
2.04	.4793	.9793	.0207	.0498
2.05	.4798	.9798	.0202	.0488
2.06	.4803	.9803	.0197	.0478
2.07	.4808	.9808	.0192	.0468
2.08	.4812	.9812	.0188	.0459
2.09	.4817	.9817	.0183	.0449

Table III. Areas and Ordinates of the Normal Curve in Terms of $Z = (X - \mu)/\sigma$—Continued

(1) Z	(2) A Area from Mean to Z	(3) B Area in Larger Portion	(4) C Area in Smaller Portion	(5) y Ordinate at Z
2.10	.4821	.9821	.0179	.0440
2.11	.4826	.9826	.0174	.0431
2.12	.4830	.9830	.0170	.0422
2.13	.4834	.9834	.0166	.0413
2.14	.4838	.9838	.0162	.0404
2.15	.4842	.9842	.0158	.0396
2.16	.4846	.9846	.0154	.0387
2.17	.4850	.9850	.0150	.0379
2.18	.4854	.9854	.0146	.0371
2.19	.4857	.9857	.0143	.0363
2.20	.4861	.9861	.0139	.0355
2.21	.4864	.9864	.0136	.0347
2.22	.4868	.9868	.0132	.0339
2.23	.4871	.9871	.0129	.0332
2.24	.4875	.9875	.0125	.0325
2.25	.4878	.9878	.0122	.0317
2.26	.4881	.9881	.0119	.0310
2.27	.4884	.9884	.0116	.0303
2.28	.4887	.9887	.0113	.0297
2.29	.4890	.9890	.0110	.0290
2.30	.4893	.9893	.0107	.0283
2.31	.4896	.9896	.0104	.0277
2.32	.4898	.9898	.0102	.0270
2.33	.4901	.9901	.0099	.0264
2.34	.4904	.9904	.0096	.0258
2.35	.4906	.9906	.0094	.0252
2.36	.4909	.9909	.0091	.0246
2.37	.4911	.9911	.0089	.0241
2.38	.4913	.9913	.0087	.0235
2.39	.4916	.9916	.0084	.0229
2.40	.4918	.9918	.0082	.0224
2.41	.4920	.9920	.0080	.0219
2.42	.4922	.9922	.0078	.0213
2.43	.4925	.9925	.0075	.0208
2.44	.4927	.9927	.0073	.0203

Table III. Areas and Ordinates of the Normal Curve in Terms of $Z = (X - \mu)/\sigma$—Continued

(1) Z	(2) A Area from Mean to Z	(3) B Area in Larger Portion	(4) C Area in Smaller Portion	(5) y Ordinate at Z
2.45	.4929	.9929	.0071	.0198
2.46	.4931	.9931	.0069	.0194
2.47	.4932	.9932	.0068	.0189
2.48	.4934	.9934	.0066	.0184
2.49	.4936	.9936	.0064	.0180
2.50	.4938	.9938	.0062	.0175
2.51	.4940	.9940	.0060	.0171
2.52	.4941	.9941	.0059	.0167
2.53	.4943	.9943	.0057	.0163
2.54	.4945	.9945	.0055	.0158
2.55	.4946	.9946	.0054	.0154
2.56	.4948	.9948	.0052	.0151
2.57	.4949	.9949	.0051	.0147
2.58	.4951	.9951	.0049	.0143
2.59	.4952	.9952	.0048	.0139
2.60	.4953	.9953	.0047	.0136
2.61	.4955	.9955	.0045	.0132
2.62	.4956	.9956	.0044	.0129
2.63	.4957	.9957	.0043	.0126
2.64	.4959	.9959	.0041	.0122
2.65	.4960	.9960	.0040	.0119
2.66	.4961	.9961	.0039	.0116
2.67	.4962	.9962	.0038	.0113
2.68	.4963	.9963	.0037	.0110
2.69	.4964	.9964	.0036	.0107
2.70	.4965	.9965	.0035	.0104
2.71	.4966	.9966	.0034	.0101
2.72	.4967	.9967	.0033	.0099
2.73	.4968	.9968	.0032	.0096
2.74	.4969	.9969	.0031	.0093
2.75	.4970	.9970	.0030	.0091
2.76	.4971	.9971	.0029	.0088
2.77	.4972	.9972	.0028	.0086
2.78	.4973	.9973	.0027	.0084
2.79	.4974	.9974	.0026	.0081

Table III. Areas and Ordinates of the Normal Curve in Terms of $Z = (X - \mu)/\sigma$—Continued

(1) Z	(2) A Area from Mean to Z	(3) B Area in Larger Portion	(4) C Area in Smaller Portion	(5) y Ordinate at Z
2.80	.4974	.9974	.0026	.0079
2.81	.4975	.9975	.0025	.0077
2.82	.4976	.9976	.0024	.0075
2.83	.4977	.9977	.0023	.0073
2.84	.4977	.9977	.0023	.0071
2.85	.4978	.9978	.0022	.0069
2.86	.4979	.9979	.0021	.0067
2.87	.4979	.9979	.0021	.0065
2.88	.4980	.9980	.0020	.0063
2.89	.4981	.9981	.0019	.0061
2.90	.4981	.9981	.0019	.0060
2.91	.4982	.9982	.0018	.0058
2.92	.4982	.9982	.0018	.0056
2.93	.4983	.9983	.0017	.0055
2.94	.4984	.9984	.0016	.0053
2.95	.4984	.9984	.0016	.0051
2.96	.4985	.9985	.0015	.0050
2.97	.4985	.9985	.0015	.0048
2.98	.4986	.9986	.0014	.0047
2.99	.4986	.9986	.0014	.0046
3.00	.4987	.9987	.0013	.0044
3.01	.4987	.9987	.0013	.0043
3.02	.4987	.9987	.0013	.0042
3.03	.4988	.9988	.0012	.0040
3.04	.4988	.9988	.0012	.0039
3.05	.4989	.9989	.0011	.0038
3.06	.4989	.9989	.0011	.0037
3.07	.4989	.9989	.0011	.0036
3.08	.4990	.9990	.0010	.0035
3.09	.4990	.9990	.0010	.0034
3.10	.4990	.9990	.0010	.0033
3.11	.4991	.9991	.0009	.0032
3.12	.4991	.9991	.0009	.0031
3.13	.4991	.9991	.0009	.0030
3.14	.4992	.9992	.0008	.0029

Table III. Areas and Ordinates of the Normal Curve in Terms of $Z = (X - \mu)/\sigma$—Concluded

(1) Z	(2) *A* Area from Mean to *Z*	(3) *B* Area in Larger Portion	(4) *C* Area in Smaller Portion	(5) *y* Ordinate at *Z*
3.15	.4992	.9992	.0008	.0028
3.16	.4992	.9992	.0008	.0027
3.17	.4992	.9992	.0008	.0026
3.18	.4993	.9993	.0007	.0025
3.19	.4993	.9993	.0007	.0025
3.20	.4993	.9993	.0007	.0024
3.21	.4993	.9993	.0007	.0023
3.22	.4994	.9994	.0006	.0022
3.23	.4994	.9994	.0006	.0022
3.24	.4994	.9994	.0006	.0021
3.30	.4995	.9995	.0005	.0017
3.40	.4997	.9997	.0003	.0012
3.50	.4998	.9998	.0002	.0009
3.60	.4998	.9998	.0002	.0006
3.70	.4999	.9999	.0001	.0004

Table IV. Table of χ^2*

Degrees of Freedom df	P = .99	.98	.95	.90	.80	.70	.50	.30	.20	.10	.05	.02	.01
1	.000157	.000628	.00393	.0158	.0642	.148	.455	1.074	1.642	2.706	3.841	5.412	6.635
2	.0201	.0404	.103	.211	.446	.713	1.386	2.408	3.219	4.605	5.991	7.824	9.210
3	.115	.185	.352	.584	1.005	1.424	2.366	3.665	4.642	6.251	7.815	9.837	11.341
4	.297	.429	.711	1.064	1.649	2.195	3.357	4.878	5.989	7.779	9.488	11.668	13.277
5	.554	.752	1.145	1.610	2.343	3.000	4.351	6.064	7.289	9.236	11.070	13.388	15.086
6	.872	1.134	1.635	2.204	3.070	3.828	5.348	7.231	8.558	10.645	12.592	15.033	16.812
7	1.239	1.564	2.167	2.833	3.822	4.671	6.346	8.383	9.803	12.017	14.067	16.622	18.475
8	1.646	2.032	2.733	3.490	4.594	5.527	7.344	9.524	11.030	13.362	15.507	18.168	20.090
9	2.088	2.532	3.325	4.168	5.380	6.393	8.343	10.656	12.242	14.684	16.919	19.679	21.666
10	2.558	3.059	3.940	4.865	6.179	7.267	9.342	11.781	13.442	15.987	18.307	21.161	23.209
11	3.053	3.609	4.575	5.578	6.989	8.148	10.341	12.899	14.631	17.275	19.675	22.618	24.725
12	3.571	4.178	5.226	6.304	7.807	9.034	11.340	14.011	15.812	18.549	21.026	24.054	26.217
13	4.107	4.765	5.892	7.042	8.634	9.926	12.340	15.119	16.985	19.812	22.362	25.472	27.688
14	4.660	5.368	6.571	7.790	9.467	10.821	13.339	16.222	18.151	21.064	23.685	26.873	29.141
15	5.229	5.985	7.261	8.547	10.307	11.721	14.339	17.322	19.311	22.307	24.996	28.259	30.578
16	5.812	6.614	7.962	9.312	11.152	12.624	15.338	18.418	20.465	23.542	26.296	29.633	32.000
17	6.408	7.255	8.672	10.085	12.002	13.531	16.338	19.511	21.615	24.769	27.587	30.995	33.409
18	7.015	7.906	9.390	10.865	12.857	14.440	17.338	20.601	22.760	25.989	28.869	32.346	34.805
19	7.633	8.567	10.117	11.651	13.716	15.352	18.338	21.689	23.900	27.204	30.144	33.687	36.191
20	8.260	9.237	10.851	12.443	14.578	16.266	19.337	22.775	25.038	28.412	31.410	35.020	37.566
21	8.897	9.915	11.591	13.240	15.445	17.182	20.337	23.858	26.171	29.615	32.671	36.343	38.932
22	9.542	10.600	12.338	14.041	16.314	18.101	21.337	24.939	27.301	30.813	33.924	37.659	40.289
23	10.196	11.293	13.091	14.848	17.187	19.021	22.337	26.018	28.429	32.007	35.172	38.968	41.638
24	10.856	11.992	13.848	15.659	18.062	19.943	23.337	27.096	29.553	33.196	36.415	40.270	42.980
25	11.524	12.697	14.611	16.473	18.940	20.867	24.337	28.172	30.675	34.382	37.652	41.566	44.314
26	12.198	13.409	15.379	17.292	19.820	21.792	25.336	29.246	31.795	35.563	38.885	42.856	45.642
27	12.879	14.125	16.151	18.114	20.703	22.719	26.336	30.319	32.912	36.741	40.113	44.140	46.963
28	13.565	14.847	16.928	18.939	21.588	23.647	27.336	31.391	34.027	37.916	41.337	45.419	48.278
29	14.256	15.574	17.708	19.768	22.475	24.577	28.336	32.461	35.139	39.087	42.557	46.693	49.588
30	14.953	16.306	18.493	20.599	23.364	25.508	29.336	33.530	36.250	40.256	43.773	47.962	50.892

* Table IV is reprinted from Table III of Fisher: *Statistical Methods for Research Workers*, Oliver & Boyd Ltd., Edinburgh, by permission of the author and publishers.

For larger values of df, the expression $\sqrt{2\chi^2} - \sqrt{2(df) - 1}$ may be used as a normal deviate with unit standard error.

Table V. Table of t*

df \ P =	.450	.400	.350	.300	.250	.200	.150	.100	.050	.025	.010	.005
1	.158	.325	.510	.727	1.000	1.376	1.963	3.078	6.314	12.706	31.821	63.657
2	.142	.289	.445	.617	.816	1.061	1.386	1.886	2.920	4.303	6.965	9.925
3	.137	.277	.424	.584	.765	.978	1.250	1.638	2.353	3.182	4.541	5.841
4	.134	.271	.414	.569	.741	.941	1.190	1.533	2.132	2.776	3.747	4.604
5	.132	.267	.408	.559	.727	.920	1.156	1.476	2.015	2.571	3.365	4.032
6	.131	.265	.404	.553	.718	.906	1.134	1.440	1.943	2.447	3.143	3.707
7	.130	.263	.402	.549	.711	.896	1.119	1.415	1.895	2.365	2.998	3.499
8	.130	.262	.399	.546	.706	.889	1.108	1.397	1.860	2.306	2.896	3.355
9	.129	.261	.398	.543	.703	.883	1.100	1.383	1.833	2.262	2.821	3.250
10	.129	.260	.397	.542	.700	.879	1.093	1.372	1.812	2.228	2.764	3.169
11	.129	.260	.396	.540	.697	.876	1.088	1.363	1.796	2.201	2.718	3.106
12	.128	.259	.395	.539	.695	.873	1.083	1.356	1.782	2.179	2.681	3.055
13	.128	.259	.394	.538	.694	.870	1.079	1.350	1.771	2.160	2.650	3.012
14	.128	.258	.393	.537	.692	.868	1.076	1.345	1.761	2.145	2.624	2.977
15	.128	.258	.393	.536	.691	.866	1.074	1.341	1.753	2.131	2.602	2.947
16	.128	.258	.392	.535	.690	.865	1.071	1.337	1.746	2.120	2.583	2.921
17	.128	.257	.392	.534	.689	.863	1.069	1.333	1.740	2.110	2.567	2.898
18	.127	.257	.392	.534	.688	.862	1.067	1.330	1.734	2.101	2.552	2.878
19	.127	.257	.391	.533	.688	.861	1.066	1.328	1.729	2.093	2.539	2.861
20	.127	.257	.391	.533	.687	.860	1.064	1.325	1.725	2.086	2.528	2.845
21	.127	.257	.391	.532	.686	.859	1.063	1.323	1.721	2.080	2.518	2.831
22	.127	.256	.390	.532	.686	.858	1.061	1.321	1.717	2.074	2.508	2.819
23	.127	.256	.390	.532	.685	.858	1.060	1.319	1.714	2.069	2.500	2.807
24	.127	.256	.390	.531	.685	.857	1.059	1.318	1.711	2.064	2.492	2.797
25	.127	.256	.390	.531	.684	.856	1.058	1.316	1.708	2.060	2.485	2.787
26	.127	.256	.390	.531	.684	.856	1.058	1.315	1.706	2.056	2.479	2.779
27	.127	.256	.389	.531	.684	.855	1.057	1.314	1.703	2.052	2.473	2.771
28	.127	.256	.389	.530	.683	.855	1.056	1.313	1.701	2.048	2.467	2.763
29	.127	.256	.389	.530	.683	.854	1.055	1.311	1.699	2.045	2.462	2.756
30	.127	.256	.389	.530	.683	.854	1.055	1.310	1.697	2.042	2.457	2.750
∞	.12566	.25335	.38532	.52440	.67449	.84162	1.03643	1.28155	1.64485	1.95996	2.32634	2.57582

Additional Values of t at the .025 and .005 Levels of Significance†

df	.025	.005	df	.025	.005	df	.025	.005
32	2.037	2.739	55	2.005	2.668	125	1.979	2.616
34	2.032	2.728	60	2.000	2.660	150	1.976	2.609
36	2.027	2.718	65	1.998	2.653	175	1.974	2.605
38	2.025	2.711	70	1.994	2.648	200	1.972	2.601
40	2.021	2.704	75	1.992	2.643	300	1.968	2.592
42	2.017	2.696	80	1.990	2.638	400	1.966	2.588
44	2.015	2.691	85	1.989	2.635	500	1.965	2.586
46	2.012	2.685	90	1.987	2.632	1000	1.962	2.581
48	2.010	2.681	95	1.986	2.629	∞	1.960	2.576
50	2.008	2.678	100	1.984	2.626			

* Table V is reprinted from Table IV of Fisher: *Statistical Methods for Research* [illegible]*rkers*, Oliver & Boyd Ltd., Edinburgh, by permission of the author and publishers.

† Additional entries were taken from Snedecor: *Statistical Methods*, Io[illegible]ate College Press, Ames, Iowa, by permission of the author and publisher. Values for 75, 85, 95, [illegible] degrees of freedom were obtained by linear interpolation.

The probabilities given are for a one-sided test.

Table VI. Values of the Correlation Coefficient for Different Levels of Significance*

df	*P* = .050	.025	.010	.005
1	.988	.997	.9995	.9999
2	.900	.950	.980	.990
3	.805	.878	.934	.959
4	.729	.811	.882	.917
5	.669	.754	.833	.874
6	.622	.707	.789	.834
7	.582	.666	.750	.798
8	.549	.632	.716	.765
9	.521	.602	.685	.735
10	.497	.576	.658	.708
11	.476	.553	.634	.684
12	.458	.532	.612	.661
13	.441	.514	.592	.641
14	.426	.497	.574	.623
15	.412	.482	.558	.606
16	.400	.468	.542	.590
17	.389	.456	.528	.575
18	.378	.444	.516	.561
19	.369	.433	.503	.549
20	.360	.423	.492	.537
21	.352	.413	.482	.526
22	.344	.404	.472	.515
23	.337	.396	.462	.505
24	.330	.388	.453	.496
25	.323	.381	.445	.487
26	.317	.374	.437	.479
27	.311	.367	.430	.471
28	.306	.361	.423	.463
29	.301	.355	.416	.456
30	.296	.349	.409	.449
35	.275	.325	.381	.418
40	.257	.304	.358	.393
45	.243	.288	.338	.372
50	.231	.273	.322	.354
60	.211	.250	.295	.325
70	.195	.232	.274	.302
80	.183	.217	.256	.283
90	.173	.205	.242	.267
100	.164	.195	.230	.254

Additional values of *r* at the .025 and .005 Levels of Significance

df	.0[illegible]	.005	*df*	.025	.005	*df*	.025	.005
3[illegible]	.339	.436	48	.279	.361	150	.159	.208
34	.329	.424	55	.261	.338	175	.148	.193
36	.320	[illegible]413	65	.241	.313	200	.138	.181
38	.312	[illegible]3	75	.224	.292	300	.113	.148
42	.297	[illegible]	85	.211	.275	400	.098	.128
44	.291	[illegible]	95	.200	.260	500	.088	.115
46	.284	.3[illegible]	125	.174	.228	1,000	.062	.081

* Table VI is repr[illegible] Oliver & Boyd Ltd., Edinb[illegible] [illegible]n Table V.A. of R. A. Fisher, *Statistical Methods for Research Workers*,
Additional entries [illegible] [illegible]ermission of the author and publishers.
The probabilities gi[illegible] [illegible]ated using the table of *t*.
[illegible] one-sided test.

Table VII. Table of z' Values for r*

r	z'	r	z'	r	z'	r	z'	r	z'
.000	.000	.200	.203	.400	.424	.600	.693	.800	1.099
.005	.005	.205	.208	.405	.430	.605	.701	.805	1.113
.010	.010	.210	.213	.410	.436	.610	.709	.810	1.127
.015	.015	.215	.218	.415	.442	.615	.717	.815	1.142
.020	.020	.220	.224	.420	.448	.620	.725	.820	1.157
.025	.025	.225	.229	.425	.454	.625	.733	.825	1.172
.030	.030	.230	.234	.430	.460	.630	.741	.830	1.188
.035	.035	.235	.239	.435	.466	.635	.750	.835	1.204
.040	.040	.240	.245	.440	.472	.640	.758	.840	1.221
.045	.045	.245	.250	.445	.478	.645	.767	.845	1.238
.050	.050	.250	.255	.450	.485	.650	.775	.850	1.256
.055	.055	.255	.261	.455	.491	.655	.784	.855	1.274
.060	.060	.260	.266	.460	.497	.660	.793	.860	1.293
.065	.065	.265	.271	.465	.504	.665	.802	.865	1.313
.070	.070	.270	.277	.470	.510	.670	.811	.870	1.333
.075	.075	.275	.282	.475	.517	.675	.820	.875	1.354
.080	.080	.280	.288	.480	.523	.680	.829	.880	1.376
.085	.085	.285	.293	.485	.530	.685	.838	.885	1.398
.090	.090	.290	.299	.490	.536	.690	.848	.890	1.422
.095	.095	.295	.304	.495	.543	.695	.858	.895	1.447
.100	.100	.300	.310	.500	.549	.700	.867	.900	1.472
.105	.105	.305	.315	.505	.556	.705	.877	.905	1.499
.110	.110	.310	.321	.510	.563	.710	.887	.910	1.528
.115	.116	.315	.326	.515	.570	.715	.897	.915	1.557
.120	.121	.320	.332	.520	.576	.720	.908	.920	1.589
.125	.126	.325	.337	.525	.583	.725	.918	.925	1.623
.130	.131	.330	.343	.530	.590	.730	.929	.930	1.658
.135	.136	.335	.348	.535	.597	.735	.940	.935	1.697
.140	.141	.340	.354	.540	.604	.740	.950	.940	1.738
.145	.146	.345	.360	.545	.611	.745	.962	.945	1.783
.150	.151	.350	.365	.550	.618	.750	.973	.950	1.832
.155	.156	.355	.371	.555	.626	.755	.984	.955	1.886
.160	.161	.360	.377	.560	.633	.760	.996	.960	1.946
.165	.167	.365	.383	.565	.640	.765	1.008	.965	2.014
.170	.172	.370	.388	.570	.648	.770	1.020	.970	2.092
.175	.177	.375	.394	.575	.655	.775	1.033	.975	2.185
.180	.182	.380	.400	.580	.662	.780	1.045	.980	2.298
.185	.187	.385	.406	.585	.670	.785	1.058	.985	2.443
.190	.192	.390	.412	.590	.678	.790	1.071	.990	2.647
.195	.198	.395	.418	.595	.685	.795	1.085	.995	2.994

*Table VII was constructed by F. P. Kilpatrick and D. A. Buchanan.

Table VIII. The 5 (Roman Type) and 1 (Boldface Type) Percent Points for the Distribution of F*

n_2	n_1 degrees of freedom (for greater mean square)																						
	1	2	3	4	5	6	7	8	9	10	11	12	14	16	20	24	30	40	50	75	100	200	500	∞
1	161	200	216	225	230	234	237	239	241	242	243	244	245	246	248	249	250	251	252	253	253	254	254	254
	4,052	**4,999**	**5,403**	**5,625**	**5,764**	**5,859**	**5,928**	**5,981**	**6,022**	**6,056**	**6,082**	**6,106**	**6,142**	**6,169**	**6,208**	**6,234**	**6,258**	**6,286**	**6,302**	**6,323**	**6,334**	**6,352**	**6,361**	**6,366**
2	18.51	19.00	19.16	19.25	19.30	19.33	19.36	19.37	19.38	19.39	19.40	19.41	19.42	19.43	19.44	19.45	19.46	19.47	19.47	19.48	19.49	19.49	19.50	19.50
	98.49	**99.00**	**99.17**	**99.25**	**99.30**	**99.33**	**99.34**	**99.36**	**99.38**	**99.40**	**99.41**	**99.42**	**99.43**	**99.44**	**99.45**	**99.46**	**99.47**	**99.48**	**99.48**	**99.49**	**99.49**	**99.49**	**99.50**	**99.50**
3	10.13	9.55	9.28	9.12	9.01	8.94	8.88	8.84	8.81	8.78	8.76	8.74	8.71	8.69	8.66	8.64	8.62	8.60	8.58	8.57	8.56	8.54	8.54	8.53
	34.12	**30.82**	**29.46**	**28.71**	**28.24**	**27.91**	**27.67**	**27.49**	**27.34**	**27.23**	**27.13**	**27.05**	**26.92**	**26.83**	**26.69**	**26.60**	**26.50**	**26.41**	**26.35**	**26.27**	**26.23**	**26.18**	**26.14**	**26.12**
4	7.71	6.94	6.59	6.39	6.26	6.16	6.09	6.04	6.00	5.96	5.93	5.91	5.87	5.84	5.80	5.77	5.74	5.71	5.70	5.68	5.66	5.65	5.64	5.63
	21.20	**18.00**	**16.69**	**15.98**	**15.52**	**15.21**	**14.98**	**14.80**	**14.66**	**14.54**	**14.45**	**14.37**	**14.24**	**14.15**	**14.02**	**13.93**	**13.83**	**13.74**	**13.69**	**13.61**	**13.57**	**13.52**	**13.48**	**13.46**
5	6.61	5.79	5.41	5.19	5.05	4.95	4.88	4.82	4.78	4.74	4.70	4.68	4.64	4.60	4.56	4.53	4.50	4.46	4.44	4.42	4.40	4.38	4.37	4.36
	16.26	**13.27**	**12.06**	**11.39**	**10.97**	**10.67**	**10.45**	**10.27**	**10.15**	**10.05**	**9.96**	**9.89**	**9.77**	**9.68**	**9.55**	**9.47**	**9.38**	**9.29**	**9.24**	**9.17**	**9.13**	**9.07**	**9.04**	**9.02**
6	5.99	5.14	4.76	4.53	4.39	4.28	4.21	4.15	4.10	4.06	4.03	4.00	3.96	3.92	3.87	3.84	3.81	3.77	3.75	3.72	3.71	3.69	3.68	3.67
	13.74	**10.92**	**9.78**	**9.15**	**8.75**	**8.47**	**8.26**	**8.10**	**7.98**	**7.87**	**7.79**	**7.72**	**7.60**	**7.52**	**7.39**	**7.31**	**7.23**	**7.14**	**7.09**	**7.02**	**6.99**	**6.94**	**6.90**	**6.88**
7	5.59	4.74	4.35	4.12	3.97	3.87	3.79	3.73	3.68	3.63	3.60	3.57	3.52	3.49	3.44	3.41	3.38	3.34	3.32	3.29	3.28	3.25	3.24	3.23
	12.25	**9.55**	**8.45**	**7.85**	**7.46**	**7.19**	**7.00**	**6.84**	**6.71**	**6.62**	**6.54**	**6.47**	**6.35**	**6.27**	**6.15**	**6.07**	**5.98**	**5.90**	**5.85**	**5.78**	**5.75**	**5.70**	**5.67**	**5.65**
8	5.32	4.46	4.07	3.84	3.69	3.58	3.50	3.44	3.39	3.34	3.31	3.28	3.23	3.20	3.15	3.12	3.08	3.05	3.03	3.00	2.98	2.96	2.94	2.93
	11.26	**8.65**	**7.59**	**7.01**	**6.63**	**6.37**	**6.19**	**6.03**	**5.91**	**5.82**	**5.74**	**5.67**	**5.56**	**5.48**	**5.36**	**5.28**	**5.20**	**5.11**	**5.06**	**5.00**	**4.96**	**4.91**	**4.88**	**4.86**
9	5.12	4.26	3.86	3.63	3.48	3.37	3.29	3.23	3.18	3.13	3.10	3.07	3.02	2.98	2.93	2.90	2.86	2.82	2.80	2.77	2.76	2.73	2.72	2.71
	10.56	**8.02**	**6.99**	**6.42**	**6.06**	**5.80**	**5.62**	**5.47**	**5.35**	**5.26**	**5.18**	**5.11**	**5.00**	**4.92**	**4.80**	**4.73**	**4.64**	**4.56**	**4.51**	**4.45**	**4.41**	**4.36**	**4.33**	**4.31**
10	4.96	4.10	3.71	3.48	3.33	3.22	3.14	3.07	3.02	2.97	2.94	2.91	2.86	2.82	2.77	2.74	2.70	2.67	2.64	2.61	2.59	2.56	2.55	2.54
	10.04	**7.56**	**6.55**	**5.99**	**5.64**	**5.39**	**5.21**	**5.06**	**4.95**	**4.85**	**4.78**	**4.71**	**4.60**	**4.52**	**4.41**	**4.33**	**4.25**	**4.17**	**4.12**	**4.05**	**4.01**	**3.96**	**3.93**	**3.91**
11	4.84	3.98	3.59	3.36	3.20	3.09	3.01	2.95	2.90	2.86	2.82	2.79	2.74	2.70	2.65	2.61	2.57	2.53	2.50	2.47	2.45	2.42	2.41	2.40
	9.65	**7.20**	**6.22**	**5.67**	**5.32**	**5.07**	**4.88**	**4.74**	**4.63**	**4.54**	**4.46**	**4.40**	**4.29**	**4.21**	**4.10**	**4.02**	**3.94**	**3.86**	**3.80**	**3.74**	**3.70**	**3.66**	**3.62**	**3.60**
12	4.75	3.88	3.49	3.26	3.11	3.00	2.92	2.85	2.80	2.76	2.72	2.69	2.64	2.60	2.54	2.50	2.46	2.42	2.40	2.36	2.35	2.32	2.31	2.30
	9.33	**6.93**	**5.95**	**5.41**	**5.06**	**4.82**	**4.65**	**4.50**	**4.39**	**4.30**	**4.22**	**4.16**	**4.05**	**3.98**	**3.86**	**3.78**	**3.70**	**3.61**	**3.56**	**3.49**	**3.46**	**3.41**	**3.38**	**3.36**
13	4.67	3.80	3.41	3.18	3.02	2.92	2.84	2.77	2.72	2.67	2.63	2.60	2.55	2.51	2.46	2.42	2.38	2.34	2.32	2.28	2.26	2.24	2.22	2.21
	9.07	**6.70**	**5.74**	**5.20**	**4.86**	**4.62**	**4.44**	**4.30**	**4.19**	**4.10**	**4.02**	**3.96**	**3.85**	**3.78**	**3.67**	**3.59**	**3.51**	**3.42**	**3.37**	**3.30**	**3.27**	**3.21**	**3.18**	**3.16**

* Table VIII is reproduced from Snedecor: *Statistical Methods*, Iowa State College Press, Ames, Iowa, by permission of the author and publisher.

Table VIII. The 5 (Roman Type) and 1 (Boldface Type) Percent Points for the Distribution of F^*—Continued

n_2	n_1 degrees of freedom (for greater mean square)																							
	1	2	3	4	5	6	7	8	9	10	11	12	14	16	20	24	30	40	50	75	100	200	500	∞
14	4.60	3.74	3.34	3.11	2.96	2.85	2.77	2.70	2.65	2.60	2.56	2.53	2.48	2.44	2.39	2.35	2.31	2.27	2.24	2.21	2.19	2.16	2.14	2.13
	8.86	**6.51**	**5.56**	**5.03**	**4.69**	**4.46**	**4.28**	**4.14**	**4.03**	**3.94**	**3.86**	**3.80**	**3.70**	**3.62**	**3.51**	**3.43**	**3.34**	**3.26**	**3.21**	**3.14**	**3.11**	**3.06**	**3.02**	**3.00**
15	4.54	3.68	3.29	3.06	2.90	2.79	2.70	2.64	2.59	2.55	2.51	2.48	2.43	2.39	2.33	2.29	2.25	2.21	2.18	2.15	2.12	2.10	2.08	2.07
	8.68	**6.36**	**5.42**	**4.89**	**4.56**	**4.32**	**4.14**	**4.00**	**3.89**	**3.80**	**3.73**	**3.67**	**3.56**	**3.48**	**3.36**	**3.29**	**3.20**	**3.12**	**3.07**	**3.00**	**2.97**	**2.92**	**2.89**	**2.87**
16	4.49	3.63	3.24	3.01	2.85	2.74	2.66	2.59	2.54	2.49	2.45	2.42	2.37	2.33	2.28	2.24	2.20	2.16	2.13	2.09	2.07	2.04	2.02	2.01
	8.53	**6.23**	**5.29**	**4.77**	**4.44**	**4.20**	**4.03**	**3.89**	**3.78**	**3.69**	**3.61**	**3.55**	**3.45**	**3.37**	**3.25**	**3.18**	**3.10**	**3.01**	**2.96**	**2.89**	**2.86**	**2.80**	**2.77**	**2.75**
17	4.45	3.59	3.20	2.96	2.81	2.70	2.62	2.55	2.50	2.45	2.41	2.38	2.33	2.29	2.23	2.19	2.15	2.11	2.08	2.04	2.02	1.99	1.97	1.96
	8.40	**6.11**	**5.18**	**4.67**	**4.34**	**4.10**	**3.93**	**3.79**	**3.68**	**3.59**	**3.52**	**3.45**	**3.35**	**3.27**	**3.16**	**3.08**	**3.00**	**2.92**	**2.86**	**2.79**	**2.76**	**2.70**	**2.67**	**2.65**
18	4.41	3.55	3.16	2.93	2.77	2.66	2.58	2.51	2.46	2.41	2.37	2.34	2.29	2.25	2.19	2.15	2.11	2.07	2.04	2.00	1.98	1.95	1.93	1.92
	8.28	**6.01**	**5.09**	**4.58**	**4.25**	**4.01**	**3.85**	**3.71**	**3.60**	**3.51**	**3.44**	**3.37**	**3.27**	**3.19**	**3.07**	**3.00**	**2.91**	**2.83**	**2.78**	**2.71**	**2.68**	**2.62**	**2.59**	**2.57**
19	4.38	3.52	3.13	2.90	2.74	2.63	2.55	2.48	2.43	2.38	2.34	2.31	2.26	2.21	2.15	2.11	2.07	2.02	2.00	1.96	1.94	1.91	1.90	1.88
	8.18	**5.93**	**5.01**	**4.50**	**4.17**	**3.94**	**3.77**	**3.63**	**3.52**	**3.43**	**3.36**	**3.30**	**3.19**	**3.12**	**3.00**	**2.92**	**2.84**	**2.76**	**2.70**	**2.63**	**2.60**	**2.54**	**2.51**	**2.49**
20	4.35	3.49	3.10	2.87	2.71	2.60	2.52	2.45	2.40	2.35	2.31	2.28	2.23	2.18	2.12	2.08	2.04	1.99	1.96	1.92	1.90	1.87	1.85	1.84
	8.10	**5.85**	**4.94**	**4.43**	**4.10**	**3.87**	**3.71**	**3.56**	**3.45**	**3.37**	**3.30**	**3.23**	**3.13**	**3.05**	**2.94**	**2.86**	**2.77**	**2.69**	**2.63**	**2.56**	**2.53**	**2.47**	**2.44**	**2.42**
21	4.32	3.47	3.07	2.84	2.68	2.57	2.49	2.42	2.37	2.32	2.28	2.25	2.20	2.15	2.09	2.05	2.00	1.96	1.93	1.89	1.87	1.84	1.82	1.81
	8.02	**5.78**	**4.87**	**4.37**	**4.04**	**3.81**	**3.65**	**3.51**	**3.40**	**3.31**	**3.24**	**3.17**	**3.07**	**2.99**	**2.88**	**2.80**	**2.72**	**2.63**	**2.58**	**2.51**	**2.47**	**2.42**	**2.38**	**2.36**
22	4.30	3.44	3.05	2.82	2.66	2.55	2.47	2.40	2.35	2.30	2.26	2.23	2.18	2.13	2.07	2.03	1.98	1.93	1.91	1.87	1.84	1.81	1.80	1.78
	7.94	**5.72**	**4.82**	**4.31**	**3.99**	**3.76**	**3.59**	**3.45**	**3.35**	**3.26**	**3.18**	**3.12**	**3.02**	**2.94**	**2.83**	**2.75**	**2.67**	**2.58**	**2.53**	**2.46**	**2.42**	**2.37**	**2.33**	**2.31**
23	4.28	3.42	3.03	2.80	2.64	2.53	2.45	2.38	2.32	2.28	2.24	2.20	2.14	2.10	2.04	2.00	1.96	1.91	1.88	1.84	1.82	1.79	1.77	1.76
	7.88	**5.66**	**4.76**	**4.26**	**3.94**	**3.71**	**3.54**	**3.41**	**3.30**	**3.21**	**3.14**	**3.07**	**2.97**	**2.89**	**2.78**	**2.70**	**2.62**	**2.53**	**2.48**	**2.41**	**2.37**	**2.32**	**2.28**	**2.26**
24	4.26	3.40	3.01	2.78	2.62	2.51	2.43	2.36	2.30	2.26	2.22	2.18	2.13	2.09	2.02	1.98	1.94	1.89	1.86	1.82	1.80	1.76	1.74	1.73
	7.82	**5.61**	**4.72**	**4.22**	**3.90**	**3.67**	**3.50**	**3.36**	**3.25**	**3.17**	**3.09**	**3.03**	**2.93**	**2.85**	**2.74**	**2.66**	**2.58**	**2.49**	**2.44**	**2.36**	**2.33**	**2.27**	**2.23**	**2.21**
25	4.24	3.38	2.99	2.76	2.60	2.49	2.41	2.34	2.28	2.24	2.20	2.16	2.11	2.06	2.00	1.96	1.92	1.87	1.84	1.80	1.77	1.74	1.72	1.71
	7.77	**5.57**	**4.68**	**4.18**	**3.86**	**3.63**	**3.46**	**3.32**	**3.21**	**3.13**	**3.05**	**2.99**	**2.89**	**2.81**	**2.70**	**2.62**	**2.54**	**2.45**	**2.40**	**2.32**	**2.29**	**2.23**	**2.19**	**2.17**
26	4.22	3.37	2.98	2.74	2.59	2.47	2.39	2.32	2.27	2.22	2.18	2.15	2.10	2.05	1.99	1.95	1.90	1.85	1.82	1.78	1.76	1.72	1.70	1.69
	7.72	**5.53**	**4.64**	**4.14**	**3.82**	**3.59**	**3.42**	**3.29**	**3.17**	**3.09**	**3.02**	**2.96**	**2.86**	**2.77**	**2.66**	**2.58**	**2.50**	**2.41**	**2.36**	**2.28**	**2.25**	**2.19**	**2.15**	**2.13**

Table VIII. The 5 (Roman Type) and 1 (Boldface Type) Percent Points for the Distribution of F*—Continued

n_2	n_1 degrees of freedom (for greater mean square)																							
	1	2	3	4	5	6	7	8	9	10	11	12	14	16	20	24	30	40	50	75	100	200	500	∞
27	4.21	3.35	2.96	2.73	2.57	2.46	2.37	2.30	2.25	2.20	2.16	2.13	2.08	2.03	1.97	1.93	1.88	1.84	1.80	1.76	1.74	1.71	1.68	1.67
	7.68	**5.49**	**4.60**	**4.11**	**3.79**	**3.56**	**3.39**	**3.26**	**3.14**	**3.06**	**2.98**	**2.93**	**2.83**	**2.74**	**2.63**	**2.55**	**2.47**	**2.38**	**2.33**	**2.25**	**2.21**	**2.16**	**2.12**	**2.10**
28	4.20	3.34	2.95	2.71	2.56	2.44	2.36	2.29	2.24	2.19	2.15	2.12	2.06	2.02	1.96	1.91	1.87	1.81	1.78	1.75	1.72	1.69	1.67	1.65
	7.64	**5.45**	**4.57**	**4.07**	**3.76**	**3.53**	**3.36**	**3.23**	**3.11**	**3.03**	**2.95**	**2.90**	**2.80**	**2.71**	**2.60**	**2.52**	**2.44**	**2.35**	**2.30**	**2.22**	**2.18**	**2.13**	**2.09**	**2.06**
29	4.18	3.33	2.93	2.70	2.54	2.43	2.35	2.28	2.22	2.18	2.14	2.10	2.05	2.00	1.94	1.90	1.85	1.80	1.77	1.73	1.71	1.68	1.65	1.64
	7.60	**5.42**	**4.54**	**4.04**	**3.73**	**3.50**	**3.33**	**3.20**	**3.08**	**3.00**	**2.92**	**2.87**	**2.77**	**2.68**	**2.57**	**2.49**	**2.41**	**2.32**	**2.27**	**2.19**	**2.15**	**2.10**	**2.06**	**2.03**
30	4.17	3.32	2.92	2.69	2.53	2.42	2.34	2.27	2.21	2.16	2.12	2.09	2.04	1.99	1.93	1.89	1.84	1.79	1.76	1.72	1.69	1.66	1.64	1.62
	7.56	**5.39**	**4.51**	**4.02**	**3.70**	**3.47**	**3.30**	**3.17**	**3.06**	**2.98**	**2.90**	**2.84**	**2.74**	**2.66**	**2.55**	**2.47**	**2.38**	**2.29**	**2.24**	**2.16**	**2.13**	**2.07**	**2.03**	**2.01**
32	4.15	3.30	2.90	2.67	2.51	2.40	2.32	2.25	2.19	2.14	2.10	2.07	2.02	1.97	1.91	1.86	1.82	1.76	1.74	1.69	1.67	1.64	1.61	1.59
	7.50	**5.34**	**4.46**	**3.97**	**3.66**	**3.42**	**3.25**	**3.12**	**3.01**	**2.94**	**2.86**	**2.80**	**2.70**	**2.62**	**2.51**	**2.42**	**2.34**	**2.25**	**2.20**	**2.12**	**2.08**	**2.02**	**1.98**	**1.96**
34	4.13	3.28	2.88	2.65	2.49	2.38	2.30	2.23	2.17	2.12	2.08	2.05	2.00	1.95	1.89	1.84	1.80	1.74	1.71	1.67	1.64	1.61	1.59	1.57
	7.44	**5.29**	**4.42**	**3.93**	**3.61**	**3.38**	**3.21**	**3.08**	**2.97**	**2.89**	**2.82**	**2.76**	**2.66**	**2.58**	**2.47**	**2.38**	**2.30**	**2.21**	**2.15**	**2.08**	**2.04**	**1.98**	**1.94**	**1.91**
36	4.11	3.26	2.86	2.63	2.48	2.36	2.28	2.21	2.15	2.10	2.06	2.03	1.98	1.93	1.87	1.82	1.78	1.72	1.69	1.65	1.62	1.59	1.56	1.55
	7.39	**5.25**	**4.38**	**3.89**	**3.58**	**3.35**	**3.18**	**3.04**	**2.94**	**2.86**	**2.78**	**2.72**	**2.62**	**2.54**	**2.43**	**2.35**	**2.26**	**2.17**	**2.12**	**2.04**	**2.00**	**1.94**	**1.90**	**1.87**
38	4.10	3.25	2.85	2.62	2.46	2.35	2.26	2.19	2.14	2.09	2.05	2.02	1.96	1.92	1.85	1.80	1.76	1.71	1.67	1.63	1.60	1.57	1.54	1.53
	7.35	**5.21**	**4.34**	**3.86**	**3.54**	**3.32**	**3.15**	**3.02**	**2.91**	**2.82**	**2.75**	**2.69**	**2.59**	**2.51**	**2.40**	**2.32**	**2.22**	**2.14**	**2.08**	**2.00**	**1.97**	**1.90**	**1.86**	**1.84**
40	4.08	3.23	2.84	2.61	2.45	2.34	2.25	2.18	2.12	2.07	2.04	2.00	1.95	1.90	1.84	1.79	1.74	1.69	1.66	1.61	1.59	1.55	1.53	1.51
	7.31	**5.18**	**4.31**	**3.83**	**3.51**	**3.29**	**3.12**	**2.99**	**2.88**	**2.80**	**2.73**	**2.66**	**2.56**	**2.49**	**2.37**	**2.29**	**2.20**	**2.11**	**2.05**	**1.97**	**1.94**	**1.88**	**1.84**	**1.81**
42	4.07	3.22	2.83	2.59	2.44	2.32	2.24	2.17	2.11	2.06	2.02	1.99	1.94	1.89	1.82	1.78	1.73	1.68	1.64	1.60	1.57	1.54	1.51	1.49
	7.27	**5.15**	**4.29**	**3.80**	**3.49**	**3.26**	**3.10**	**2.96**	**2.86**	**2.77**	**2.70**	**2.64**	**2.54**	**2.46**	**2.35**	**2.26**	**2.17**	**2.08**	**2.02**	**1.94**	**1.91**	**1.85**	**1.80**	**1.78**
44	4.06	3.21	2.82	2.58	2.43	2.31	2.23	2.16	2.10	2.05	2.01	1.98	1.92	1.88	1.81	1.76	1.72	1.66	1.63	1.58	1.56	1.52	1.50	1.48
	7.24	**5.12**	**4.26**	**3.78**	**3.46**	**3.24**	**3.07**	**2.94**	**2.84**	**2.75**	**2.68**	**2.62**	**2.52**	**2.44**	**2.32**	**2.24**	**2.15**	**2.06**	**2.00**	**1.92**	**1.88**	**1.82**	**1.78**	**1.75**
46	4.05	3.20	2.81	2.57	2.42	2.30	2.22	2.14	2.09	2.04	2.00	1.97	1.91	1.87	1.80	1.75	1.71	1.65	1.62	1.57	1.54	1.51	1.48	1.46
	7.21	**5.10**	**4.24**	**3.76**	**3.44**	**3.22**	**3.05**	**2.92**	**2.82**	**2.73**	**2.66**	**2.60**	**2.50**	**2.42**	**2.30**	**2.22**	**2.13**	**2.04**	**1.98**	**1.90**	**1.86**	**1.80**	**1.76**	**1.72**
48	4.04	3.19	2.80	2.56	2.41	2.30	2.21	2.14	2.08	2.03	1.99	1.96	1.90	1.86	1.79	1.74	1.70	1.64	1.61	1.56	1.53	1.50	1.47	1.45
	7.19	**5.08**	**4.22**	**3.74**	**3.42**	**3.20**	**3.04**	**2.90**	**2.80**	**2.71**	**2.64**	**2.58**	**2.48**	**2.40**	**2.28**	**2.20**	**2.11**	**2.02**	**1.96**	**1.88**	**1.84**	**1.78**	**1.73**	**1.70**

* Table VIII is reproduced from Snedecor: *Statistical Methods*, Iowa State College Press, Ames, Iowa, by permission of the author and publisher.

Table VIII. The 5 (Roman Type) and 1 (Boldface Type) Percent Points for the Distribution of F*—Concluded

n_2	n_1 degrees of freedom (for greater mean square)																							
	1	2	3	4	5	6	7	8	9	10	11	12	14	16	20	24	30	40	50	75	100	200	500	∞
50	4.03	3.18	2.79	2.56	2.40	2.29	2.20	2.13	2.07	2.02	1.98	1.95	1.90	1.85	1.78	1.74	1.69	1.63	1.60	1.55	1.52	1.48	1.46	1.44
	7.17	**5.06**	**4.20**	**3.72**	**3.41**	**3.18**	**3.02**	**2.88**	**2.78**	**2.70**	**2.62**	**2.56**	**2.46**	**2.39**	**2.26**	**2.18**	**2.10**	**2.00**	**1.94**	**1.86**	**1.82**	**1.76**	**1.71**	**1.68**
55	4.02	3.17	2.78	2.54	2.38	2.27	2.18	2.11	2.05	2.00	1.97	1.93	1.88	1.83	1.76	1.72	1.67	1.61	1.58	1.52	1.50	1.46	1.43	1.41
	7.12	**5.01**	**4.16**	**3.68**	**3.37**	**3.15**	**2.98**	**2.85**	**2.75**	**2.66**	**2.59**	**2.53**	**2.43**	**2.35**	**2.23**	**2.15**	**2.06**	**1.96**	**1.90**	**1.82**	**1.78**	**1.71**	**1.66**	**1.64**
60	4.00	3.15	2.76	2.52	2.37	2.25	2.17	2.10	2.04	1.99	1.95	1.92	1.86	1.81	1.75	1.70	1.65	1.59	1.56	1.50	1.48	1.44	1.41	1.39
	7.08	**4.98**	**4.13**	**3.65**	**3.34**	**3.12**	**2.95**	**2.82**	**2.72**	**2.63**	**2.56**	**2.50**	**2.40**	**2.32**	**2.20**	**2.12**	**2.03**	**1.93**	**1.87**	**1.79**	**1.74**	**1.68**	**1.63**	**1.60**
65	3.99	3.14	2.75	2.51	2.36	2.24	2.15	2.08	2.02	1.98	1.94	1.90	1.85	1.80	1.73	1.68	1.63	1.57	1.54	1.49	1.46	1.42	1.39	1.37
	7.04	**4.95**	**4.10**	**3.62**	**3.31**	**3.09**	**2.93**	**2.79**	**2.70**	**2.61**	**2.54**	**2.47**	**2.37**	**2.30**	**2.18**	**2.09**	**2.00**	**1.90**	**1.84**	**1.76**	**1.71**	**1.64**	**1.60**	**1.56**
70	3.98	3.13	2.74	2.50	2.35	2.23	2.14	2.07	2.01	1.97	1.93	1.89	1.84	1.79	1.72	1.67	1.62	1.56	1.53	1.47	1.45	1.40	1.37	1.35
	7.01	**4.92**	**4.08**	**3.60**	**3.29**	**3.07**	**2.91**	**2.77**	**2.67**	**2.59**	**2.51**	**2.45**	**2.35**	**2.28**	**2.15**	**2.07**	**1.98**	**1.88**	**1.82**	**1.74**	**1.69**	**1.62**	**1.56**	**1.53**
80	3.96	3.11	2.72	2.48	2.33	2.21	2.12	2.05	1.99	1.95	1.91	1.88	1.82	1.77	1.70	1.65	1.60	1.54	1.51	1.45	1.42	1.38	1.35	1.32
	6.96	**4.88**	**4.04**	**3.56**	**3.25**	**3.04**	**2.87**	**2.74**	**2.64**	**2.55**	**2.48**	**2.41**	**2.32**	**2.24**	**2.11**	**2.03**	**1.94**	**1.84**	**1.78**	**1.70**	**1.65**	**1.57**	**1.52**	**1.49**
100	3.94	3.09	2.70	2.46	2.30	2.19	2.10	2.03	1.97	1.92	1.88	1.85	1.79	1.75	1.68	1.63	1.57	1.51	1.48	1.42	1.39	1.34	1.30	1.28
	6.90	**4.82**	**3.98**	**3.51**	**3.20**	**2.99**	**2.82**	**2.69**	**2.59**	**2.51**	**2.43**	**2.36**	**2.26**	**2.19**	**2.06**	**1.98**	**1.89**	**1.79**	**1.73**	**1.64**	**1.59**	**1.51**	**1.46**	**1.43**
125	3.92	3.07	2.68	2.44	2.29	2.17	2.08	2.01	1.95	1.90	1.86	1.83	1.77	1.72	1.65	1.60	1.55	1.49	1.45	1.39	1.36	1.31	1.27	1.25
	6.84	**4.78**	**3.94**	**3.47**	**3.17**	**2.95**	**2.79**	**2.65**	**2.56**	**2.47**	**2.40**	**2.33**	**2.23**	**2.15**	**2.03**	**1.94**	**1.85**	**1.75**	**1.68**	**1.59**	**1.54**	**1.46**	**1.40**	**1.37**
150	3.91	3.06	2.67	2.43	2.27	2.16	2.07	2.00	1.94	1.89	1.85	1.82	1.76	1.71	1.64	1.59	1.54	1.47	1.44	1.37	1.34	1.29	1.25	1.22
	6.81	**4.75**	**3.91**	**3.44**	**3.14**	**2.92**	**2.76**	**2.62**	**2.53**	**2.44**	**2.37**	**2.30**	**2.20**	**2.12**	**2.00**	**1.91**	**1.83**	**1.72**	**1.66**	**1.56**	**1.51**	**1.43**	**1.37**	**1.33**
200	3.89	3.04	2.65	2.41	2.26	2.14	2.05	1.98	1.92	1.87	1.83	1.80	1.74	1.69	1.62	1.57	1.52	1.45	1.42	1.35	1.32	1.26	1.22	1.19
	6.76	**4.71**	**3.88**	**3.41**	**3.11**	**2.90**	**2.73**	**2.60**	**2.50**	**2.41**	**2.34**	**2.28**	**2.17**	**2.09**	**1.97**	**1.88**	**1.79**	**1.69**	**1.62**	**1.53**	**1.48**	**1.39**	**1.33**	**1.28**
400	3.86	3.02	2.62	2.39	2.23	2.12	2.03	1.96	1.90	1.85	1.81	1.78	1.72	1.67	1.60	1.54	1.49	1.42	1.38	1.32	1.28	1.22	1.16	1.13
	6.70	**4.66**	**3.83**	**3.36**	**3.06**	**2.85**	**2.69**	**2.55**	**2.46**	**2.37**	**2.29**	**2.23**	**2.12**	**2.04**	**1.92**	**1.84**	**1.74**	**1.64**	**1.57**	**1.47**	**1.42**	**1.32**	**1.24**	**1.19**
1000	3.85	3.00	2.61	2.38	2.22	2.10	2.02	1.95	1.89	1.84	1.80	1.76	1.70	1.65	1.58	1.53	1.47	1.41	1.36	1.30	1.26	1.19	1.13	1.08
	6.66	**4.62**	**3.80**	**3.34**	**3.04**	**2.82**	**2.66**	**2.53**	**2.43**	**2.34**	**2.26**	**2.20**	**2.09**	**2.01**	**1.89**	**1.81**	**1.71**	**1.61**	**1.54**	**1.44**	**1.38**	**1.28**	**1.19**	**1.11**
∞	3.84	2.99	2.60	2.37	2.21	2.09	2.01	1.94	1.88	1.83	1.79	1.75	1.69	1.64	1.57	1.52	1.46	1.40	1.35	1.28	1.24	1.17	1.11	1.00
	6.64	**4.60**	**3.78**	**3.32**	**3.02**	**2.80**	**2.64**	**2.51**	**2.41**	**2.32**	**2.24**	**2.18**	**2.07**	**1.99**	**1.87**	**1.79**	**1.69**	**1.59**	**1.52**	**1.41**	**1.36**	**1.25**	**1.15**	**1.00**

* Table VIII is reproduced from Snedecor: *Statistical Methods*, Iowa State College Press, Ames, Iowa, by permission of the author and publisher.

Table IX. The 25, 10, 2.5, and 0.5 Percent Points for the Distribution of F*

n_2	P	n_1 Degrees of Freedom (for greater mean square)																		
		1	2	3	4	5	6	7	8	9	10	12	15	20	24	30	40	60	120	∞
1	.250	5.83	7.50	8.20	8.58	8.82	8.98	9.10	9.19	9.26	9.32	9.41	9.49	9.58	9.63	9.67	9.71	9.76	9.80	9.85
	.100	39.86	49.50	53.59	55.83	57.24	58.20	58.91	59.44	59.86	60.20	60.70	61.22	61.74	62.00	62.26	62.53	62.79	63.06	63.33
	.025	648	800	864	900	922	937	948	957	963	969	977	985	993	997	1,001	1,006	1,010	1,014	1,018
	.005	16,211	20,000	21,615	22,500	23,056	23,437	23,715	23,925	24,091	24,224	24,426	24,630	24,836	24,940	25,044	25,148	25,253	25,359	25,465
2	.250	2.57	3.00	3.15	3.23	3.28	3.31	3.34	3.35	3.37	3.38	3.39	3.41	3.43	3.43	3.44	3.45	3.46	3.47	3.48
	.100	8.53	9.00	9.16	9.24	9.29	9.33	9.35	9.37	9.38	9.39	9.41	9.42	9.44	9.45	9.46	9.47	9.47	9.48	9.49
	.025	38.51	39.00	39.16	39.25	39.30	39.33	39.36	39.37	39.39	39.40	39.42	39.43	39.45	39.46	39.46	39.47	39.48	39.49	39.50
	.005	198	199	199	199	199	199	199	199	199	199	199	199	199	199	199	199	199	199	200
3	.250	2.02	2.28	2.36	2.39	2.41	2.42	2.43	2.44	2.44	2.44	2.45	2.46	2.46	2.46	2.46	2.47	2.47	2.47	2.47
	.100	5.54	5.46	5.39	5.34	5.31	5.28	5.27	5.25	5.24	5.23	5.22	5.20	5.18	5.18	5.17	5.16	5.15	5.14	5.13
	.025	17.44	16.04	15.44	15.10	14.88	14.74	14.62	14.54	14.47	14.42	14.34	14.25	14.17	14.12	14.08	14.04	13.99	13.95	13.90
	.005	55.55	49.80	47.47	46.20	45.39	44.84	44.43	44.13	43.88	43.69	43.39	43.08	42.78	42.62	42.47	42.31	42.15	41.99	41.83
4	.250	1.81	2.00	2.05	2.06	2.07	2.08	2.08	2.08	2.08	2.08	2.08	2.08	2.08	2.08	2.08	2.08	2.08	2.08	2.08
	.100	4.54	4.32	4.19	4.11	4.05	4.01	3.98	3.95	3.94	3.92	3.90	3.87	3.84	3.83	3.82	3.80	3.79	3.78	3.76
	.025	12.22	10.65	9.98	9.60	9.36	9.20	9.07	8.98	8.90	8.84	8.75	8.66	8.56	8.51	8.46	8.41	8.36	8.31	8.26
	.005	31.33	26.28	24.26	23.16	22.46	21.98	21.62	21.35	21.14	20.97	20.70	20.44	20.17	20.03	19.89	19.75	19.61	19.47	19.32
5	.250	1.69	1.85	1.88	1.89	1.89	1.89	1.89	1.89	1.89	1.89	1.89	1.89	1.88	1.88	1.88	1.88	1.87	1.87	1.87
	.100	4.06	3.78	3.62	3.52	3.45	3.40	3.37	3.34	3.32	3.30	3.27	3.24	3.21	3.19	3.17	3.16	3.14	3.12	3.10
	.025	10.01	8.43	7.76	7.39	7.15	6.98	6.85	6.76	6.68	6.62	6.52	6.43	6.33	6.28	6.23	6.18	6.12	6.07	6.02
	.005	22.78	18.31	16.53	15.56	14.94	14.51	14.20	13.96	13.77	13.62	13.38	13.15	12.90	12.78	12.66	12.53	12.40	12.27	12.14
6	.250	1.62	1.76	1.78	1.79	1.79	1.78	1.78	1.78	1.77	1.77	1.77	1.76	1.76	1.75	1.75	1.75	1.74	1.74	1.74
	.100	3.78	3.46	3.29	3.18	3.11	3.05	3.01	2.98	2.96	2.94	2.90	2.87	2.84	2.82	2.80	2.78	2.76	2.74	2.72
	.025	8.81	7.26	6.60	6.23	5.99	5.82	5.70	5.60	5.52	5.46	5.37	5.27	5.17	5.12	5.07	5.01	4.96	4.90	4.85
	.005	18.64	14.54	12.92	12.03	11.46	11.07	10.79	10.57	10.39	10.25	10.03	9.81	9.59	9.47	9.36	9.24	9.12	9.00	8.88
7	.250	1.57	1.70	1.72	1.72	1.71	1.71	1.70	1.70	1.69	1.69	1.68	1.68	1.67	1.67	1.66	1.66	1.65	1.65	1.65
	.100	3.59	3.26	3.07	2.96	2.88	2.83	2.78	2.75	2.72	2.70	2.67	2.63	2.59	2.58	2.56	2.54	2.51	2.49	2.47
	.025	8.07	6.54	5.89	5.52	5.29	5.12	4.99	4.90	4.82	4.76	4.67	4.57	4.47	4.42	4.36	4.31	4.25	4.20	4.14
	.005	16.24	12.40	10.88	10.05	9.52	9.16	8.89	8.68	8.51	8.38	8.18	7.97	7.75	7.64	7.53	7.42	7.31	7.19	7.08
8	.250	1.54	1.66	1.67	1.66	1.66	1.65	1.64	1.64	1.64	1.63	1.62	1.62	1.61	1.60	1.60	1.59	1.59	1.58	1.58
	.100	3.46	3.11	2.92	2.81	2.73	2.67	2.62	2.59	2.56	2.54	2.50	2.46	2.42	2.40	2.38	2.36	2.34	2.32	2.29
	.025	7.57	6.06	5.42	5.05	4.82	4.65	4.53	4.43	4.36	4.30	4.20	4.10	4.00	3.95	3.89	3.84	3.78	3.73	3.67
	.005	14.69	11.04	9.60	8.81	8.30	7.95	7.69	7.50	7.34	7.21	7.01	6.81	6.61	6.50	6.40	6.29	6.18	6.06	5.95

* Table IX is reprinted from Maxine Merrington and Catherine M. Thompson: Tables of percentage points of the inverted beta (F) distribution, *Biometrika*, 1943, **33**, 73–78, by permission of the authors and *Biometrika*.

Table IX. The 25, 10, 2.5, and 0.5 Percent Points for the Distribution of F*—Continued

n_2	P	n_1 Degrees of Freedom (for greater mean square) 1	2	3	4	5	6	7	8	9	10	12	15	20	24	30	40	60	120	∞
9	.250	1.51	1.62	1.63	1.63	1.62	1.61	1.60	1.60	1.59	1.59	1.58	1.57	1.56	1.56	1.55	1.54	1.54	1.53	1.53
	.100	3.36	3.01	2.81	2.69	2.61	2.55	2.51	2.47	2.44	2.42	2.38	2.34	2.30	2.28	2.25	2.23	2.21	2.18	2.16
	.025	7.21	5.71	5.08	4.72	4.48	4.32	4.20	4.10	4.03	3.96	3.87	3.77	3.67	3.61	3.56	3.51	3.45	3.39	3.33
	.005	13.61	10.11	8.72	7.96	7.47	7.13	6.88	6.69	6.54	6.42	6.23	6.03	5.83	5.73	5.62	5.52	5.41	5.30	5.19
10	.250	1.49	1.60	1.60	1.59	1.59	1.58	1.57	1.56	1.56	1.55	1.54	1.53	1.52	1.52	1.51	1.51	1.50	1.49	1.48
	.100	3.28	2.92	2.73	2.61	2.52	2.46	2.41	2.38	2.35	2.32	2.28	2.24	2.20	2.18	2.16	2.13	2.11	2.08	2.06
	.025	6.94	5.46	4.83	4.47	4.24	4.07	3.95	3.85	3.78	3.72	3.62	3.52	3.42	3.37	3.31	3.26	3.20	3.14	3.08
	.005	12.83	9.43	8.08	7.34	6.87	6.54	6.30	6.12	5.97	5.85	5.66	5.47	5.27	5.17	5.07	4.97	4.86	4.75	4.64
11	.250	1.47	1.58	1.58	1.57	1.56	1.55	1.54	1.53	1.53	1.52	1.51	1.50	1.49	1.49	1.48	1.47	1.47	1.46	1.45
	.100	3.23	2.86	2.66	2.54	2.45	2.39	2.34	2.30	2.27	2.25	2.21	2.17	2.12	2.10	2.08	2.05	2.03	2.00	1.97
	.025	6.72	5.26	4.63	4.28	4.04	3.88	3.76	3.66	3.59	3.53	3.43	3.33	3.23	3.17	3.12	3.06	3.00	2.94	2.88
	.005	12.23	8.91	7.60	6.88	6.42	6.10	5.86	5.68	5.54	5.42	5.24	5.05	4.86	4.76	4.65	4.55	4.44	4.34	4.23
12	.250	1.46	1.56	1.56	1.55	1.54	1.53	1.52	1.51	1.51	1.50	1.49	1.48	1.47	1.46	1.45	1.45	1.44	1.43	1.42
	.100	3.18	2.81	2.61	2.48	2.39	2.33	2.28	2.24	2.21	2.19	2.15	2.10	2.06	2.04	2.01	1.99	1.96	1.93	1.90
	.025	6.55	5.10	4.47	4.12	3.89	3.73	3.61	3.51	3.44	3.37	3.28	3.18	3.07	3.02	2.96	2.91	2.85	2.79	2.72
	.005	11.75	8.51	7.23	6.52	6.07	5.76	5.52	5.35	5.20	5.09	4.91	4.72	4.53	4.43	4.33	4.23	4.12	4.01	3.90
13	.250	1.45	1.55	1.55	1.53	1.52	1.51	1.50	1.49	1.49	1.48	1.47	1.46	1.45	1.44	1.43	1.42	1.42	1.41	1.40
	.100	3.14	2.76	2.56	2.43	2.35	2.28	2.23	2.20	2.16	2.14	2.10	2.05	2.01	1.98	1.96	1.93	1.90	1.88	1.85
	.025	6.41	4.97	4.35	4.00	3.77	3.60	3.48	3.39	3.31	3.25	3.15	3.05	2.95	2.89	2.84	2.78	2.72	2.66	2.60
	.005	11.37	8.19	6.93	6.23	5.79	5.48	5.25	5.08	4.94	4.82	4.64	4.46	4.27	4.17	4.07	3.97	3.87	3.76	3.65
14	.250	1.44	1.53	1.53	1.52	1.51	1.50	1.49	1.48	1.47	1.46	1.45	1.44	1.43	1.42	1.41	1.41	1.40	1.39	1.38
	.100	3.10	2.73	2.52	2.39	2.31	2.24	2.19	2.15	2.12	2.10	2.05	2.01	1.96	1.94	1.91	1.89	1.86	1.83	1.80
	.025	6.30	4.86	4.24	3.89	3.66	3.50	3.38	3.29	3.21	3.15	3.05	2.95	2.84	2.79	2.73	2.67	2.61	2.55	2.49
	.005	11.06	7.92	6.68	6.00	5.56	5.26	5.03	4.86	4.72	4.60	4.43	4.25	4.06	3.96	3.86	3.76	3.66	3.55	3.44
15	.250	1.43	1.52	1.52	1.51	1.49	1.48	1.47	1.46	1.46	1.45	1.44	1.43	1.41	1.41	1.40	1.39	1.38	1.37	1.36
	.100	3.07	2.70	2.49	2.36	2.27	2.21	2.16	2.12	2.09	2.06	2.02	1.97	1.92	1.90	1.87	1.85	1.82	1.79	1.76
	.025	6.20	4.76	4.15	3.80	3.58	3.41	3.29	3.20	3.12	3.06	2.96	2.86	2.76	2.70	2.64	2.58	2.52	2.46	2.40
	.005	10.80	7.70	6.48	5.80	5.37	5.07	4.85	4.67	4.54	4.42	4.25	4.07	3.88	3.79	3.69	3.58	3.48	3.37	3.26
16	.250	1.42	1.51	1.51	1.50	1.48	1.47	1.46	1.45	1.44	1.44	1.43	1.41	1.40	1.39	1.38	1.37	1.36	1.35	1.34
	.100	3.05	2.67	2.46	2.33	2.24	2.18	2.13	2.09	2.06	2.03	1.99	1.94	1.89	1.87	1.84	1.81	1.78	1.75	1.72
	.025	6.12	4.69	4.08	3.73	3.50	3.34	3.22	3.12	3.05	2.99	2.89	2.79	2.68	2.63	2.57	2.51	2.45	2.38	2.32
	.005	10.58	7.51	6.30	5.64	5.21	4.91	4.69	4.52	4.38	4.27	4.10	3.92	3.73	3.64	3.54	3.44	3.33	3.22	3.11
17	.250	1.42	1.51	1.50	1.49	1.47	1.46	1.45	1.44	1.43	1.43	1.41	1.40	1.39	1.38	1.37	1.36	1.35	1.34	1.33
	.100	3.03	2.64	2.44	2.31	2.22	2.15	2.10	2.06	2.03	2.00	1.96	1.91	1.86	1.84	1.81	1.78	1.75	1.72	1.69
	.025	6.04	4.62	4.01	3.66	3.44	3.28	3.16	3.06	2.98	2.92	2.82	2.72	2.62	2.56	2.50	2.44	2.38	2.32	2.25
	.005	10.38	7.35	6.16	5.50	5.07	4.78	4.56	4.39	4.25	4.14	3.97	3.79	3.61	3.51	3.41	3.31	3.21	3.10	2.98

* Table IX is reprinted from Maxine Merrington and Catherine M. Thompson: Tables of percentage points of the inverted beta (F) distribution, *Biometrika*, 1943, **33**, 73–78, by permission of the authors and *Biometrika*.

Table IX. The 25, 10, 2.5, and 0.5 Percent Points for the Distribution of F^*—Continued

n_2	P	n_1 Degrees of Freedom (for greater mean square)																		
		1	2	3	4	5	6	7	8	9	10	12	15	20	24	30	40	60	120	∞
18	.250	1.41	1.50	1.49	1.48	1.46	1.45	1.44	1.43	1.42	1.42	1.40	1.39	1.38	1.37	1.36	1.35	1.34	1.33	1.32
	.100	3.01	2.62	2.42	2.29	2.20	2.13	2.08	2.04	2.00	1.98	1.93	1.89	1.84	1.81	1.78	1.75	1.72	1.69	1.66
	.025	5.98	4.56	3.95	3.61	3.38	3.22	3.10	3.01	2.93	2.87	2.77	2.67	2.56	2.50	2.44	2.38	2.32	2.26	2.19
	.005	10.22	7.21	6.03	5.37	4.96	4.66	4.44	4.28	4.14	4.03	3.86	3.68	3.50	3.40	3.30	3.20	3.10	2.99	2.87
19	.250	1.41	1.49	1.49	1.47	1.46	1.44	4.43	1.42	1.41	1.41	1.40	1.38	1.37	1.36	1.35	1.34	1.33	1.32	1.30
	.100	2.99	2.61	2.40	2.27	2.18	2.11	2.06	2.02	1.98	1.96	1.91	1.86	1.81	1.79	1.76	1.73	1.70	1.67	1.63
	.025	5.92	4.51	3.90	3.56	3.33	3.17	3.05	2.96	2.88	2.82	2.72	2.62	2.51	2.45	2.39	2.33	2.27	2.20	2.13
	.005	10.07	7.09	5.92	5.27	4.85	4.56	4.34	4.18	4.04	3.93	3.76	3.59	3.40	3.31	3.21	3.11	3.00	2.89	2.78
20	.250	1.40	1.49	1.48	1.47	1.45	1.44	1.43	1.42	1.41	1.40	1.39	1.37	1.36	1.35	1.34	1.33	1.32	1.31	1.29
	.100	2.97	2.59	2.38	2.25	2.16	2.09	2.04	2.00	1.96	1.94	1.89	1.84	1.79	1.77	1.74	1.71	1.68	1.64	1.61
	.025	5.87	4.46	3.86	3.51	3.29	3.13	3.01	2.91	2.84	2.77	2.68	2.57	2.46	2.41	2.35	2.29	2.22	2.16	2.09
	.005	9.94	6.99	5.82	5.17	4.76	4.47	4.26	4.09	3.96	3.85	3.68	3.50	3.32	3.22	3.12	3.02	2.92	2.81	2.69
21	.250	1.40	1.48	1.48	1.46	1.44	1.43	1.42	1.41	1.40	1.39	1.38	1.37	1.35	1.34	1.33	1.32	1.31	1.30	1.28
	.100	2.96	2.57	2.36	2.23	2.14	2.08	2.02	1.98	1.95	1.92	1.88	1.83	1.78	1.75	1.72	1.69	1.66	1.62	1.59
	.025	5.83	4.42	3.82	3.48	3.25	3.09	2.97	2.87	2.80	2.73	2.64	2.53	2.42	2.37	2.31	2.25	2.18	2.11	2.04
	.005	9.83	6.89	5.73	5.09	4.68	4.39	4.18	4.01	3.88	3.77	3.60	3.43	3.24	3.15	3.05	2.95	2.84	2.73	2.61
22	.250	1.40	1.48	1.47	1.45	1.44	1.42	1.41	1.40	1.39	1.39	1.37	1.36	1.34	1.33	1.32	1.31	1.30	1.29	1.28
	.100	2.95	2.56	2.35	2.22	2.13	2.06	2.01	1.97	1.93	1.90	1.86	1.81	1.76	1.73	1.70	1.67	1.64	1.60	1.57
	.025	5.79	4.38	3.78	3.44	3.22	3.05	2.93	2.84	2.76	2.70	2.60	2.50	2.39	2.33	2.27	2.21	2.14	2.08	2.00
	.005	9.73	6.81	5.65	5.02	4.61	4.32	4.11	3.94	3.81	3.70	3.54	3.36	3.18	3.08	2.98	2.88	2.77	2.66	2.55
23	.250	1.39	1.47	1.47	1.45	1.43	1.42	1.41	1.40	1.39	1.38	1.37	1.35	1.34	1.33	1.32	1.31	1.30	1.28	1.27
	.100	2.94	2.55	2.34	2.21	2.11	2.05	1.99	1.95	1.92	1.89	1.84	1.80	1.74	1.72	1.69	1.66	1.62	1.59	1.55
	.025	5.75	4.35	3.75	3.41	3.18	3.02	2.90	2.81	2.73	2.67	2.57	2.47	2.36	2.30	2.24	2.18	2.11	2.04	1.97
	.005	9.63	6.73	5.58	4.95	4.54	4.26	4.05	3.88	3.75	3.64	3.47	3.30	3.12	3.02	2.92	2.82	2.71	2.60	2.48
24	.250	1.39	1.47	1.46	1.44	1.43	1.41	1.40	1.39	1.38	1.38	1.36	1.35	1.33	1.32	1.31	1.30	1.29	1.28	1.26
	.100	2.93	2.54	2.33	2.19	2.10	2.04	1.98	1.94	1.91	1.88	1.83	1.78	1.73	1.70	1.67	1.64	1.61	1.57	1.53
	.025	5.72	4.32	3.72	3.38	3.15	2.99	2.87	2.78	2.70	2.64	2.54	2.44	2.33	2.27	2.21	2.15	2.08	2.01	1.94
	.005	9.55	6.66	5.52	4.89	4.49	4.20	3.99	3.83	3.69	3.59	3.42	3.25	3.06	2.97	2.87	2.77	2.66	2.55	2.43
25	.250	1.39	1.47	1.46	1.44	1.42	1.41	1.40	1.39	1.38	1.37	1.36	1.34	1.33	1.32	1.31	1.29	1.28	1.27	1.25
	.100	2.92	2.53	2.32	2.18	2.09	2.02	1.97	1.93	1.89	1.87	1.82	1.77	1.72	1.69	1.66	1.63	1.59	1.56	1.52
	.025	5.69	4.29	3.69	3.35	3.13	2.97	2.85	2.75	2.68	2.61	2.51	2.41	2.30	2.24	2.18	2.12	2.05	1.98	1.91
	.005	9.48	6.60	5.46	4.84	4.43	4.15	3.94	3.78	3.64	3.54	3.37	3.20	3.01	2.92	2.82	2.72	2.61	2.50	2.38
26	.250	1.38	1.46	1.45	1.44	1.42	1.41	1.39	1.38	1.37	1.37	1.35	1.34	1.32	1.31	1.30	1.29	1.28	1.26	1.25
	.100	2.91	2.52	2.31	2.17	2.08	2.01	1.96	1.92	1.88	1.86	1.81	1.76	1.71	1.68	1.65	1.61	1.58	1.54	1.50
	.025	5.66	4.27	3.67	3.33	3.10	2.94	2.82	2.73	2.65	2.59	2.49	2.39	2.28	2.22	2.16	2.09	2.03	1.95	1.88
	.005	9.41	6.54	5.41	4.79	4.38	4.10	3.89	3.73	3.60	3.49	3.33	3.15	2.97	2.87	2.77	2.67	2.56	2.45	2.33

* Table IX is reprinted from Maxine Merrington and Catherine M. Thompson: Tables of percentage points of the inverted beta (F) distribution, *Biometrika*, 1943, **33**, 73–78, by permission of the authors and *Biometrika*.

Table IX. The 25, 10, 2.5, and 0.5 Percent Points for the Distribution of F*—Concluded

n_2	P	n_1 Degrees of Freedom (for greater mean square)																		
		1	2	3	4	5	6	7	8	9	10	12	15	20	24	30	40	60	120	∞
27	.250	1.38	1.46	1.45	1.43	1.42	1.40	1.39	1.38	1.37	1.36	1.35	1.33	1.32	1.31	1.30	1.28	1.27	1.26	1.24
	.100	2.90	2.51	2.30	2.17	2.07	2.00	1.95	1.91	1.87	1.85	1.80	1.75	1.70	1.67	1.64	1.60	1.57	1.53	1.49
	.025	5.63	4.24	3.65	3.31	3.08	2.92	2.80	2.71	2.63	2.57	2.47	2.36	2.25	2.19	2.13	2.07	2.00	1.93	1.85
	.005	9.34	6.49	5.36	4.74	4.34	4.06	3.85	3.69	3.56	3.45	3.28	3.11	2.93	2.83	2.73	2.63	2.52	2.41	2.29
28	.250	1.38	1.46	1.45	1.43	1.41	1.40	1.39	1.38	1.37	1.36	1.34	1.33	1.31	1.30	1.29	1.28	1.27	1.25	1.24
	.100	2.89	2.50	2.29	2.16	2.06	2.00	1.94	1.90	1.87	1.84	1.79	1.74	1.69	1.66	1.63	1.59	1.56	1.52	1.48
	.025	5.61	4.22	3.63	3.29	3.06	2.90	2.78	2.69	2.61	2.55	2.45	2.34	2.23	2.17	2.11	2.05	1.98	1.91	1.83
	.005	9.28	6.44	5.32	4.70	4.30	4.02	3.81	3.65	3.52	3.41	3.25	3.07	2.89	2.79	2.69	2.59	2.48	2.37	2.25
29	.250	1.38	1.45	1.45	1.43	1.41	1.40	1.38	1.37	1.36	1.35	1.34	1.32	1.31	1.30	1.29	1.27	1.26	1.25	1.23
	.100	2.89	2.50	2.28	2.15	2.06	1.99	1.93	1.89	1.86	1.83	1.78	1.73	1.68	1.65	1.62	1.58	1.55	1.51	1.47
	.025	5.59	4.20	3.61	3.27	3.04	2.88	2.76	2.67	2.59	2.53	2.43	2.32	2.21	2.15	2.09	2.03	1.96	1.89	1.81
	.005	9.23	6.40	5.28	4.66	4.26	3.98	3.77	3.61	3.48	3.38	3.21	3.04	2.86	2.76	2.66	2.56	2.45	2.33	2.21
30	.250	1.38	1.45	1.44	1.42	1.41	1.39	1.38	1.37	1.36	1.35	1.34	1.32	1.30	1.29	1.28	1.27	1.26	1.24	1.23
	.100	2.88	2.49	2.28	2.14	2.05	1.98	1.93	1.88	1.85	1.82	1.77	1.72	1.67	1.64	1.61	1.57	1.54	1.50	1.46
	.025	5.57	4.18	3.59	3.25	3.03	2.87	2.75	2.65	2.57	2.51	2.41	2.31	2.20	2.14	2.07	2.01	1.94	1.87	1.79
	.005	9.18	6.35	5.24	4.62	4.23	3.95	3.74	3.58	3.45	3.34	3.18	3.01	2.82	2.73	2.63	2.52	2.42	2.30	2.18
40	.250	1.36	1.44	1.42	1.40	1.39	1.37	1.36	1.35	1.34	1.33	1.31	1.30	1.28	1.26	1.25	1.24	1.22	1.21	1.19
	.100	2.84	2.44	2.23	2.09	2.00	1.93	1.87	1.83	1.79	1.76	1.71	1.66	1.61	1.57	1.54	1.51	1.47	1.42	1.38
	.025	5.42	4.05	3.46	3.13	2.90	2.74	2.62	2.53	2.45	2.39	2.29	2.18	2.07	2.01	1.94	1.88	1.80	1.72	1.64
	.005	8.83	6.07	4.98	4.37	3.99	3.71	3.51	3.35	3.22	3.12	2.95	2.78	2.60	2.50	2.40	2.30	2.18	2.06	1.93
60	.250	1.35	1.42	1.41	1.38	1.37	1.35	1.33	1.32	1.31	1.30	1.29	1.27	1.25	1.24	1.22	1.21	1.19	1.17	1.15
	.100	2.79	2.39	2.18	2.04	1.95	1.87	1.82	1.77	1.74	1.71	1.66	1.60	1.54	1.51	1.48	1.44	1.40	1.35	1.29
	.025	5.29	3.93	3.34	3.01	2.79	2.63	2.51	2.41	2.33	2.27	2.17	2.06	1.94	1.88	1.82	1.74	1.67	1.58	1.48
	.005	8.49	5.80	4.73	4.14	3.76	3.49	3.29	3.13	3.01	2.90	2.74	2.57	2.39	2.29	2.19	2.08	1.96	1.83	1.69
120	.250	1.34	1.40	1.39	1.37	1.35	1.33	1.31	1.30	1.29	1.28	1.26	1.24	1.22	1.21	1.19	1.18	1.16	1.13	1.10
	.100	2.75	2.35	2.13	1.99	1.90	1.82	1.77	1.72	1.68	1.65	1.60	1.54	1.48	1.45	1.41	1.37	1.32	1.26	1.19
	.025	5.15	3.80	3.23	2.89	2.67	2.52	2.39	2.30	2.22	2.16	2.05	1.94	1.82	1.76	1.69	1.61	1.53	1.43	1.31
	.005	8.18	5.54	4.50	3.92	3.55	3.28	3.09	2.93	2.81	2.71	2.54	2.37	2.19	2.09	1.98	1.87	1.75	1.61	1.43
∞	.250	1.32	1.39	1.37	1.35	1.33	1.31	1.29	1.28	1.27	1.25	1.24	1.22	1.19	1.18	1.16	1.14	1.12	1.08	1.00
	.100	2.71	2.30	2.08	1.94	1.85	1.77	1.72	1.67	1.63	1.60	1.55	1.49	1.42	1.38	1.34	1.30	1.24	1.17	1.00
	.025	5.02	3.69	3.12	2.79	2.57	2.41	2.29	2.19	2.11	2.05	1.94	1.83	1.71	1.64	1.57	1.48	1.39	1.27	1.00
	.005	7.88	5.30	4.28	3.72	3.35	3.09	2.90	2.74	2.62	2.52	2.36	2.19	2.00	1.90	1.79	1.67	1.53	1.36	1.00

* Table IX is reprinted from Maxine Merrington and Catherine M. Thompson: Tables of percentage points of the inverted beta (F) distribution, *Biometrika*, 1943, **33,** 73–78, by permission of the authors and *Biometrika*.

Table Xa. Significant Studentized Ranges for Duncan's New Multiple Range Test with $\alpha = .10$*

df \ k	2	3	4	5	6	7	8	9	10	11	12	13	14	15	16	17	18	19
2	4.130																	
3	3.328	3.330																
4	3.015	3.074	3.081															
5	2.850	2.934	2.964	2.970														
6	2.748	2.846	2.890	2.908	2.911													
7	2.680	2.785	2.838	2.864	2.876	2.878												
8	2.630	2.742	2.800	2.832	2.849	2.857	2.858											
9	2.592	2.708	2.771	2.808	2.829	2.840	2.845	2.847										
10	2.563	2.682	2.748	2.788	2.813	2.827	2.835	2.839	2.839									
11	2.540	2.660	2.730	2.772	2.799	2.817	2.827	2.833	2.835	2.835								
12	2.521	2.643	2.714	2.759	2.789	2.808	2.821	2.828	2.832	2.833	2.833							
13	2.505	2.628	2.701	2.748	2.779	2.800	2.815	2.824	2.829	2.832	2.832	2.832						
14	2.491	2.616	2.690	2.739	2.771	2.794	2.810	2.820	2.827	2.831	2.832	2.833	2.833					
15	2.479	2.605	2.681	2.731	2.765	2.789	2.805	2.817	2.825	2.830	2.833	2.834	2.834	2.834				
16	2.469	2.596	2.673	2.723	2.759	2.784	2.802	2.815	2.824	2.829	2.833	2.835	2.836	2.836	2.836			
17	2.460	2.588	2.665	2.717	2.753	2.780	2.798	2.812	2.822	2.829	2.834	2.836	2.838	2.838	2.838	2.838		
18	2.452	2.580	2.659	2.712	2.749	2.776	2.796	2.810	2.821	2.828	2.834	2.838	2.840	2.840	2.840	2.840	2.840	
19	2.445	2.574	2.653	2.707	2.745	2.773	2.793	2.808	2.820	2.828	2.834	2.839	2.841	2.842	2.843	2.843	2.843	2.843
20	2.439	2.568	2.648	2.702	2.741	2.770	2.791	2.807	2.819	2.828	2.834	2.839	2.843	2.845	2.845	2.845	2.845	2.845
24	2.420	2.550	2.632	2.688	2.729	2.760	2.783	2.801	2.816	2.827	2.835	2.842	2.848	2.851	2.854	2.856	2.857	2.857
30	2.400	2.532	2.615	2.674	2.717	2.750	2.776	2.796	2.813	2.826	2.837	2.846	2.853	2.859	2.863	2.867	2.869	2.871
40	2.381	2.514	2.600	2.660	2.705	2.741	2.769	2.791	2.810	2.825	2.838	2.849	2.858	2.866	2.873	2.878	2.883	2.887
60	2.363	2.497	2.584	2.646	2.694	2.731	2.761	2.786	2.807	2.825	2.839	2.853	2.864	2.874	2.883	2.890	2.897	2.903
120	2.344	2.479	2.568	2.632	2.682	2.722	2.754	2.781	2.804	2.824	2.842	2.857	2.871	2.883	2.893	2.903	2.912	2.920
∞	2.326	2.462	2.552	2.619	2.670	2.712	2.746	2.776	2.801	2.824	2.844	2.861	2.877	2.892	2.905	2.918	2.929	2.939

* The entries in this table were tabulated and made available by H. Leon Harter prior to their publication. See H. Leon Harter: Critical values for Duncan's new multiple range test. *Biometrics*, 1960, **16**, 671–685.

Table Xb. Significant Studentized Ranges for Duncan's New Multiple Range Test with α = .05*

df \ k	2	3	4	5	6	7	8	9	10	11	12	13	14	15	16	17	18	19
2	6.085																	
3	4.501	4.516																
4	3.927	4.013	4.033															
5	3.635	3.749	3.797	3.814														
6	3.461	3.587	3.649	3.680	3.694													
7	3.344	3.477	3.548	3.588	3.611	3.622												
8	3.261	3.399	3.475	3.521	3.549	3.566	3.575											
9	3.199	3.339	3.420	3.470	3.502	3.523	3.536	3.544										
10	3.151	3.293	3.376	3.430	3.465	3.489	3.505	3.516	3.522									
11	3.113	3.256	3.342	3.397	3.435	3.462	3.480	3.493	3.501	3.506								
12	3.082	3.225	3.313	3.370	3.410	3.439	3.459	3.474	3.484	3.491	3.496							
13	3.055	3.200	3.289	3.348	3.389	3.419	3.442	3.458	3.470	3.478	3.484	3.488						
14	3.033	3.178	3.268	3.329	3.372	3.403	3.426	3.444	3.457	3.467	3.474	3.479	3.482					
15	3.014	3.160	3.250	3.312	3.356	3.389	3.413	3.432	3.446	3.457	3.465	3.471	3.476	3.478				
16	2.998	3.144	3.235	3.298	3.343	3.376	3.402	3.422	3.437	3.449	3.458	3.465	3.470	3.473	3.477			
17	2.984	3.130	3.222	3.285	3.331	3.366	3.392	3.412	3.429	3.441	3.451	3.459	3.465	3.469	3.473	3.475		
18	2.971	3.118	3.210	3.274	3.321	3.356	3.383	3.405	3.421	3.435	3.445	3.454	3.460	3.465	3.470	3.472	3.474	
19	2.960	3.107	3.199	3.264	3.311	3.347	3.375	3.397	3.415	3.429	3.440	3.449	3.456	3.462	3.467	3.470	3.472	3.473
20	2.950	3.097	3.190	3.255	3.303	3.339	3.368	3.391	3.409	3.424	3.436	3.445	3.453	3.459	3.464	3.467	3.470	3.472
24	2.919	3.066	3.160	3.226	3.276	3.315	3.345	3.370	3.390	3.406	3.420	3.432	3.441	3.449	3.456	3.461	3.465	3.469
30	2.888	3.035	3.131	3.199	3.250	3.290	3.322	3.349	3.371	3.389	3.405	3.418	3.430	3.439	3.447	3.454	3.460	3.466
40	2.858	3.006	3.102	3.171	3.224	3.266	3.300	3.328	3.352	3.373	3.390	3.405	3.418	3.429	3.439	3.448	3.456	3.463
60	2.829	2.976	3.073	3.143	3.198	3.241	3.277	3.307	3.333	3.355	3.374	3.391	3.406	3.419	3.431	3.442	3.451	3.460
120	2.800	2.947	3.045	3.116	3.172	3.217	3.254	3.287	3.314	3.337	3.359	3.377	3.394	3.409	3.423	3.435	3.446	3.457
∞	2.772	2.918	3.017	3.089	3.146	3.193	3.232	3.265	3.294	3.320	3.343	3.363	3.382	3.399	3.414	3.428	3.442	3.454

* The entries in this table were tabulated and made available by H. Leon Harter prior to their publication. See H. Leon Harter: Critical values for Duncan's new multiple range test. *Biometrics*, 1960, **16**, 671–685.

Table Xc. Significant Studentized Ranges for Duncan's New Multiple Range Test with $\alpha = .01$*

df \ k	2	3	4	5	6	7	8	9	10	11	12	13	14	15	16	17	18	19
2	14.04																	
3	8.261	8.321																
4	6.512	6.677	6.740															
5	5.702	5.893	5.989	6.040														
6	5.243	5.439	5.549	5.614	5.655													
7	4.949	5.145	5.260	5.334	5.383	5.416												
8	4.746	4.939	5.057	5.135	5.189	5.227	5.256											
9	4.596	4.787	4.906	4.986	5.043	5.086	5.118	5.142										
10	4.482	4.671	4.790	4.871	4.931	4.975	5.010	5.037	5.058									
11	4.392	4.579	4.697	4.780	4.841	4.887	4.924	4.952	4.975	4.994								
12	4.320	4.504	4.622	4.706	4.767	4.815	4.852	4.883	4.907	4.927	4.944							
13	4.260	4.442	4.560	4.644	4.706	4.755	4.793	4.824	4.850	4.872	4.889	4.904						
14	4.210	4.391	4.508	4.591	4.654	4.704	4.743	4.775	4.802	4.824	4.843	4.859	4.872					
15	4.168	4.347	4.463	4.547	4.610	4.660	4.700	4.733	4.760	4.783	4.803	4.820	4.834	4.846				
16	4.131	4.309	4.425	4.509	4.572	4.622	4.663	4.696	4.724	4.748	4.768	4.786	4.800	4.813	4.825			
17	4.099	4.275	4.391	4.475	4.539	4.589	4.630	4.664	4.693	4.717	4.738	4.756	4.771	4.785	4.797	4.807		
18	4.071	4.246	4.362	4.445	4.509	4.560	4.601	4.635	4.664	4.689	4.711	4.729	4.745	4.759	4.772	4.783	4.792	
19	4.046	4.220	4.335	4.419	4.483	4.534	4.575	4.610	4.639	4.665	4.686	4.705	4.722	4.736	4.749	4.761	4.771	4.780
20	4.024	4.197	4.312	4.395	4.459	4.510	4.552	4.587	4.617	4.642	4.664	4.684	4.701	4.716	4.729	4.741	4.751	4.761
24	3.956	4.126	4.239	4.322	4.386	4.437	4.480	4.516	4.546	4.573	4.596	4.616	4.634	4.651	4.665	4.678	4.690	4.700
30	3.889	4.056	4.168	4.250	4.314	4.366	4.409	4.445	4.477	4.504	4.528	4.550	4.569	4.586	4.601	4.615	4.628	4.640
40	3.825	3.988	4.098	4.180	4.244	4.296	4.339	4.376	4.408	4.436	4.461	4.483	4.503	4.521	4.537	4.553	4.566	4.579
60	3.762	3.922	4.031	4.111	4.174	4.226	4.270	4.307	4.340	4.368	4.394	4.417	4.438	4.456	4.474	4.490	4.504	4.518
120	3.702	3.858	3.965	4.044	4.107	4.158	4.202	4.239	4.272	4.301	4.327	4.351	4.372	4.392	4.410	4.426	4.442	4.456
∞	3.643	3.796	3.900	3.978	4.040	4.091	4.135	4.172	4.205	4.235	4.261	4.285	4.307	4.327	4.345	4.363	4.379	4.394

* The entries in this table were tabulated and made available by H. Leon Harter prior to their publication. See H. Leon Harter: Critical values for Duncan's new multiple range test. *Biometrics*, 1960, **16**, 671–685.

Table Xd. Significant Studentized Ranges for Duncan's New Multiple Range Test with $\alpha = .005$*

df \ k	2	3	4	5	6	7	8	9	10	11	12	13	14	15	16	17	18	19
2	19.93																	
3	10.55	10.63																
4	7.916	8.126	8.210															
5	6.751	6.980	7.100	7.167														
6	6.105	6.334	6.466	6.547	6.600													
7	5.699	5.922	6.057	6.145	6.207	6.250												
8	5.420	5.638	5.773	5.864	5.930	5.978	6.014											
9	5.218	5.430	5.565	5.657	5.725	5.776	5.815	5.846										
10	5.065	5.273	5.405	5.498	5.567	5.620	5.662	5.695	5.722									
11	4.945	5.149	5.280	5.372	5.442	5.496	5.539	5.574	5.603	5.626								
12	4.849	5.048	5.178	5.270	5.341	5.396	5.439	5.475	5.505	5.531	5.552							
13	4.770	4.966	5.094	5.186	5.256	5.312	5.356	5.393	5.424	5.450	5.472	5.492						
14	4.704	4.897	5.023	5.116	5.185	5.241	5.286	5.324	5.355	5.382	5.405	5.425	5.442					
15	4.647	4.838	4.964	5.055	5.125	5.181	5.226	5.264	5.297	5.324	5.348	5.368	5.386	5.402				
16	4.599	4.787	4.912	5.003	5.073	5.129	5.175	5.213	5.245	5.273	5.298	5.319	5.338	5.354	5.368			
17	4.557	4.744	4.867	4.958	5.027	5.084	5.130	5.168	5.201	5.229	5.254	5.275	5.295	5.311	5.327	5.340		
18	4.521	4.705	4.828	4.918	4.987	5.043	5.090	5.129	5.162	5.190	5.215	5.237	5.256	5.274	5.289	5.303	5.316	
19	4.488	4.671	4.793	4.883	4.952	5.008	5.054	5.093	5.127	5.156	5.181	5.203	5.222	5.240	5.256	5.270	5.283	5.295
20	4.460	4.641	4.762	4.851	4.920	4.976	5.022	5.061	5.095	5.124	5.150	5.172	5.193	5.210	5.226	5.241	5.254	5.266
24	4.371	4.547	4.666	4.753	4.822	4.877	4.924	4.963	4.997	5.027	5.053	5.076	5.097	5.116	5.133	5.148	5.162	5.175
30	4.285	4.456	4.572	4.658	4.726	4.781	4.827	4.867	4.901	4.931	4.958	4.981	5.003	5.022	5.040	5.056	5.071	5.085
40	4.202	4.369	4.482	4.566	4.632	4.687	4.733	4.772	4.806	4.837	4.864	4.888	4.910	4.930	4.948	4.965	4.980	4.995
60	4.122	4.284	4.394	4.476	4.541	4.595	4.640	4.679	4.713	4.744	4.771	4.796	4.818	4.838	4.857	4.874	4.890	4.905
120	4.045	4.201	4.308	4.388	4.452	4.505	4.550	4.588	4.622	4.652	4.679	4.704	4.726	4.747	4.766	4.784	4.800	4.815
∞	3.970	4.121	4.225	4.303	4.365	4.417	4.461	4.499	4.532	4.562	4.589	4.614	4.636	4.657	4.676	4.694	4.710	4.726

* The entries in this table were tabulated and made available by H. Leon Harter prior to their publication. See H. Leon Harter: Critical values for Duncan's new multiple range test. *Biometrics*, 1960, **16**, 671–685.

Table Xe. Significant Studentized Ranges for Duncan's New Multiple Range Test with $\alpha = .001$*

df \ k	2	3	4	5	6	7	8	9	10	11	12	13	14	15	16	17	18	19
2	44.69																	
3	18.28	18.45																
4	12.18	12.52	12.67															
5	9.714	10.05	10.24	10.35														
6	8.427	8.743	8.932	9.055	9.139													
7	7.648	7.943	8.127	8.252	8.342	8.409												
8	7.130	7.407	7.584	7.708	7.799	7.869	7.924											
9	6.762	7.024	7.195	7.316	7.407	7.478	7.535	7.582										
10	6.487	6.738	6.902	7.021	7.111	7.182	7.240	7.287	7.327									
11	6.275	6.516	6.676	6.791	6.880	6.950	7.008	7.056	7.097	7.132								
12	6.106	6.340	6.494	6.607	6.695	6.765	6.822	6.870	6.911	6.947	6.978							
13	5.970	6.195	6.346	6.457	6.543	6.612	6.670	6.718	6.759	6.795	6.826	6.854						
14	5.856	6.075	6.223	6.332	6.416	6.485	6.542	6.590	6.631	6.667	6.699	6.727	6.752					
15	5.760	5.974	6.119	6.225	6.309	6.377	6.433	6.481	6.522	6.558	6.590	6.619	6.644	6.666				
16	5.678	5.888	6.030	6.135	6.217	6.284	6.340	6.388	6.429	6.465	6.497	6.525	6.551	6.574	6.595			
17	5.608	5.813	5.953	6.056	6.138	6.204	6.260	6.307	6.348	6.384	6.416	6.444	6.470	6.493	6.514	6.533		
18	5.546	5.748	5.886	5.988	6.068	6.134	6.189	6.236	6.277	6.313	6.345	6.373	6.399	6.422	6.443	6.462	6.480	
19	5.492	5.691	5.826	5.927	6.007	6.072	6.127	6.174	6.214	6.250	6.281	6.310	6.336	6.359	6.380	6.400	6.418	6.434
20	5.444	5.640	5.774	5.873	5.952	6.017	6.071	6.117	6.158	6.193	6.225	6.254	6.279	6.303	6.324	6.344	6.362	6.379
24	5.297	5.484	5.612	5.708	5.784	5.846	5.899	5.945	5.984	6.020	6.051	6.079	6.105	6.129	6.150	6.170	6.188	6.205
30	5.156	5.335	5.457	5.549	5.622	5.682	5.734	5.778	5.817	5.851	5.882	5.910	5.935	5.958	5.980	6.000	6.018	6.036
40	5.022	5.191	5.308	5.396	5.466	5.524	5.574	5.617	5.654	5.688	5.718	5.745	5.770	5.793	5.814	5.834	5.852	5.869
60	4.894	5.055	5.166	5.249	5.317	5.372	5.420	5.461	5.498	5.530	5.559	5.586	5.610	5.632	5.653	5.672	5.690	5.707
120	4.771	4.924	5.029	5.109	5.173	5.226	5.271	5.311	5.346	5.377	5.405	5.431	5.454	5.476	5.496	5.515	5.532	5.549
∞	4.654	4.798	4.898	4.974	5.034	5.085	5.128	5.166	5.199	5.229	5.256	5.280	5.303	5.324	5.343	5.361	5.378	5.394

* The entries in this table were tabulated and made available by H. Leon Harter prior to their publication. See H. Leon Harter: Critical values for Duncan's new multiple range test. *Biometrics*, 1960, **16**, 671–685.

Table XI. Coefficients for Obtaining the Linear and Quadratic Components of the Treatment Sum of Squares When Treatments are Equally Spaced

Comparison	Number of Treatments 1	2	3	4	5	6	7	8	9	$\sum a_{.i}^2$
LINEAR	−1	0	1							2
QUADRATIC	1	−2	1							6
LINEAR	−3	−1	1	3						20
QUADRATIC	1	−1	−1	1						4
LINEAR	−2	−1	0	1	2					10
QUADRATIC	2	−1	−2	−1	2					14
LINEAR	−5	−3	−1	1	3	5				70
QUADRATIC	5	−1	−4	−4	−1	5				84
LINEAR	−3	−2	−1	0	1	2	3			28
QUADRATIC	5	0	−3	−4	−3	0	5			84
LINEAR	−7	−5	−3	−1	1	3	5	7		168
QUADRATIC	7	1	−3	−5	−5	−3	1	7		168
LINEAR	−4	−3	−2	−1	0	1	2	3	4	60
QUADRATIC	28	7	−8	−17	−20	−17	−8	7	28	2772

Table XIIa. Table of t for One-Sided Comparisons Between k Treatment Means and a Control for a Joint Confidence Coefficient of P = 95 Percent*

	k, Number Of Treatment Means (Excluding The Control)								
df	1	2	3	4	5	6	7	8	9
5	2.02	2.44	2.68	2.85	2.98	3.08	3.16	3.24	3.30
6	1.94	2.34	2.56	2.71	2.83	2.92	3.00	3.07	3.12
7	1.89	2.27	2.48	2.62	2.73	2.82	2.89	2.95	3.01
8	1.86	2.22	2.42	2.55	2.66	2.74	2.81	2.87	2.92
9	1.83	2.18	2.37	2.50	2.60	2.68	2.75	2.81	2.86
10	1.81	2.15	2.34	2.47	2.56	2.64	2.70	2.76	2.81
11	1.80	2.13	2.31	2.44	2.53	2.60	2.67	2.72	2.77
12	1.78	2.11	2.29	2.41	2.50	2.58	2.64	2.69	2.74
13	1.77	2.09	2.27	2.39	2.48	2.55	2.61	2.66	2.71
14	1.76	2.08	2.25	2.37	2.46	2.53	2.59	2.64	2.69
15	1.75	2.07	2.24	2.36	2.44	2.51	2.57	2.62	2.67
16	1.75	2.06	2.23	2.34	2.43	2.50	2.56	2.61	2.65
17	1.74	2.05	2.22	2.33	2.42	2.49	2.54	2.59	2.64
18	1.73	2.04	2.21	2.32	2.41	2.48	2.53	2.58	2.62
19	1.73	2.03	2.20	2.31	2.40	2.47	2.52	2.57	2.61
20	1.72	2.03	2.19	2.30	2.39	2.46	2.51	2.56	2.60
24	1.71	2.01	2.17	2.28	2.36	2.43	2.48	2.53	2.57
30	1.70	1.99	2.15	2.25	2.33	2.40	2.45	2.50	2.54
40	1.68	1.97	2.13	2.23	2.31	2.37	2.42	2.47	2.51
60	1.67	1.95	2.10	2.21	2.28	2.35	2.39	2.44	2.48
120	1.66	1.93	2.08	2.18	2.26	2.32	2.37	2.41	2.45
inf.	1.64	1.92	2.06	2.16	2.23	2.29	2.34	2.38	2.42

* Table XIIa is reprinted from C. W. Dunnett, A multiple comparison procedure for comparing several treatments with a control. *Journal of the American Statistical Association,* 1955, **50**, 1096–1121, by permission of the author and the editors of the *Journal of the American Statistical Association.*

Table XIIb. Table of t for One-Sided Comparisons Between k Treatment Means and a Control for a Joint Confidence Coefficient of $P = 99$ Percent*

	k, Number Of Treatment Means (Excluding The Control)								
df	1	2	3	4	5	6	7	8	9
5	3.37	3.90	4.21	4.43	4.60	4.73	4.85	4.94	5.03
6	3.14	3.61	3.88	4.07	4.21	4.33	4.43	4.51	4.59
7	3.00	3.42	3.66	3.83	3.96	4.07	4.15	4.23	4.30
8	2.90	3.29	3.51	3.67	3.79	3.88	3.96	4.03	4.09
9	2.82	3.19	3.40	3.55	3.66	3.75	3.82	3.89	3.94
10	2.76	3.11	3.31	3.45	3.56	3.64	3.71	3.78	3.83
11	2.72	3.06	3.25	3.38	3.48	3.56	3.63	3.69	3.74
12	2.68	3.01	3.19	3.32	3.42	3.50	3.56	3.62	3.67
13	2.65	2.97	3.15	3.27	3.37	3.44	3.51	3.56	3.61
14	2.62	2.94	3.11	3.23	3.32	3.40	3.46	3.51	3.56
15	2.60	2.91	3.08	3.20	3.29	3.36	3.42	3.47	3.52
16	2.58	2.88	3.05	3.17	3.26	3.33	3.39	3.44	3.48
17	2.57	2.86	3.03	3.14	3.23	3.30	3.36	3.41	3.45
18	2.55	2.84	3.01	3.12	3.21	3.27	3.33	3.38	3.42
19	2.54	2.83	2.99	3.10	3.18	3.25	3.31	3.36	3.40
20	2.53	2.81	2.97	3.08	3.17	3.23	3.29	3.34	3.38
24	2.49	2.77	2.92	3.03	3.11	3.17	3.22	3.27	3.31
30	2.46	2.72	2.87	2.97	3.05	3.11	3.16	3.21	3.24
40	2.42	2.68	2.82	2.92	2.99	3.05	3.10	3.14	3.18
60	2.39	2.64	2.78	2.87	2.94	3.00	3.04	3.08	3.12
120	2.36	2.60	2.73	2.82	2.89	2.94	2.99	3.03	3.06
inf.	2.33	2.56	2.68	2.77	2.84	2.89	2.93	2.97	3.00

* Table XIIb is reprinted from C. W. Dunnett, A multiple comparison procedure for comparing several treatments with a control. *Journal of the American Statistical Association*, 1955, **50**, 1096–1121, by permission of the author and the editors of the *Journal of the American Statistical Association*.

Table XIIc. Table of t for Two-Sided Comparisons Between k Treatment Means and a Control for a Joint Confidence Coefficient of P = 95 Percent*

	k, Number Of Treatment Means (Excluding The Control)								
df	1	2	3	4	5	6	7	8	9
5	2.57	3.03	3.39	3.66	3.88	4.06	4.22	4.36	4.49
6	2.45	2.86	3.18	3.41	3.60	3.75	3.88	4.00	4.11
7	2.36	2.75	3.04	3.24	3.41	3.54	3.66	3.76	3.86
8	2.31	2.67	2.94	3.13	3.28	3.40	3.51	3.60	3.68
9	2.26	2.61	2.86	3.04	3.18	3.29	3.39	3.48	3.55
10	2.23	2.57	2.81	2.97	3.11	3.21	3.31	3.39	3.46
11	2.20	2.53	2.76	2.92	3.05	3.15	3.24	3.31	3.38
12	2.18	2.50	2.72	2.88	3.00	3.10	3.18	3.25	3.32
13	2.16	2.48	2.69	2.84	2.96	3.06	3.14	3.21	3.27
14	2.14	2.46	2.67	2.81	2.93	3.02	3.10	3.17	3.23
15	2.13	2.44	2.64	2.79	2.90	2.99	3.07	3.13	3.19
16	2.12	2.42	2.63	2.77	2.88	2.96	3.04	3.10	3.16
17	2.11	2.41	2.61	2.75	2.85	2.94	3.01	3.08	3.13
17	2.10	2.40	2.59	2.73	2.84	2.92	2.99	3.05	3.11
19	2.09	2.39	2.58	2.72	2.82	2.90	2.97	3.04	3.09
20	2.09	2.38	2.57	2.70	2.81	2.89	2.96	3.02	3.07
24	2.06	2.35	2.53	2.66	2.76	2.84	2.91	2.96	3.01
30	2.04	2.32	2.50	2.62	2.72	2.79	2.86	2.91	2.96
40	2.02	2.29	2.47	2.58	2.67	2.75	2.81	2.86	2.90
60	2.00	2.27	2.43	2.55	2.63	2.70	2.76	2.81	2.85
120	1.98	2.24	2.40	2.51	2.59	2.66	2.71	2.76	2.80
inf.	1.96	2.21	2.37	2.47	2.55	2.62	2.67	2.71	2.75

* Table XIIc is reprinted from C. W. Dunnett, A multiple comparison procedure for comparing several treatments with a control. *Journal of the American Statistical Association*, 1955, **50**, 1096–1121, by permission of the author and the editors of the *Journal of the American Statistical Association*.

Table XIId. Table of t for Two-Sided Comparisons Between k Treatment Means and a Control for a Joint Confidence Coefficient of $P = 99$ Percent*

	k, Number Of Treatment Means (Excluding The Control)								
df	1	2	3	4	5	6	7	8	9
5	4.03	4.63	5.09	5.44	5.73	5.97	6.18	6.36	6.53
6	3.71	4.22	4.60	4.88	5.11	5.30	5.47	5.61	5.74
7	3.50	3.95	4.28	4.52	4.71	4.87	5.01	5.13	5.24
8	3.36	3.77	4.06	4.27	4.44	4.58	4.70	4.81	4.90
9	3.25	3.63	3.90	4.09	4.24	4.37	4.48	4.57	4.65
10	3.17	3.53	3.78	3.95	4.10	4.21	4.31	4.40	4.47
11	3.11	3.45	3.68	3.85	3.98	4.09	4.18	4.26	4.33
12	3.05	3.39	3.61	3.76	3.89	3.99	4.08	4.15	4.22
13	3.01	3.33	3.54	3.69	3.81	3.91	3.99	4.06	4.13
14	2.98	3.29	3.49	3.64		3.84	3.92	3.99	4.05
					3.75				
15	2.95	3.25	3.45	3.59	3.70	3.79	3.86	3.93	3.99
16	2.92	3.22	3.41	3.55	3.65	3.74	3.82	3.88	3.93
17	2.90	3.19	3.38	3.51	3.62	3.70	3.77	3.83	3.89
18	2.88	3.17	3.35	3.48	3.58	3.67	3.74	3.80	3.85
19	2.86	3.15	3.33	3.46	3.55	3.64	3.70	3.76	3.81
20	2.85	3.13	3.31	3.43	3.53	3.61	3.67	3.73	3.78
24	2.80	3.07	3.24	3.36	3.45	3.52	3.58	3.64	3.69
30	2.75	3.01	3.17	3.28	3.37	3.44	3.50	3.55	3.59
40	2.70	2.95	3.10	3.21	3.29	3.36	3.41	3.46	3.50
60	2.66	2.90	3.04	3.14	3.22	3.28	3.33	3.38	3.42
120	2.62	2.84	2.98	3.08	3.15	3.21	3.25	3.30	3.33
inf.	2.58	2.79	2.92	3.01	3.08	3.14	3.18	3.22	3.25

* Table XIId is reprinted from C. W. Dunnett, A multiple comparison procedure for comparing several treatments with a control. *Journal of the American Statistical Association*, 1955, **50**, 1096–1121, by permission of the author and the editors of the *Journal of the American Statistical Association*.

Table XIII. Table of Four-Place Logarithms*

N	0	1	2	3	4	5	6	7	8	9	1 2 3	4 5 6	7 8 9
1.0	.0000	.0043	.0086	.0128	.0170	.0212	.0253	.0294	.0334	.0374	4 8 12	17 21 25	29 33 37
1.1	.0414	.0453	.0492	.0531	.0569	.0607	.0645	.0682	.0719	.0755	4 8 11	15 19 23	26 30 34
1.2	.0792	.0828	.0864	.0899	.0934	.0969	.1004	.1038	.1072	.1106	3 7 10	14 17 21	24 28 31
1.3	.1139	.1173	.1206	.1239	.1271	.1303	.1335	.1367	.1399	.1430	3 6 10	13 16 19	23 26 29
1.4	.1461	.1492	.1523	.1553	.1584	.1614	.1644	.1673	.1703	.1732	3 6 9	12 15 18	21 24 27
1.5	.1761	.1790	.1818	.1847	.1875	.1903	.1931	.1959	.1987	.2014	3 6 8	11 14 17	20 22 25
1.6	.2041	.2068	.2095	.2122	.2148	.2175	.2201	.2227	.2253	.2279	3 5 8	11 13 16	18 21 24
1.7	.2304	.2330	.2355	.2380	.2405	.2430	.2455	.2480	.2504	.2529	2 5 7	10 12 15	17 20 22
1.8	.2553	.2577	.2601	.2625	.2648	.2672	.2695	.2718	.2742	.2765	2 5 7	9 12 14	16 19 21
1.9	.2788	.2810	.2833	.2856	.2878	.2900	.2923	.2945	.2967	.2989	2 4 7	9 11 13	16 18 20
2.0	.3010	.3032	.3054	.3075	.3096	.3118	.3139	.3160	.3181	.3201	2 4 6	8 11 13	15 17 19
2.1	.3222	.3243	.3263	.3284	.3304	.3324	.3345	.3365	.3385	.3404	2 4 6	8 10 12	14 16 18
2.2	.3424	.3444	.3464	.3483	.3502	.3522	.3541	.3560	.3579	.3598	2 4 6	8 10 12	14 15 17
2.3	.3617	.3636	.3655	.3674	.3692	.3711	.3729	.3747	.3766	.3784	2 4 6	7 9 11	13 15 17
2.4	.3802	.3820	.3838	.3856	.3874	.3892	.3909	.3927	.3945	.3962	2 4 5	7 9 11	12 14 16
2.5	.3979	.3997	.4014	.4031	.4048	.4065	.4082	.4099	.4116	.4133	2 3 5	7 9 10	12 14 15
2.6	.4150	.4166	.4183	.4200	.4216	.4232	.4249	.4265	.4281	.4298	2 3 5	7 8 10	11 13 15
2.7	.4314	.4330	.4346	.4362	.4378	.4393	.4409	.4425	.4440	.4456	2 3 5	6 8 9	11 13 14
2.8	.4472	.4487	.4502	.4518	.4533	.4548	.4564	.4579	.4594	.4609	2 3 5	6 8 9	11 12 14
2.9	.4624	.4639	.4654	.4669	.4683	.4698	.4713	.4728	.4742	.4757	1 3 4	6 7 9	10 12 13
3.0	.4771	.4786	.4800	.4814	.4829	.4843	.4857	.4871	.4886	.4900	1 3 4	6 7 9	10 11 13
3.1	.4914	.4928	.4942	.4955	.4969	.4983	.4997	.5011	.5024	.5038	1 3 4	6 7 8	10 11 12
3.2	.5051	.5065	.5079	.5092	.5105	.5119	.5132	.5145	.5159	.5172	1 3 4	5 7 8	9 11 12
3.3	.5185	.5198	.5211	.5224	.5237	.5250	.5263	.5276	.5289	.5302	1 3 4	5 6 8	9 10 12
3.4	.5315	.5328	.5340	.5353	.5366	.5378	.5391	.5403	.5416	.5428	1 3 4	5 6 8	9 10 11
3.5	.5441	.5453	.5465	.5478	.5490	.5502	.5514	.5527	.5539	.5551	1 2 4	5 6 7	9 10 11
3.6	.5563	.5575	.5587	.5599	.5611	.5623	.5635	.5647	.5658	.5670	1 2 4	5 6 7	8 10 11
3.7	.5682	.5694	.5705	.5717	.5729	.5740	.5752	.5763	.5775	.5786	1 2 3	5 6 7	8 9 10
3.8	.5798	.5809	.5821	.5832	.5843	.5855	.5866	.5877	.5888	.5899	1 2 3	5 6 7	8 9 10
3.9	.5911	.5922	.5933	.5944	.5955	.5966	.5977	.5988	.5999	.6010	1 2 3	4 5 7	8 9 10
4.0	.6021	.6031	.6042	.6053	.6064	.6075	.6085	.6096	.6107	.6117	1 2 3	4 5 6	8 9 10
4.1	.6128	.6138	.6149	.6160	.6170	.6180	.6191	.6201	.6212	.6222	1 2 3	4 5 6	7 8 9
4.2	.6232	.6243	.6253	.6263	.6274	.6284	.6294	.6304	.6314	.6325	1 2 3	4 5 6	7 8 9
4.3	.6335	.6345	.6355	.6365	.6375	.6385	.6395	.6405	.6415	.6425	1 2 3	4 5 6	7 8 9
4.4	.6435	.6444	.6454	.6464	.6474	.6484	.6493	.6503	.6513	.6522	1 2 3	4 5 6	7 8 9
4.5	.6532	.6542	.6551	.6561	.6571	.6580	.6590	.6599	.6609	.6618	1 2 3	4 5 6	7 8 9
4.6	.6628	.6637	.6646	.6656	.6665	.6675	.6684	.6693	.6702	.6712	1 2 3	4 5 6	7 7 8
4.7	.6721	.6730	.6739	.6749	.6758	.6767	.6776	.6785	.6794	.6803	1 2 3	4 5 5	6 7 8
4.8	.6812	.6821	.6830	.6839	.6848	.6857	.6866	.6875	.6884	.6893	1 2 3	4 4 5	6 7 8
4.9	.6902	.6911	.6920	.6928	.6937	.6946	.6955	.6964	.6972	.6981	1 2 3	4 4 5	6 7 8
5.0	.6990	.6998	.7007	.7016	.7024	.7033	.7042	.7050	.7059	.7067	1 2 3	3 4 5	6 7 8
5.1	.7076	.7084	.7093	.7101	.7110	.7118	.7126	.7135	.7143	.7152	1 2 3	3 4 5	6 7 8
5.2	.7160	.7168	.7177	.7185	.7193	.7202	.7210	.7218	.7226	.7235	1 2 2	3 4 5	6 7 7
5.3	.7243	.7251	.7259	.7267	.7275	.7284	.7292	.7300	.7308	.7316	1 2 2	3 4 5	6 6 7
5.4	.7324	.7332	.7340	.7348	.7356	.7364	.7372	.7380	.7388	.7396	1 2 2	3 4 5	6 6 7

* Table XIII is reprinted from D. E. Smith, W. D. Reeve, and E. L. Morss: *Elementary Mathematical Tables*, Ginn and Company, by permission of the authors and publishers.

To obtain the mantissa for a four-digit number, find in the body of the table the mantissa for the first three digits and then, neglecting the decimal point temporarily, add the number in the proportional-parts table at the right which is on the same line as the mantissa already obtained and in the column corresponding to the fourth digit.

Table XIII. Table of Four-Place Logarithms*—Concluded

N	0	1	2	3	4	5	6	7	8	9	1	2	3	4	5	6	7	8	9
5.5	.7404	.7412	.7419	.7427	.7435	.7443	.7451	.7459	.7466	.7474	1	2	2	3	4	5	5	6	7
5.6	.7482	.7490	.7497	.7505	.7513	.7520	.7528	.7536	.7543	.7551	1	2	2	3	4	5	5	6	7
5.7	.7559	.7566	.7574	.7582	.7589	.7597	.7604	.7612	.7619	.7627	1	2	2	3	4	5	5	6	7
5.8	.7634	.7642	.7649	.7657	.7664	.7672	.7679	.7686	.7694	.7701	1	1	2	3	4	4	5	6	7
5.9	.7709	.7716	.7723	.7731	.7738	.7745	.7752	.7760	.7767	.7774	1	1	2	3	4	4	5	6	7
6.0	.7782	.7789	.7796	.7803	.7810	.7818	.7825	.7832	.7839	.7846	1	1	2	3	4	4	5	6	6
6.1	.7853	.7860	.7868	.7875	.7882	.7889	.7896	.7903	.7910	.7917	1	1	2	3	4	4	5	6	6
6.2	.7924	.7931	.7938	.7945	.7952	.7959	.7966	.7973	.7980	.7987	1	1	2	3	3	4	5	6	6
6.3	.7993	.8000	.8007	.8014	.8021	.8028	.8035	.8041	.8048	.8055	1	1	2	3	3	4	5	5	6
6.4	.8062	.8069	.8075	.8082	.8089	.8096	.8102	.8109	.8116	.8122	1	1	2	3	3	4	5	5	6
6.5	.8129	.8136	.8142	.8149	.8156	.8162	.8169	.8176	.8182	.8189	1	1	2	3	3	4	5	5	6
6.6	.8195	.8202	.8209	.8215	.8222	.8228	.8235	.8241	.8248	.8254	1	1	2	3	3	4	5	5	6
6.7	.8261	.8267	.8274	.8280	.8287	.8293	.8299	.8306	.8312	.8319	1	1	2	3	3	4	5	5	6
6.8	.8325	.8331	.8338	.8344	.8351	.8357	.8363	.8370	.8376	.8382	1	1	2	3	3	4	4	5	6
6.9	.8388	.8395	.8401	.8407	.8414	.8420	.8426	.8432	.8439	.8445	1	1	2	2	3	4	4	5	6
7.0	.8451	.8457	.8463	.8470	.8476	.8482	.8488	.8494	.8500	.8506	1	1	2	2	3	4	4	5	6
7.1	.8513	.8519	.8525	.8531	.8537	.8543	.8549	.8555	.8561	.8567	1	1	2	2	3	4	4	5	5
7.2	.8573	.8579	.8585	.8591	.8597	.8603	.8609	.8615	.8621	.8627	1	1	2	2	3	4	4	5	5
7.3	.8633	.8639	.8645	.8651	.8657	.8663	.8669	.8675	.8681	.8686	1	1	2	2	3	4	4	5	5
7.4	.8692	.8698	.8704	.8710	.8716	.8722	.8727	.8733	.8739	.8745	1	1	2	2	3	4	4	5	5
7.5	.8751	.8756	.8762	.8768	.8774	.8779	.8785	.8791	.8797	.8802	1	1	2	2	3	3	4	5	5
7.6	.8808	.8814	.8820	.8825	.8831	.8837	.8842	.8848	.8854	.8859	1	1	2	2	3	3	4	5	5
7.7	.8865	.8871	.8876	.8882	.8887	.8893	.8899	.8904	.8910	.8915	1	1	2	2	3	3	4	4	5
7.8	.8921	.8927	.8932	.8938	.8943	.8949	.8954	.8960	.8965	.8971	1	1	2	2	3	3	4	4	5
7.9	.8976	.8982	.8987	.8993	.8998	.9004	.9009	.9015	.9020	.9025	1	1	2	2	3	3	4	4	5
8.0	.9031	.9036	.9042	.9047	.9053	.9058	.9063	.9069	.9074	.9079	1	1	2	2	3	3	4	4	5
8.1	.9085	.9090	.9096	.9101	.9106	.9112	.9117	.9122	.9128	.9133	1	1	2	2	3	3	4	4	5
8.2	.9138	.9143	.9149	.9154	.9159	.9165	.9170	.9175	.9180	.9186	1	1	2	2	3	3	4	4	5
8.3	.9191	.9196	.9201	.9206	.9212	.9217	.9222	.9227	.9232	.9238	1	1	2	2	3	3	4	4	5
8.4	.9243	.9248	.9253	.9258	.9263	.9269	.9274	.9279	.9284	.9289	1	1	2	2	3	3	4	4	5
8.5	.9294	.9299	.9304	.9309	.9315	.9320	.9325	.9330	.9335	.9340	1	1	2	2	3	3	4	4	5
8.6	.9345	.9350	.9355	.9360	.9365	.9370	.9375	.9380	.9385	.9390	1	1	2	2	3	3	4	4	5
8.7	.9395	.9400	.9405	.9410	.9415	.9420	.9425	.9430	.9435	.9440	0	1	1	2	2	3	3	4	4
8.8	.9445	.9450	.9455	.9460	.9465	.9469	.9474	.9479	.9484	.9489	0	1	1	2	2	3	3	4	4
8.9	.9494	.9499	.9504	.9509	.9513	.9518	.9523	.9528	.9533	.9538	0	1	1	2	2	3	3	4	4
9.0	.9542	.9547	.9552	.9557	.9562	.9566	.9571	.9576	.9581	.9586	0	1	1	2	2	3	3	4	4
9.1	.9590	.9595	.9600	.9605	.9609	.9614	.9619	.9624	.9628	.9633	0	1	1	2	2	3	3	4	4
9.2	.9638	.9643	.9647	.9652	.9657	.9661	.9666	.9671	.9675	.9680	0	1	1	2	2	3	3	4	4
9.3	.9685	.9689	.9694	.9699	.9703	.9708	.9713	.9717	.9722	.9727	0	1	1	2	2	3	3	4	4
9.4	.9731	.9736	.9741	.9745	.9750	.9754	.9759	.9763	.9768	.9773	0	1	1	2	2	3	3	4	4
9.5	.9777	.9782	.9786	.9791	.9795	.9800	.9805	.9809	.9814	.9818	0	1	1	2	2	3	3	4	4
9.6	.9823	.9827	.9832	.9836	.9841	.9845	.9850	.9854	.9859	.9863	0	1	1	2	2	3	3	4	4
9.7	.9868	.9872	.9877	.9881	.9886	.9890	.9894	.9899	.9903	.9908	0	1	1	2	2	3	3	4	4
9.8	.9912	.9917	.9921	.9926	.9930	.9934	.9939	.9943	.9948	.9952	0	1	1	2	2	3	3	4	4
9.9	.9956	.9961	.9965	.9969	.9974	.9978	.9983	.9987	.9991	.9996	0	1	1	2	2	3	3	3	4

* Table XIII is reprinted from D. E. Smith, W. D. Reeve, and E. L. Morss: *Elementary Mathematical Tables*, Ginn and Company, by permission of the authors and publishers.

To obtain the mantissa for a four-digit number, find in the body of the table the mantissa for the first three digits and then, neglecting the decimal point temporarily, add the number in the proportional-parts table at the right which is on the same line as the mantissa already obtained and in the column corresponding to the fourth digit.

NAME INDEX

SUBJECT INDEX